Challenges and Opportunities for Aviation Stakeholders in a Post-Pandemic World

Salim Kurnaz
Suleyman Demirel University, Turkey & Kazimiero Simonaviciaus University, Lithuania

Antonio Rodrigues
Instituto Superior de Gestão, Portugal

Dorothea Bowyer
Western Sydney University, Australia

IGI Global
PUBLISHER of TIMELY KNOWLEDGE

A volume in the Advances in Business Strategy and Competitive Advantage (ABSCA) Book Series

Published in the United States of America by
IGI Global
Business Science Reference (an imprint of IGI Global)
701 E. Chocolate Avenue
Hershey PA, USA 17033
Tel: 717-533-8845
Fax: 717-533-8661
E-mail: cust@igi-global.com
Web site: http://www.igi-global.com

Library of Congress Cataloging-in-Publication Data

Names: Kurnaz, Salim, 1978- editor. | Rodrigues, António, 1973- editor. | Bowyer, Dorothea Maria, 1980- editor.
Title: Challenges and opportunities for aviation stakeholders in a post-pandemic world / edited by: Salim Kurnaz, Antonio Rodrigues, Dorothea Bowyer.
Description: Hershey, PA : Engineering Science Reference, [2023] | Includes bibliographical references and index. | Summary: "This publication aims to offer both empirical and theoretical information focused on the current situation and future expectations of aviation. The book will include research results, authoritative overview articles, high-quality analysis trends, comparative studies, and analysis of particular casesin aviation"-- Provided by publisher.
Identifiers: LCCN 2022044506 (print) | LCCN 2022044507 (ebook) | ISBN 9781668468357 (hardcover) | ISBN 9781668468364 (paperback) | ISBN 9781668468371 (ebook)
Subjects: LCSH: Aircraft industry. | COVID-19 Pandemic, 2020---Influence.
Classification: LCC HD9711.A2 C43 2023 (print) | LCC HD9711.A2 (ebook) | DDC 338.4/762913334--dc23/eng/20220926
LC record available at https://lccn.loc.gov/2022044506
LC ebook record available at https://lccn.loc.gov/2022044507

This book is published in the IGI Global book series Advances in Business Strategy and Competitive Advantage (ABSCA) (ISSN: 2327-3429; eISSN: 2327-3437)

British Cataloguing in Publication Data
A Cataloguing in Publication record for this book is available from the British Library.

For electronic access to this publication, please contact: eresources@igi-global.com.

Advances in Business Strategy and Competitive Advantage (ABSCA) Book Series

Patricia Ordóñez de Pablos
Universidad de Oviedo, Spain

ISSN:2327-3429
EISSN:2327-3437

Mission

Business entities are constantly seeking new ways through which to gain advantage over their competitors and strengthen their position within the business environment. With competition at an all-time high due to technological advancements allowing for competition on a global scale, firms continue to seek new ways through which to improve and strengthen their business processes, procedures, and profitability.

The **Advances in Business Strategy and Competitive Advantage (ABSCA) Book Series** is a timely series responding to the high demand for state-of-the-art research on how business strategies are created, implemented and re-designed to meet the demands of globalized competitive markets. With a focus on local and global challenges, business opportunities and the needs of society, the **ABSCA** encourages scientific discourse on doing business and managing information technologies for the creation of sustainable competitive advantage.

Coverage

- International Business Strategy
- Strategy Performance Management
- Strategic Management
- Tacit Knowledge
- Co-operative Strategies
- Customer-Orientation Strategy
- Adaptive Enterprise
- Globalization
- Small and Medium Enterprises
- Competitive Strategy

IGI Global is currently accepting manuscripts for publication within this series. To submit a proposal for a volume in this series, please contact our Acquisition Editors at Acquisitions@igi-global.com or visit: http://www.igi-global.com/publish/.

The Advances in Business Strategy and Competitive Advantage (ABSCA) Book Series (ISSN 2327-3429) is published by IGI Global, 701 E. Chocolate Avenue, Hershey, PA 17033-1240, USA, www.igi-global.com. This series is composed of titles available for purchase individually; each title is edited to be contextually exclusive from any other title within the series. For pricing and ordering information please visit http://www.igi-global.com/book-series/advances-business-strategy-competitive-advantage/73672. Postmaster: Send all address changes to above address.

Titles in this Series

For a list of additional titles in this series, please visit: www.igi-global.com/book-series

Opportunities and Challenges of Business 5.0 in Emerging Markets
Sumesh Dadwal (Northumbria University, UK) Pawan Kumar (Lovely Professional University, India) Rajesh Verma (Lovely Professional University, India) and Gursimranjit Singh (Lovely Professional University, ndia)
Business Science Reference • © 2023 • 315pp • H/C (ISBN: 9781668464038) • US $250.00

Rapid Innovation and Development in the Global Cannabis Market
Pfano Mashau (University of KwaZulu-Natal, South Africa) and Jabulani Nyawo (University of KwaZulu-Natal, South Arica)
Business Science Reference • © 2023 • 352pp • H/C (ISBN: 9781668463468) • US $250.00

Analyzing International Business Operations in the Post-Pandemic Era
Bryan Christiansen (The Christiansen Foundation, USA) and John D. Branch (University of Michigan, USA)
Business Science Reference • © 2023 • 309pp • H/C (ISBN: 9781668437490) • US $250.00

Handbook of Research on Business Model Innovation Through Disruption and Digitalization
Erik Stavnsager Rasmussen (University of Southern Denmark, Denmark) and Nicolaj Hannesbo Petersen (UCL University College, Denmark)
Business Science Reference • © 2023 • 400pp • H/C (ISBN: 9781668448953) • US $325.00

Intellectual Capital as a Precursor to Sustainable Corporate Social Responsibility
Bartolomé Marco-Lajara (University of Alicante, Spain) Patrocinio Zaragoza-Sáez (University of Alicante, Spain) and Javier Martínez-Falcó (University of Alicante, Spain)
Business Science Reference • © 2023 • 291pp • H/C (ISBN: 9781668468159) • US $250.00

The Transformation of Global Trade in a New World
Bartolomé Marco-Lajara (University of Alicante, Spain) Ahu Coşkun Özer (Marmara University, Turkey) and Javier Martínez Falcó (University of Alicante, Spain)
Business Science Reference • © 2023 • 308pp • H/C (ISBN: 9781668459508) • US $250.00

Exploring the Economic Opportunities and Impacts of Migrant Entrepreneurship Success Stories and Case Studies
Meena Chavan (Macquarie University, Australia) and Sheba Nandkeolyar (Multiconnexions, Australia)
Business Science Reference • © 2022 • 231pp • H/C (ISBN: 9781668449868) • US $215.00

IGI Global
PUBLISHER of TIMELY KNOWLEDGE

701 East Chocolate Avenue, Hershey, PA 17033, USA
Tel: 717-533-8845 x100 • Fax: 717-533-8661
E-Mail: cust@igi-global.com • www.igi-global.com

Table of Contents

Detailed Table of Contents

The need for human resources will expand as the air transport industry recovers in the post-pandemic era. Improving employer attractiveness will be one of the most effective strategies for enabling companies to retain existing employees while also recruiting new talent. This chapter aims to provide a scale to measure employer attractiveness in the air transport industry. For this purpose, the air transport employer attractiveness scale (AirEmpAt) is introduced, based on adapting and validating the 25-item employer attractiveness scale (EmpAt). The participants of the research were university students in aviation management departments, and 623 responses were collected from students in Turkey. Exploratory factor analysis and confirmatory factor analysis were conducted, respectively. Then, the nomological validity was ensured. In conclusion, the analyses support the validity of the AirEmpAt scale, and provide evidence of three distinct measures of employer attractiveness: functional value, psychological value, and economic value.

Air transportation demand increases all over the world every day, and airports have been serving more passengers. Furthermore, the number of passengers with disabilities who get service at the airport has been increasing recently. But the service which disabled passengers get is not enough, and these passengers are confronted with difficulties while using air transportation. The biggest issue these passengers face is not getting efficient guidance service and not benefiting from some right which other passengers do. Hence, this study aims to analyse collaboration of airport stakeholders to provide safe air transportation journey and reduce or remove difficulties for disabled passengers. The research is based on literature review, interviews with airport stakeholders, and analysis of these with AHP method.

The aviation industry is adopting a promising approach to decarbonize the aviation sector dramatically by incorporating hydrogen fuel cell technology and expects to start some commercial services as early as 2035. The authors estimate the aircraft operating cost to become less expensive eventually with this forthcoming fuel cell technology, and the total maintenance cost of hydrogen-powered aircraft to come down in the long run for all segments of the aviation sector, from short haul to long haul services. With the possible incorporation of a new superconducting electrical system, a pneumatic power system, and a fully integrated electrical drive train system, additional maintenance will be required due to their cryogenic factors and associated safety issues. This chapter focuses on the potential impacts of hydrogen fuel cell technology on aircraft maintenance reliability, cost, and safety across the wide range of commuter, regional, short, medium, and long-haul segments of the aviation industry.

Covid-19 has adversely affected many sectors, especially aviation. Because the plane is an enclosed space and spreads the virus faster by carrying passengers between countries, aviation has been affected more than other sectors. In this process, revenues of aviation enterprises have decreased, flights and therefore the amount of required labor has decreased due to domestic and international flight bans. For this reason, the human resources units of the enterprises have taken some actions against this situation. Although almost every airline company was adversely affected by the pandemic, airlines following the right policies were able to close this process with less damage. In this book chapter, what is done in the human resources departments of aviation companies in the world will be examined.

The aviation industry is currently on a path to becoming climate-neutral and various stakeholders such as airlines, airports, and regulatory bodies are setting targets to achieve this goal. The purpose of this chapter is to explain aviation's contribution to climate change and the existing efforts of the industry to reduce and mitigate climate impact. The chapter covers recent evidence suggesting that our changing climate is beginning to impact the industry, largely in a negative manner. Therefore, the industry will need to plan and adapt to its changing environment while developing new technologies and approaches. The challenges and opportunities of alternative aircraft fuel and energy source are presented, along with modifications to infrastructure that will accelerate the transition to a carbon and climate-neutral industry.

This chapter investigates the managerial practices of low-cost carriers during the pandemic enhancing its operations to remain in track. The managerial practices in the study were high competence workforce, organizational culture, risk management, and strategic human resources management. The multiple regression was applied. The results found that LCCs highlighted on risk management, followed by high competence workforce, and strategic human resource management affecting to organizational structure. Overall, there seems to be some evidence to indicate that the discrepancies and risk management plan for any circumstances of uncertainty must be implemented. While the training and development of employees should be conducted continuously especially during the period of non-operating. A further issue suggested to practice the strategic human resource management (SHRM), that aligned organization objectives fit with the current situation both internally and externally, since the SHRM would eventually affect to organizational structure as a whole.

Events that deeply affect the world and the airline industry, such as terrorist attacks or disease outbreaks, are expressed as unexpected events since they are not predicted or desired events. The responses employed by airlines are of great importance for the future of airline companies for their survival. In this context, this study aims to reveal a strategic response repertoire that airlines can employ to sustain their existence in case of facing an unexpected events for the post-Covid-19 era. Studies on strategic responses during unexpected events were collected from WoS and Sciencedirect databases. These studies were analyzed and discussed under the themes of preparedness/resilience efforts, resume, exit, adaptation, and action, determined by the authors. The study is expected to guide airline companies on how to respond to various unexpected events post-Covid-19 period.

Of all the Portuguese private airlines that for many years competed with TAP Air Portugal (the Portuguese owned state airline) only Portugália Airlines (PGA) resisted the hegemony of the Portuguese flag carrier until it was bought by TAP itself. Portugália Airlines, a private airline, managed to maintain its operation, even against the power of the state company. This case study shows how PGA operated certain routes (here the Turin example, in Italy) as a survival strategy. The approach is exploratory and based on an observational description of facts. In view of the recent events caused by COVID 19, the pandemic has brought about a change in business models, also providing the warning that anyone who does not have a clear strategy, dies at birth. The case study leaves some interesting clues to explore.

Ground handling services must be provided in accordance with the airlines' strategy in order to add value and create competitive advantage for the contracting airline, and it is therefore critical to understand how airlines shape ground handling services based on their business models. This study addresses ground handling services by focusing on the aircraft turnarounds of full-service, hybrid and low-cost airlines. It aims to extend the knowledge of the procedures, activities, and resources of aircraft turnarounds by comparing, based on qualitative research using the case study method, the characteristics of Lufthansa, Vueling and Wizz Air aircraft turnarounds. The results point out that the tasks related to unloading/unloading the holds, cleaning the cabin, and disembarking/embarking passengers, are the turnaround tasks that show the most differences between the three business models, playing an important role in the value creation strategy and in increasing the competitive advantage of airlines in ground handling services.

The pandemic had a very negative and penalizing impact on service activity sectors, including tourism and transport, to which Portugal was not immune. TAP Air Portugal, like Ryanair and easyJet, felt these adverse effects in their operations, forcing them to readjust their operating network. The present investigation aims to assess the effects of the pandemic on these airlines at Lisbon and Porto airports between 2019 (pre-pandemic) and 2022. In methodological terms, the analysis will be exploratory, descriptive, and mostly qualitative, based on the collection of information on the departure flights operated from these airports by the selected airlines in selecting dates. The results obtained show that TAP was the airline company most penalized given its international operation with many countries and their borders closed. In Porto, the results of this company were even more severe, reducing more than 50% of the airports served.

This chapter intends to investigate the factors influencing the green purchase intention of passengers in the context of the airline industry and to develop a hierarchical interdependencies model among these factors. To this end, consensus-based interviews with academic experts were conducted. Drawing on a total of 8 factors defined, interpretive structural modeling (ISM) created a hierarchical model. The Matrice d'Impacts Croisés Multiplication Appliquée á un Classement (MICMAC) analysis was additionally applied to validate the ISM model and determine the driving and dependence power of each factor. In terms of influencing green purchasing intention, the results demonstrated that perceived consumer effectiveness, subjective norm, environmental knowledge, environmental awareness, and environmental concern were at the root of the hierarchical model, and green image was at the top. In order to manage passengers' green purchase intentions, policymakers and marketers could benefit from the ISM model designed to comprehend the interdependencies between factors.

The purpose of this study is to understand and identify sustainability measures applicable in the aviation industry. Further, the study also explores the enablers of sustainability in the aviation sector. The study uses a systematic literature review, published in Scopus-indexed journals. The study addresses the complexity of sustainability in the sector and identifies key indicators based on comprehensive and valid data from different stakeholders. The study focuses separately on the three pillars of sustainability—social economic, and environmental sustainability—and identifies sustainability indicators for each pillar from the views of the triple bottom line. The study finds that there is a lack of sustainability knowledge and awareness in the aviation sector despite its resistant growth and expansion. The results show that the aviation industry highlights the importance of sustainability indicators that value equitable development in the pursuit of business goals and environmental and economic efficiency.

Preface

The COVID-19 pandemic that emerged in 2019 has profoundly affected many aspects of everyday life; the pandemic changed the world we live and work in. According to the International Air Transport Association (IATA), the pandemic caused by COVID-19 is the greatest impact aviation has had since World War II. IATA, even in its most optimistic estimates, does not expect air traffic to recover to 2019 levels until 2024. Commercial airlines were forced to reinvent themselves to keep their operations at economically viable levels.

Now that it seems that the pandemic situation has stabilized and hope has been reborn, we have to think about the future.

Sustainability must be a key lever to improve the efficiency of the fleet and operations, accelerating the recovery from the impact of the pandemic. In this way, airlines must realistically consider the potential for reducing mature technologies, including more efficient aircraft models or the use of SAF fuel, which would make it possible, together with the implementation of the Single European Sky, to reach 36% of reduction of current emission levels. Greater ambition will require collaboration with institutions and other industries to develop mechanisms that encourage and promote the adoption of other technologies that will be mature in the long term, such as hydrogen aircraft or large-scale SAF production.

The development of a differential experience based on the integration of services, multimodality and digitization for the post-COVID traveler must reinforce the purpose and digital transformation programs, accelerating the integration of new technologies for data analysis and the creation of multimodal and multiservice platforms. This process must promote a differential offer during the traveler's travel journey, based on anticipating changes in demand and making decisions in real time about the experience perceived by the traveler. In short, airlines must transform their vision of the passenger towards that of a traveler who accesses all the services and information he needs for his trip through the airline itself.

The strategy must enable the financial discipline that mitigates the impact of the pandemic and promotes the development of new businesses, preventing the recovery of its financial situation from extending throughout the decade. The strategy must advance strict financial control and seize opportunities to make the cost model more flexible, while promoting the generation of additional income with new pricing models, as well as with the exploitation of new businesses, segments or market niches. It is key that the strategy generates the economic resources necessary to develop the investments that allow the airline business to be reinvented.

The recovery of the commercial aviation sector is undoubtedly a vital and strategic aspect at a global level. On the one hand, it is a key sector in the mobility of citizens and, on the other, it represents a vital link in the value chain of the tourism sector, without forgetting its contribution to the industrial sector, among others, such as the development and maintenance of air powered. An efficient, sustainable and interconnected transport system is crucial to laying the foundations for economic recovery as well as contributing to the well-being of citizens and the competitiveness of companies.

Of no less concern are the initiatives aviation companies must take to understand their cyber exposures, protect critical assets, and develop and maintain robust cyber risk management strategies.

Collaboration among academics working in various fields and disciplines is not new, and in Higher Education, this is the foundation for generating, sharing, and disseminating knowledge. This book's origin has an exciting beginning as the idea evolved during the COVID-19 pandemic. Academics on opposite sides of the world, passionate about research in the air transportation industry, witnessed the dramatic stand-still of the entire sector and questioned the long way to recovery for the aviation industry. The three editors collaborated on this global book project to exemplify what was happening to air transport stakeholders worldwide; the industry that was the hardest struck during the COVID-19 pandemic. As such, there have been a variety of government assistance programs in various countries. However, while there have been many studies into the going concerns of businesses, there is a lack of reported research into COVID-19's impact on the aviation sector, in which the financial health of organizations is investigated while examining viable outsourcing and offshoring practices in a post-pandemic world.

Did COVID-19 result in fundamental changes in the structure of the airline industry? Did this prompt regulatory change in various countries, and if so, in what form? Is it possible to maintain pre-Covid levels of competition in a once flourishing aviation sector? How will competition regulators respond as competitors might look to cooperate or consolidate due to the demise of several aviation stakeholders?

This textbook's primary motivation was to share stories with interested readers, give a voice to the post-COVID-19 aviation actors the winners and losers, and showcase new emerging business models that air transport stakeholders considered a steppingstone to recovery.

The financial impact of COVID-19 has been huge on the aviation industry. The aviation industry is a key enabler of closely linked economic activities in each country. The pandemic affected the aviation and tourism sectors the most, with border closures used by health regulators, leading to considerable passenger and freight demand declines. The imposed travel restrictions stopped the interconnectivity among nations. Certain aviation stakeholders ceased to exist, and others, up until today, are trying to regain their customers. The dramatic drop in demand for passenger air transport (and freight, to a lesser extent) due to the COVID-19 pandemic and containment measures threatened the viability of many actors in the air transport sector and the rest of the aviation industry, with many jobs at stake. While the aviation industry has often been a target of government policies, the COVID-19 crisis has precipitated a new suite of loans, loan guarantees, wage subsidies, and equity injections, raising concerns about competition and the efficient use of public resources. There is thus a need to report research on the current situation, future expectations, and possible change scenarios in the aviation industry.

Technological advancements have also played a profound role in the pre, during, and post-COVID restructuring of the aviation industry and the various privatization and liberalization movements across the globe. Without a doubt, the air-transport stakeholders' experiences will vary in the years to come, and it is thus essential to report on the present status quo and to understand the past. This book, therefore, focuses on contemporary studies addressing the effects of economic crises, pandemics, digitalization, and war environment on the aviation industry.

The book includes research results, authoritative overview articles, high-quality analysis trends, comparative studies, and analysis of particular cases in aviation. It draws attention to the aviation industry's current situation and future expectations by focusing on stakeholders, gender issues in aviation, e-aviation, and various other industry trends in the context of full-service network carriers, low-cost airlines, charter, regional, cargo, and hybrid airlines; aviation education and management, highlighting the use of digital technology in the aviation sector and the various communication technologies available.

We hope that this book will reach its target audience of researchers, scholars, academics, post-graduate and undergraduate students, and professionals in the areas of philosophy, ethics, sociology, development studies, economics, political science, cultural studies, the entire humanities, and social sciences family, as well as the general public. Academics, researchers, and aviation students will find this book helpful in furthering their research exposure to pertinent topics and assisting in further promoting their research efforts in this field.

Lastly, the recent conflict between Russia and Ukraine, and its significant impact in the geopolitical and economic spheres, could result in great uncertainty about the evolution of the sector and its implications. The current inflationary crisis also requires firm resilience so that the sector can maintain its growth trajectory by responding to the challenges that the aeronautical industry is facing.

We would like to thank the authors of all chapters for their invaluable efforts and contributions to our book. The geographical diversity of the authors has certainly enriched the discussion of Challenges and Opportunities for Aviation Stakeholders in a Post-Pandemic, so without their efforts it would be impossible to finish this book.

ORGANIZATION OF THE BOOK

The book is organized into thirteen chapters. A brief description of each of the chapters follows:

Chapter 1 focuses on attracting future employees. The section aims to provide a scale to measure employer attractiveness in the air transport industry. For this purpose, the Air Transport Employer Attractiveness Scale (AirEmpAt) is introduced, based on adapting and validating the 25-item Employer Attractiveness Scale (EmpAt).

Chapter 2 aims to analyze collaboration of airport stakeholders to provide safe air transportation journey and reduce or remove difficulties for disabled passenger. The research is based on literature review, interviews with airport stakeholders and analysis of these with AHP method.

Chapter 3 focuses on the potential impacts of hydrogen fuel cell technology on aircraft maintenance reliability, cost, and safety across the wide range of commuter, regional, short, medium, and long-haul segments of the aviation industry.

Chapter 4 focused on the aims and functions of human resources, and especially the selection and placement part in the aviation sector after the pandemic. In the chapter, recruitment and firing processes are examined. Furthermore, the changes in the performance of the sector employees, during the pandemic process, are discussed in connection with the concerns they experience.

Chapter 5 tried to explain aviation's contribution to climate change and the existing efforts of the industry to reduce and mitigate climate impact. The chapter covers recent evidence suggesting that our changing climate is beginning to impact the industry, largely in a negative manner. Therefore, the industry will need to plan and adapt to its changing environment while developing new technologies and approaches. The challenges and opportunities of alternative aircraft fuel and energy source are presented, along with modifications to infrastructure that will accelerate the transition to a carbon and climate-neutral industry.

Chapter 6 investigates the managerial practices of low-cost carriers during the pandemic enhancing its operations to remain in track. The managerial practices in the study were high competence workforce, organizational culture, risk management, and strategic human resources management. The multiple regression was applied. The results found that LCCs highlighted on risk management, followed by high competence workforce, and strategic human resource management affecting to organizational structure.

Chapter 7 aims to reveal a strategic response repertoire that airlines can employ to sustain their existence in case of facing unexpected events for the post-Covid-19 era. Studies on strategic responses during unexpected events were collected from WoS and Sciencedirect databases. These studies were analyzed and discussed under the themes of preparedness/resilience efforts, resume, exit, adaptation, and action determined by the authors. The study is expected to guide airline companies on how to respond to various unexpected events post-Covid-19 period.

Chapter 8 shows how PGA operated certain routes as a survival strategy. The approach is exploratory and based on an observational description of facts. In view of the recent events caused by COVID 19, the pandemic has brought about a change in business models, also providing the warning that anyone who does not have a clear strategy, dies at birth. The case study leaves some interesting clues to explore.

Chapter 9 addresses ground handling services by focusing on the aircraft turnarounds of Full-Service, Hybrid and Low-Cost airlines. It aims to extend the knowledge of the procedures, activities and resources of aircraft turnarounds by comparing, based on qualitative research using the case study method, the characteristics of Lufthansa, Vueling and Wizz Air aircraft turnarounds. The results point out that the tasks related to loading/unloading the holds, cleaning the cabin and disembarking/embarking passengers, are the turnaround tasks that show the most differences between the three business models, playing an important role in the value creation strategy and in increasing the competitive advantage of airlines in ground handling services.

Chapter 10 aims to assess the effects of the pandemic on these airlines at Lisbon and Porto airports between 2019 (pre-pandemic) and 2022. In methodological terms, the analysis was exploratory, descriptive and mostly qualitative, based on the collection of information on the departure flights operated from these airports by the selected airlines in selecting dates. The results obtained shows that TAP was the company most penalized airline given its international operation with many countries and their borders closed. In Porto, the results of this company were even more severe, reducing more than 50% of the airports served.

Chapter 11 intends to investigate the factors influencing the green purchase intention of passengers in the context of the airline industry and to develop a hierarchical interdependencies model among these factors. To this end, consensus-based interviews with academic experts were conducted. The results demonstrated that perceived consumer effectiveness, subjective norm, environmental knowledge, environmental awareness, and environmental concern were at the root of the hierarchical model, and green image was at the top.

Chapter 12 tried to understand and identify sustainability measures applicable in the aviation industry. Further, the study also explores the enablers of sustainability in the aviation sector. The study uses a systematic literature review, published in Scopus-indexed journals. The study addresses the complexity of sustainability in the sector and identifies key indicators based on comprehensive and valid data from different stakeholders.

Chapter 13 examines the occupational commitment levels of the employee in the aviation industry during the post-pandemic period and offers suggestions for the future. Participants consist of airport employees operating in Turkey. According to the results of the research, the occupational commitment levels of the participants was high.

Salim Kurnaz
Süleyman Demirel University, Turkey & Kazimiero Simonaviciaus University, Lithuania

António Augusto Baptista Rodrigues
ISG-Business and Economics School, Portugal

Maria Dorothea Bowyer
Western Sydney University, Australia

REFERENCES

Deloitte. (2022). La guía de la transformación de las aerolíneas en la postpandemia [The guide to the transformation of airlines in the post-pandemic]. *DeLoitte*. https://www2.deloitte.com/es/es/pages/operations/articles/despegue-sector-aereo.html

International Air Transport Association. (2020). Annual Report 2020, https://www.iata.org/ contentassets/c81222d96c9a4e0bb 4ff6ced0126f0bb/iata-annual-review-2020.pdf, .

International Air Transport Association. (2021). *Annual Report 2021*. IATA. https://www.iata.org/contentassets/c81222d96c9a4e0bb4ff6ced0126f0bb/iata-annual-review-2021.pdf

International Civil Aviation Organization (2020). Effects of Novel Coronavirus (COVID-19) on Civil Aviation: Economic Impact Analysis. *ICAO Reports*, 1-45

Zhang, X. A., Fan, H., Qi, R. Z., Zheng, W., Zheng, K., Gong, J. H., Fang, L.-Q., & Liu, W. (2020). Importing coronavirus disease 2019 (covid-19) into China after international air travel. *Travel Medicine and Infectious Disease*, *1-5*, 101620. doi:10.1016/j.tmaid.2020.101620 PMID:32222421

Acknowledgment

The editors would like to acknowledge the help of all the people involved in this project and, more specifically, to the authors and reviewers that took part in the review process. Without their support, this book would not have become a reality.

First, the editors would like to thank each one of the authors for their contributions. Our sincere gratitude goes to the chapters authors who contributed their time and expertise to this book. The geographical diversity of the authors has certainly enriched the discussion of Challenges and Opportunities for Aviation Stakeholders in a Post-Pandemic, so without their efforts it would be impossible to finish this book.

Second, the editors wish to acknowledge the valuable contributions of the reviewers regarding the improvement of quality, coherence, and content presentation of chapters. Most of the authors also served as referees; we highly appreciate their double task.

Salim Kurnaz
Süleyman Demirel University, Turkey & Kazimiero Simonaviciaus University, Lithuania

António Augusto Baptista Rodrigues
ISG-Business and Economics School, Portugal

Maria Dorothea Bowyer
Western Sydney University, Australia

Chapter 1
Adaptation and Validation of the Employer Attractiveness Scale for the Air Transport Industry:
Attracting Future Employees

Emircan Ozdemir
https://orcid.org/0000-0002-1383-4712
Eskisehir Technical University, Turkey

Umran Under
Eskisehir Technical University, Turkey

K. Gulnaz Bulbul
Eskisehir Technical University, Turkey

Ender Gerede
Eskisehir Technical University, Turkey

ABSTRACT

The need for human resources will expand as the air transport industry recovers in the post-pandemic era. Improving employer attractiveness will be one of the most effective strategies for enabling companies to retain existing employees while also recruiting new talent. This chapter aims to provide a scale to measure employer attractiveness in the air transport industry. For this purpose, the air transport employer attractiveness scale (AirEmpAt) is introduced, based on adapting and validating the 25-item employer attractiveness scale (EmpAt). The participants of the research were university students in aviation management departments, and 623 responses were collected from students in Turkey. Exploratory factor analysis and confirmatory factor analysis were conducted, respectively. Then, the nomological validity was ensured. In conclusion, the analyses support the validity of the AirEmpAt scale, and provide evidence of three distinct measures of employer attractiveness: functional value, psychological value, and economic value.

DOI: 10.4018/978-1-6684-6835-7.ch001

INTRODUCTION

The air transport industry involves a wide range of companies which are airlines, airports, air cargo carriers, ground handling companies, and other facilitating organizations such as air traffic management organizations, maintenance, and repair organizations (MROs), and catering companies (Schmitt & Gollnick, 2016; Tabares & Mora-Camino, 2017). The primary goal of this extensive and complex system is to provide air transport service, which requires the cooperation of organizations specialized in various businesses. In addition to specialization in operations, human capital is one of the primary pillars in achieving a sustainable competitive advantage for companies in the air transport industry (Berthon et al., 2005; Harvey & Turnbull, 2020). Due to the service-intensive nature of the air transport industry, employees play a crucial role in the success of companies. In this context, it is evident that the specialization of air transport companies in operations directly depends on the competencies of the employees.

On the other hand, service quality has a significant effect on business performance in service businesses (Ramayah et al., 2011). As in all service businesses, it is the same for air transport companies. The ultimate goal of companies in the air transport industry is to provide a service that meets the passengers' needs and expectations. Many studies in this field have presented that providing the service that best meets passenger needs and expectations will positively affect service quality and, subsequently, customer satisfaction (Farooq et al., 2018; F. T. Shah et al., 2020; Wang et al., 2011). Because the service delivery process involves simultaneous production and consumption, employees have critical importance in ensuring service quality. As a result, the air transport industry needs to employ qualified employees to fulfill the expectations of passengers.

Recent studies reveal the importance of internal marketing for service businesses (Gounaris, 2008). The internal marketing concept acknowledges that the initial market of a company is its employees (George, 1990). When the situation is considered from this perspective, the internal products that companies offer to their employees are jobs (Berthon et al., 2005). This also makes employees the initial customers. Therefore, when jobs are ensured to motivate employees and meet their wants and needs, the level of service quality offered by the company will also improve. In addition, the efforts to be carried out in the context of internal marketing will ensure that the company is preferable by qualified personnel. On the other hand, the sum of communication efforts aiming at current and potential employees that the company is a desirable place to work is defined as employer brand, which is closely associated with the internal marketing concept (Lloyd, 2002). As Slåtten, Lien and Svenkerud stated (2019), internal market-oriented culture and organizational attractiveness are associated with turnover intentions, employee engagement, and service quality provision. In this context, companies place high importance on efforts to build strong employer brands to gain advantages, such as lower employee acquisition costs, improved employee relations, increased employee retention, and the ability to offer lower salaries for employees at the same level than companies with weaker employer brands (Berthon et al., 2005; Leekha Chhabra & Sharma, 2014).

When an employer is perceived as more attractive, the brand equity of the employer will also be stronger. In other words, the attractiveness of an employer for potential employees is the total value of its employer brand (Kissel & Büttgen, 2015). For this reason, the concept of employer attractiveness, which is closely related to the concept of employer brand, can be considered an antecedent of employer brand and is defined as "the envisioned benefits that a potential employee sees in working for a specific organization" (Berthon et al., 2005). According to another similar definition, the interest of job-seekers to be employed by an organization is called employer attractiveness (Ritz & Waldner, 2011). When the employer branding frameworks proposed by Leekha Chhabra and Sharma (2014) and Backhaus and Tikoo (2004) are examined by following the same rationale, both of them discuss employer attractiveness as a part of the employer branding process. As a result, in studies on the antecedents of employer branding, the measurement of the concept of employer attractiveness becomes prominent. The Employer Attractiveness Scale (EmpAt) developed by Berthon et al. (2005) to measure employer attractiveness is one of the most prominent scales in this field. However, as in other studies in the field of marketing, studies of employer attractiveness corroborating results of previous researches remain rare (Benraïss-Noailles & Viot, 2021). This shortcoming also exists for the air transport industry, where being favored by the right employees provides a considerable strategic advantage for the companies.

As the service industry, including the air transport industry, is labor-intensive, it is obvious that employees have a direct impact on the quality of service delivery. Improving the quality of the service delivery process is a significant factor for companies in terms of competition and global marketing (Alzaydi et al., 2018). Employment of staff with the necessary qualifications and skills is the key determinant of business success for air transport companies. The air transport industry is one of the major global employers and supported 87.7 million jobs globally in 2019. 11.3 million jobs were provided in aviation organizations such as airlines, airports, and air navigation services (ATAG, 2020). With such a high demand for human resources, specific issues arise. Because of high labor costs in the air transport industry, the turnover rate is a critical issue to deal with (Chen, 2006). In this context, the search for effective strategies that enable businesses to retain existing employees while also recruiting new talents emphasizes the employer brand concept. (Ambler & Barrow, 1996; Dabirian, Berthon, et al., 2019).

On the other hand, the air transport industry faced the devastating effects of the COVID-19 pandemic in 2019, since the last significant passenger traffic decline it last experienced during the financial crisis in 2008 (ICAO, 2022). When the effects of COVID-19 are examined, this pandemic was perhaps the biggest challenge to the air transport industry of all time and resulted in closed borders, and many airlines came to a standstill (Tanrıverdi et al., 2020). Even some airlines went bankrupt in this era and the number of bankrupt airlines is projected to increase as the effects of the pandemic still continue (Nižetić, 2020). Therefore, many staff employed by the air transport industry faced pay cuts and layoffs (Deveci et al., 2022). However, with the implementation of vaccines, therapeutics, and tests, the COVID-19 pandemic has been brought under control and has begun to be recognized as an epidemic in many regions of the world, including Europe and the United States (Charumilind et al., 2022), the air transport industry has started to recover as well as many other industries. The average recovery time for the air transport industry is estimated to take 2.4 years, and the industry is expected to reach pre-COVID-19 passenger traffic numbers by the late-2022 (Gudmundsson et al., 2021). Passenger traffic statistics also show that the loss of the air transport industry, which experienced a 60.1% decrease in 2020 compared to the previous year (ICAO, 2021), will largely catch up with the 2019 figures by the end of 2022 (ICAO, 2022). Therefore, the need for human resources in the air transport industry, which decreased significantly during the COVID-19 period, will increase in the post-COVID-19 period. In this process, being preferred again by both new and existing qualified employees will become an important competitive advantage for air transport companies. For this reason, air transport companies need to define the expectations of their employees accurately and build strong employer brands that will help them to attract the best employees. At this point, it is important for the air transport companies to monitor how attractive they are from the perspective of their potential employees. Although various measurement tools have been developed on employer attractiveness in many different business-domains, there is no valid tool for the air transport industry. Considering that different business-domains or industries have different elements of organizational culture, employee expectations also differ between business-domains or industries (Chatman & Jehn, 1994; Gupta, 2011). Therefore, it is emphasized in this research that the necessity of adapting an employer attractiveness scale, which was not previously available for the air transport industry. Thus, this scale will be a useful tool that enables air transport companies to be successful in employer branding practices.

This chapter aims to adapt and validate the EmpAt scale for measuring employer attractiveness from the perspective of potential employees of the air transport industry while analyzing psychometric properties. The employer attractiveness was examined by focusing specifically air transport industry from the perspective of potential employers. In this context, the EmpAt scale was applied to university students studying in aviation management and civil air transport management departments, who are the future employees of the air transport industry. In order to confirm the factorial structure of the questionnaire, the study of Berthon et al. (2005) was taken into consideration, and several factor analyzes were carried out in this context. As a result of the analyses conducted, the air transport employer attractiveness scale (AirEmpAt) was adapted to and validated for the air transport industry by considering the peculiar characteristics of this industry. In this way, the requirements for the companies in the air transport industry to be preferred by the best potential employees have been put forward. It is anticipated that the use of the proposed AirEmpAt scale by air transport companies to measure how attractive they are in the minds of their potential employees will provide them important insights to be preferred by qualified human resources. In terms of implementation, it is expected that the air transport companies, which will attract and employ qualified personnel, will gain competitive advantages during the recovery phase of the industry in the post-COVID-19 period.

BACKGROUND

Today, the impact of employees on the success of businesses is now widely recognized in all business sectors. As a result, businesses endeavor to increase employees' satisfaction just as much as they do for their customers. In fact, businesses develop brands for their employees (employer brand) as well as the brand they develop for their customers (consumer brand). The primary goal of employer brand practices is to help businesses attract both current and potential employees. Because companies are competing for the top employees in today's job market and are aware of that employer brand practices will provide them a competitive advantage. In order to maximize the advantages of attracting qualified potential employees and hiring them, companies build and maintain a wide range of applications, such as internal marketing, employer brand, employer attractiveness, and HR marketing. Firstly, these concepts are discussed in this section. Next, the concept of employer attractiveness is emphasized. Finally, employment figures in the aviation industry are addressed for the pre-COVID-19 and post-COVID-19 periods, with a focus on the relationship between employer attractiveness and post-COVID-19 employment expectations.

Definition and Measures of the Employer Attractiveness

The concept of internal marketing was used for the first time in the marketing literature in the late 1970s, and it was emphasized that businesses must first ensure the satisfaction of their employees to achieve customer satisfaction (Berry et al., 1976). Since the concept of internal marketing exists in the marketing literature, it has been discussed that businesses should recruit the best employees to provide the best product/service to their customers and to be preferred by the best employee candidates. At this point, the concept of employer brand has been defined as "the package of functional, economic, and psychological benefits provided by employment, and identified with the employing company", which corresponds to the concept of customer branding in terms of employees (Ambler & Barrow, 1996). Employer attractiveness, an antecedent of employer brand equity, is another concept closely related to the concept of employer branding (Berthon et al., 2005). However, in the literature, employer brand and employer attractiveness terms are frequently used interchangeably, which may confuse (Tkalac Verčič et al., 2021). As a result, while previous work on evaluating employer attractiveness is discussed, studies on employer branding are also included where the distinction between concepts is unclear and used interchangeably. Ambler and Barrow (1996) addressed the employer brand concept for the first time in the literature. They carried out a qualitative study with executives from 27 different companies in the United Kingdom. According to them, the dimensions of the employer brand were functional, economic, and psychological. They also defined the concept as the benefits provided to employees by employers through these dimensions. It can be seen that the employer brand dimensions they determined were the same with the benefits provided to customers by the conventional brands in external markets. Based on the inductive outputs of Ambler and Barrow (1996), Berthon et al. (2005) conducted a study to determine the dimensions of employer attractiveness in a sample of Australian university students. They gave an operational definition of employer attractiveness as the benefits that a company provides to potential employees. They developed the EmpAt scale to reveal the dimensions of employer attractiveness, highlighting the distinction between the concepts of employer attractiveness and employer brand. Five dimensions were discovered under the employer attractiveness structure: Interest value, social value, economic value, development value, and application value. This is one of the most well-known studies on employer branding in the literature.

There are also several alternative scales based on the study developed by Berthon et al. (2005) to measure employer attractiveness. Roy (2008) determined employer attractiveness dimensions by collecting data from university students in India. He developed an 8-factor employer attractiveness scale, including ethical value, psychological value, and career opportunities dimensions, in addition to dimensions of the EmpAt. In another study conducted by Schlager et al. (2011), the employer brand dimensions were determined as economic value, development value, social value, diversity value, and reputation value based on data collected from employees in 12 different countries. An exploratory factor analysis on employer brand dimensions was conducted by Sivertzen, Nilsen, and Olafsen (2013) on university students in Norway, and it was reported that three dimensions, which are psychological value, innovation value, and application value, explain the employer brand structure. In the Chinese private sector, Jiang and Iles (2011) used the concept of employer-based brand equity, which corresponds to employer attractiveness, and determined its dimensions as economic value, developmental value, social value, interest value, and brand trust. Dabirian, Berthon, et al. (2019) studied a group of students in North America and came up with eight value propositions for employer branding in the IT industry. Some studies in the literature used the structure proposed by Berthon et al. (2005) for measuring employer brands in different cultures and samples while keeping the same structure (Benraïss-Noailles & Viot, 2021; Reis & Braga, 2016). The EmpAt scale was commonly used in studies examining employer attractiveness within the context of employer branding. Although the EmpAt scale was used in different studies, it is obvious that the original structure may vary among cultures.

Apart from the EmpAt scale, other unique structures have been identified in some research on employer attractiveness and employer branding. Biswas and Suar (2016) defined antecedents of employer branding as realistic job previews, perceived organizational support, equity in reward administration, perceived organizational prestige, organizational trust, the leadership of top management, psychological contract obligations, and corporate social responsibility influence by collecting data from Indian company executives. Shrivastava and Shukla (2021) developed the Employer Brand (EmBran) scale to assess employer brand, including the dimensions of good human resource (HR) practices, business impression and work conditions, financial compensation, work-life balance, passive culture, and standard HR policy. The EmBran scale was developed by collecting data from both potential and existing employees in India. Arachchige and Robertson (2011) evaluated employer attractiveness from the perspective of Sri Lankan university graduates and determined eight dimensions of employer attractiveness as corporate environment, job structure, social commitment, social environment, relationships, personal growth, organizational dynamism, and enjoyment. When the literature on employer attractiveness and the employer brand is examined, it appears that quite different measurements were proposed in the previous research, which was mainly focused on employee values perceived from companies, as the EmpAt scale indicated. On the other hand, the measurements suggested by the earlier studies place a greater emphasis on organizational characteristics than the values offered to employees. The employer attractiveness measurements, which focus on the values offered to employees rather than organizational or job characteristics, reveal the structure of employer attractiveness more precisely. In the same way that the customer value proposition is critical in building the optimal brand-customer relationship in consumer brands (Aaker, 1996), the employee value proposition should be considered critical in managing the employer brand-employee relationship. The studies conducted based on the value-based approach, which is dominant in the field of employer attractiveness in the literature, are given in Table 1.

Table 1. Dimensions of the employer attractiveness in the literature

	Functional Value	Psychological Value	Economic Value	Interest Value	Social Value	Development Value	Application Value	Ethical Value	Career Opportunities	Diversity Value	Reputation Value	Innovation Value	Brand Trust	Management Value	Work/Life Balance	Brand Image
Ambler and Barrow (1996)	X	X	X													
Berthon et al. (2005)			X	X	X	X	X									
Roy (2008)		X	X	X	X	X	X	X	X							
Schlager et al. (2011)			X		X	X				X	X					
Jiang and Iles (2011)			X	X	X	X							X			
Sivertzen et al. (2013)		X					X					X				
Reis and Braga (2016)			X	X	X	X	X									
Dabirian, Berthon, et al. (2019)			X	X	X	X	X							X	X	X
Benraïss-Noailles and Viot (2021)			X	X	X	X	X									

Source: Authors' own elaboration

There are also some disputes in the literature on the measurement of employer attractiveness. Maxwell and Knox (2009) criticized that most of the studies measuring employer attractiveness or employer branding generally tend to focus on potential employees. As a result, they focused on existing employees and determined which organizational attributes make companies attractive to current employees. In order to fill the same gap that Maxwell and Knox (2009) mentioned, Tanwar and Prasad (2017) developed an employer attractiveness scale that includes training and development, ethics and corporate social responsibility, work-life balance, healthy work atmosphere, compensation, and benefits factors for both potential and current employees. Similarly, Shrivastava and Shukla (2021) conducted a study to address the same gap in the literature, focusing on both current and prospective employees.

On the other hand, Schlager et al. (2011) categorized employer attractiveness in terms of current and potential employees into two distinct structures. They defined two distinct employer attractiveness structures for current and potential employees, which are entitled as perceived and expected employer brands, respectively. Additionally, they have proved that expected employer branding affects only employer identification, and perceived employer branding affects both employer identification and employee satisfaction (Schlager et al., 2011). Given the differences in expectations and perceptions between potential and current employees, it makes more sense to distinguish groups when measuring the attractiveness of companies.

Among the employer attractiveness studies in the literature, there is also one specific study that examines former employees. Dabirian, Paschen and Kietzmann (2019) analyzed online reviews of IT companies in terms of employer attractiveness, covering both former and current employees. They focused on the praise and complaint rates for eight employer attractiveness factors in their study and made comparisons between current and former employees. They provided an effective technique for identifying the factors that lead to the improvement of employer brand in this manner.

Since this chapter aims to propose a valid scale to measure the employer attractiveness of potential employees in the air transport industry, the EmpAt scale was adapted because it focuses directly on the concept of employer attractiveness, and it is widely used in previous studies in the literature. Also, Berthon et al. (2005)'s EmpAt scale is based on Ambler and Barrow (1996)'s research, which addressed the employer brand concept in the literature for the first time. With this aspect, the EmpAt scale can theoretically be considered the most successful and used structure in order to answer how potential employees perceive the value of the employer brand. When all these conditions are evaluated together, it was decided to present the employer attractiveness scale specifically for the air transport industry by adapting and validating the original EmpAt scale. Thus, the AirEmpAt scale presented in this chapter is intended to be an instrument that meets the basic characteristics of the air transport industry and reveals which dimensions have an impact on the attractiveness of employers from the perspective of potential employees of this industry.

Importance of the Employer Attractiveness

The number of studies on developing relationships with the most important stakeholder of companies, which are employees, has increased since Berry, Hensel and Burke (1976) emphasized the concept of internal marketing in the marketing literature. Internal marketing strategies play a critical role in the external marketing success of companies, which is why the internal marketing concept is essential (Greene et al., 1994). On the other hand, as services became more prevalent in economic activities since the 1970s, businesses have been forced to adopt service-dominant logic, which drives forward offering the customer value proposition (CVP) by admitting goods and services are inextricably linked (Gummesson, 2007). Payne, Frow and Eggert (2017, p. 472) defined CVP as "A customer value proposition (CVP) is a strategic tool facilitating communication of an organization's ability to share resources and offer a superior value package to targeted customers." CVP is a strategic tool that requires being customer-centric, as stated in its definition. With the shift in marketing understanding, the benefits of the customer-centric paradigm for businesses became apparent, and it was emphasized that employees serve as customer advocates for customer-centered organizations (D. Shah et al., 2006). In this context, it has become evident that to succeed, businesses need to focus not only on their external customers but also on their internal customers, who are their employees. At this point, the value proposition concept should be offered not only to customers but also to all stakeholders and, most crucially, employees (Frow & Payne, 2011). This one offered to employees is also called the employee value proposition.

In any case, a company must first ensure the job satisfaction of its employees in order to enhance the perceived quality of its service and ensure customer satisfaction (S. P. Brown & Lam, 2008). Moreover, jobs are the internal products that businesses offer to their employees (Berthon et al., 2005). In this respect, companies should apply marketing techniques and manage their brands not only for customers but also for employees (Ambler & Barrow, 1996). On the other hand, Human Resources Management (HRM) undertakes the management of all employee related business operations such as recruiting, hiring, training, compensating, and developing. The basic functions of HRM are defined as employee selection/placement, rewards, appraisal, development, and career planning (Devanna et al., 1981).

Recruiting, which is the first function of HRM, actually means recruiting the right people for the right job. In this direction, recruiting the right employees for the organization while offering the proper jobs is the focus of HRM. On the other hand, a new field where marketing applications and HRM applications will collaborate on internal issues has appeared with the start of evaluating the employees as customers. To attract potential employees in this environment, companies then need to manage their brands from the perspective of their employees. Thus, a new marketing discipline, namely human resources marketing (HR marketing), emerged as the scope and impact of the employer brand notion became obvious (Urbancová & Hudáková, 2017). Through a combination of employer branding and recruiting marketing strategy, HR marketing aims to provide employees with a remarkable and rewarding experience. Therefore, it is inevitable for businesses to invest in developing and managing an employer brand. To do so, companies must first comprehend the expectations of both potential and current employees and determine what will attract them accurately. Responding to the expectations of the employees, which are one of the most important stakeholders of the enterprises, is indispensable for the success of a business.

Employment in the Air Transport Industry Before and After the COVID-19

COVID-19 cases, which first appeared in Wuhan, China, on December 12, 2019, continued to rise and reached 118,000 cases in 114 countries by March 2020. The World Health Organization (WHO) declared COVID-19 a pandemic on March 11, 2020 (CDC, 2022). Following this, national governments issued statements for citizens to stay at their homes to prevent the spread of COVID-19 and shut down all but essential businesses. The negative effects of the COVID-19 period were most severe in the air transport industry, tourism, and hospitality industry as a result of policies including government-ordered lockdowns and travel restrictions (Vidovic, 2022).

There have been numerous crises in the past, including the 2008 financial crisis, the oil crisis, the Iran-Iraq war, the gulf crisis, the Asian crisis, the 9/11 terrorist assault, the SARS outbreak, and others that had a detrimental impact on the air transport industry. However, compared to past crises, COVID-19 has had a much more damaging impact on the air transport industry. The COVID-19 pandemic certainly caused the largest drop in passenger traffic since 1945, when the consequences of crises on the number of passengers carried by the air transport industry are examined (see Figure 1).

Figure 1. Effects of crises on the number of passengers carried by the air transport industry
Source: ICAO (2022)

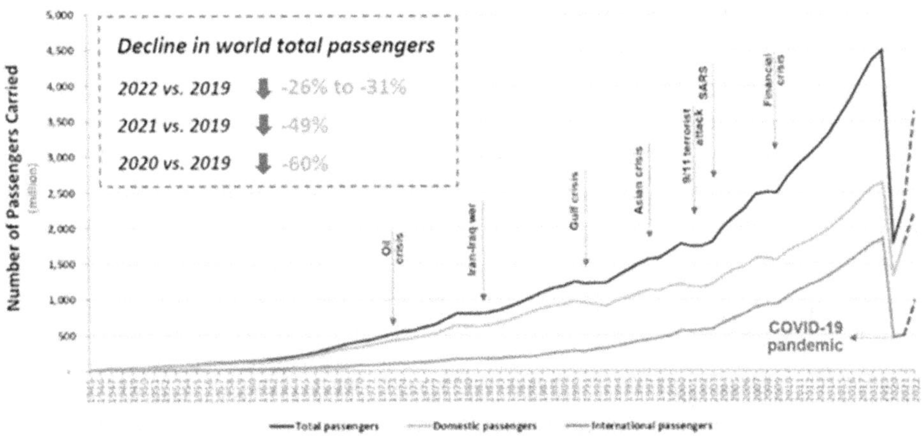

Due to COVID-19, global passenger traffic fell by 60% in 2020 compared to the previous year, marking the largest decline in global passenger numbers ever. In April 2020, 64% of the aircrafts in the global fleet of the air transport industry were towed to storage as a result of the decline in passenger traffic (ATAG, 2020). One of the reasons for the huge decline in sectoral figures is that air transport enables the efficient spread of COVID-19 contagion. Air transport companies have been forced to cut the number of their current employees as a result of the huge decline in passenger and flight traffic statistics in the aviation industry. Prior to COVID-19, the air transport industry was employing 87.7 million people in the world under normal circumstances. Because of COVID-19, that number dropped to 41.7 million (ATAG, 2020). The negative difference of 46 million is the number of unemployed workers in the air transport industry globally. When the employment statistics of the air transport industry in the European Union (EU) are examined, it becomes clear that young people are more adversely affected by this unfavorable situation and the 15-39 age group experienced the highest rate of layoffs during the COVID-19 period. Despite the fact that various businesses implemented many different applications, such as wage subsidies, job protection, employment retention subsidies, and tax incentives, were implemented for employee retention in many countries during the COVID-19 period (Eichhorst et al., 2022), due to significant damage to the air transport industry, layoffs could not be avoided.

The air transport industry is expected to recover between 2023 - 2024 in the post-COVID-19 period by recapturing the numbers of 2019 (ATAG, 2020; Gudmundsson et al., 2021). When the statistics of passengers carried in 2022 are examined, it is predicted that the air transport industry will recover up to 75% by the end of the year 2022 (ICAO, 2022). Considering the employment figures of the air transport industry decreased by 52.5% during the COVID-19 period, 46 million employees will be required again globally when the industry returns back to the circumstances of 2019 in the post-COVID-19 period (ATAG, 2020). As a result, during the recovery period, air transport companies will need to make an effort to hire qualified employees in a short time. The most important factor in overcoming this situation and gaining a competitive advantage over rivals will be to become preferred by successful potential employees. Employer attractiveness practices are, therefore, one of the most useful instruments which are available to organizations in this process. In order to successfully manage employee acquisition and retention activities in the post-COVID-19 period, air transport companies should accurately understand the expectations of their potential employees and find efficient ways to attract them in today's competitive job market. Instruments and measurements for assessing potential employees as well as techniques for building and maintaining a strong employer brand will undoubtedly become more significant during this period. From the perspective of potential employees in the air transport industry, this chapter aims to adapt and validate the employer attractiveness scale, which will provide several advantages especially in the HR marketing practices of companies.

METHODOLOGY

Participants and Procedure

The population of the research consists of junior and senior students enrolled in aviation management bachelor's degree programs and second-year students enrolled in civil air transport management associate degree programs at universities in Turkey, which represents an emerging air transport market. All of the participants are future employees of the air transport industry. University students are appropriate participants for this research because they are future job seekers and potential employees for companies (Sivertzen et al., 2013). A web-based survey was distributed to students enrolled in 4-year aviation management and 2-year civil air transport management programs at universities in Turkey between May 2021 and July 2021. The participants of this research were chosen using a non-probability convenience sampling technique. The aim was to obtain a homogeneous sample by including only students from universities that train human resources for the air transport industry. Online participation invitations were sent to the students of 33 universities via e-mail, and a total of 4000 students were invited. 658 valid responses were collected during the 3-month period during which the questionnaire was accessible, with a response rate of 16.5%. The participants were divided into 3 sub-samples according to their submission order. In this context, the first 35 participants of the survey constituted the pilot sample, the following 224 participants of the survey constituted the Sample_1 (S1), and the last 399 participants constituted the Sample_2 (S2). Responses of the S1 were used in the exploratory factor analysis (EFA), and responses of the S2 were used in the confirmatory factor analysis (CFA). The pilot sample was obtained only from the students of Eskisehir Technical University. Other samples included responses from students from 33 out of 77 universities that offer bachelor's and associate degree programs in air transport, which represents 42.86% of all universities that train human resources for the air transport industry in these programs in Turkey. In terms of gender distribution, 47.3% of S1 were female, and 52.7% were male, 45.6% of S2 were female, and 54.4% were male. According to the distribution of educational departments, 43.8% of S1 were students of associate degree programs, while 56.2% of S1 were bachelor's students; 34.3%

of S2 were students of associate degree programs, while 65.7% of S2 were bachelor's students. Finally, 39.3% of the S1 and 40.1% of the S2 had completed an internship program in an air transport company.

In the research, the cross-cultural scale adaptation process suggested by the Gjersing et al. (2010) was considered. Because of that different business-domains or industries have different organizational culture elements (Chatman & Jehn, 1994), adapting a scale into a specific industry should be considered a cross-cultural adaptation, and it requires an approach in this direction. In the first step of the procedure, the original EmpAt scale was translated into Turkish. A committee was formed to lead the process of translating the scale from the original English language to the target language and evaluating its linguistic validity. The committee was formed with the participation of four experts in the field of air transportation and five experts in the field of English education. To ensure the accuracy of the translation, a back-translation approach was used in translating the scale (Douglas & Craig, 2007). In addition, an expert committee consisting of four human resource management specialists in the aviation management discipline evaluated whether the items of the EmpAt scale used in the research had content validity appropriate to the purpose of the research (Berthon et al., 2005). Thus, the content of the items was examined carefully, and their content validity was established in terms of assessing the concept of employer attractiveness in the field of air transportation. After the cross-cultural scale adaptation procedure, the experts concluded that it was not necessary to add any new items to the scale because the existing item pool was sufficient to measure the concept of employer attractiveness in the air transport industry.

Previous cross-cultural studies have recommended retesting the measurement structure and distribution of items across dimensions when applying the scale to a new sample in a different culture or domain (Gjersing et al., 2010). In some studies where the EmpAt scale was used, the EFA was conducted within this procedure (Shrivastava & Shukla, 2021; Sivertzen et al., 2013; Tanwar & Prasad, 2017). The aforementioned studies used the EFA to determine whether they needed to add new items to the original scale (Shrivastava & Shukla, 2021; Tanwar & Prasad, 2017) or create a new measurement structure by protecting the original scale items (Sivertzen et al., 2013). Thus, it is possible to first ensure cultural validity via EFA, which also addresses the cultural elements peculiar to a business-domain, and then to establish structural validity using the CFA. Even if the translation process is well managed, there is always the possibility of syntax, ambiguity, or other problems, and the length of the items is language-dependent (Steenkamp & Maydeu-Olivares, 2021). Therefore, using the EFA in the initial phase allows researchers to determine the optimal structure for cross-cultural scale adaptation. As this chapter aims to present the employer attractiveness scale peculiar to the air transport industry, primarily using the EFA option was chosen. Considering the significant concerns associated with the process of cross-cultural adaptation, the pilot test, the EFA, and the CFA were conducted and also back-translation procedures were followed. After analyses the final instrument (AirEmpAt) was presented.

Finally, considering the implementation time of the research, it is obvious that this chapter has special importance in terms of interpreting the effects of COVID-19 on human resources in the air transport industry. These dates coincide with a time when the devastating effects of COVID-19 began to wane for the air transport industry, but the industry was still not able to recover. This period could also be defined as the period when resilience and recovery activities started to gain importance in the air transport industry. When approaching the implementation period of this research in terms of human resources, it was a period in which layoffs were still common because the passenger traffic statistics of the industry were not yet at a sufficient level. Therefore, the output of this section, the AirEmpAt scale, reflects the perceptions of potential employees of the air transport industry for the COVID-19 period and beyond.

Instruments

The 25-item EmpAt scale developed by Berthon et al. (2005) was adapted and validated to measure the employer attractiveness of potential air transport employees. The EmpAt scale includes five different measurements: Interest Value, Social Value, Economic Value, Development Value, and Application Value. These five value dimensions of employer attractiveness were derived from the functional (developmental and/or useful activities), economic (material or monetary rewards), and psychological (feelings such as belonging, direction, and purpose) benefits that conventional brand offerings provide to consumers (Ambler & Barrow, 1996). The benefits that constitute the value that a product/service offers to the consumers are also the three major aspects from which the EmpAt scale obtains its measures. The EmpAt scale assesses the perceived importance of five value dimensions of employer attractiveness from the perspective of potential employees using a 7-point Likert-type scale. There is a slight difference in reliability between 7-point and 5-point Likert-type scales, and it has been observed that quality and validity decrease as the number of response categories increases (Revilla et al., 2014). The literature suggests that a 5-point Likert scale is less confusing, and it increases the response rate (Bouranta et al., 2009). As the number of categories increases, additional information about the topic of interest is needed. Consequently, the 5-point Likert scale was used for the items of the EmpAt scale in this research because the participants were university students, and the 5-response category is typical for studies on students (1=Not at all important, 3=Moderately important, 5=Extremely important).

As a part of the adaptation and validation process, the relationship between employer attractiveness and organizational familiarity was examined to assess the nomological validity of the EmpAt scale. There was evidence in the literature that these two structures have a positive correlation (Saini et al., 2014, 2015). Accordingly, Likert scale statements measuring familiarity of the employer were also included in the survey (e.g., "To say that I am familiar with this company, I should be aware of the economic opportunities that this company offers to its employees.").

Data Analysis and Tools

The following analysis processes were used to adapt the AirEmpAt scale in line with research purposes: Evaluating the initial model fit, exploring the dimensions (factors) through the EFA, assessing the construct reliability, and conducting the CFA to assess structural validity. EFA is a statistical method for determining the relationship between variables in a dataset. A factor structure is provided in this analysis by a grouping of variables based on strong correlations (Gaskin, 2021b). Generally, an EFA is conducted prior to structural equation modeling. After the measurement model is determined, it is confirmed by conducting CFA. Evaluating the model fit, validity and reliability steps are included in the CFA. The model fit of our proposed model refers to how well it accounts for the correlations between variables in the dataset (Gaskin, 2021a). When adapting and validating a scale, EFA and CFA analyses are generally accepted analyses. In this chapter, the analyses were performed using SPSS 26 and AMOS 24.

Results

The EFA analysis for cross-cultural adaptation of the EmpAt scale was performed using the principal axis factoring method and Promax rotation. CFA analysis to assess structural validity was conducted using the maximum likelihood method for structural validity assessment. Three different data sets were collected as the pilot sample (n=35), S1 sample (n=224), and S2 (n=399) sample. The S1 was used to investigate the factorial structure, while the S2 was used to confirm the measurement model. The minimum sample size depends on the number of factors, the number of variables per factor, and the degree of communality in performing the EFA (Mundfrom et al., 2005). The sample S1 corresponded to the appropriate sample size for EFA. On the other hand, the sample S2 used in the CFA was almost 400 and adequate for conducting CFA (MacCallum et al., 1999). $\chi2/df$, RMSEA, TLI, CFI, and SRMR values were recommended to be reported while evaluating the goodness of fit of the CFA models (Dogan, 2018). Cut-off values for model fit indices accepted as $\chi2/df < 3$; RMSEA < .08; TLI > .90; CFI > .90; SRMR < .08 (Gürbüz, 2019; Hu & Bentler, 1999).

By conducting a pilot test, the intelligibility of the items was examined for the proper translation of the EmpAt scale into the target language. It was concluded that the intended meanings of the items were clear. Cronbach's α of the items was $\alpha = 0.953$, and it was seen that internal consistency was acceptable, which is above 0.70 (Nunnally, 1978). Internal consistency of the scale was found to be quite good, and progressive analyses were performed.

Exploring the Factorial Structure of the AirEmpAt

While adapting the scale of perceived employer attractiveness by potential employees of air transport companies, the sample S1 was used first. Then, the model fit of the original structure of the EmpAt was evaluated by conducting the CFA to provide findings of the initial factorial structure. First, the CFA conducted using the original five-dimensional structure of the EmpAt was not adequate and yielded a bad model fit ($\chi2/df = 2.437$, $p < 0.001$, RMSEA = 0.08, TLI = 0.873, CFI = 0.888, SRMR = 0.0547). This issue has been identified due to the cross-loadings of four items and low factor loadings. Rather than using a large number of adjusting modification indices to improve model fit, the EFA was conducted to explore the distribution of items in the target business-domain, as previously indicated. The EFA results demonstrated that the sample size was adequate (KMO = 0.943), and Bartlett's test of sphericity was significant ($p<0.001$). All communality values were above 0.4. Results showed that 22 items of employer attractiveness were distributed within three factors, with an explained variance of 53.77%. The minimum threshold of 0.3 is accepted for factor loadings, which is an acceptable cutoff considering the sample size (Yaşlıoğlu, 2017). Internal consistencies of factors were adequate, as seen in Table 2. Three items, which are "gaining career-enhancing experience", "recognition/appreciation from management", and "the employees have a good relationship with their superiors", were discarded because of cross-loadings. 22-item and 3-factor employer attractiveness scale was renamed the Air Transport Employer Attractiveness Scale (AirEmpAt), which focuses on measuring employer attractiveness only in the air transport industry. Three factors were named functional value, psychological value, and economic value, respectively. As can be observed, the measurements of the suggested AirEmpAt structure were identical to the dimensions proposed by Ambler and Barrow (1996) for employer attractiveness.

Confirming the AirEmpAt Measurement Model

According to the EFA results, the AirEmpAt scale consists of three measurements as functional, psychological, and economic values, respectively. The CFA was conducted by using maximum likelihood estimation using sample S2 to reveal the psychometric properties of the first-order measurement model. CFA results demonstrated that the measurement model yielded good model fit ($\chi2/df = 2.705$, $p < 0.001$,

RMSEA = 0.065, TLI = 0.921, CFI = 0.930, SRMR = 0.0455). The CFA findings for the AirEmpAt model are given in Figure 2. When the relations between the variables in the model were examined, all regression weights were significant ($p<0.001$), standardized regression weights > 0.60, and correlations among measurements were significant ($p<0.001$) (Gürbüz, 2019; Hair, Black, et al., 2019). There were only two standardized regression weights that fell below 0.6 (FV7=0.58; EV1=0.59). These were approved, though, because the differences were just minor. Also, there was no multicollinearity problem among constructs ($r<0.85$) (T. A. Brown, 2015).

Table 2. Results of the EFA

Factor	Item Label	Items*	Factor Loadings	Cronbach's α
Functional Value (FV)	FV1	The organization produces innovative products and services	0,882	0.904
	FV2	Innovative employer – novel work practices/forward-thinking	0,834	
	FV3	The organization produces high-quality products and services	0,737	
	FV4	Opportunity to teach others what you have learned	0,716	
	FV5	Opportunity to apply what was learned at a tertiary institution	0,635	
	FV6	Hands-on inter-departmental experience	0,601	
	FV7	The organization is customer-orientated	0,556	
	FV8	Humanitarian organization – gives back to society	0,492	
	FV9	The organization both values and makes use of your creativity	0,469	
Psychological Value (PV)	PV1	Happy work environment	0,949	0.913
	PV2	Feeling good about yourself as a result of working for this particular organization	0,841	
	PV3	Acceptance and belonging	0,697	
	PV4	Good relationship with colleagues	0,655	
	PV5	A fun working environment	0,648	
	PV6	The organization has an exciting environment	0,642	
	PV7	Feeling more self-confident as a result of working for this particular organization	0,639	
	PV8	Supportive and encouraging colleagues	0,475	
Economic Value (EV)	EV1	An above-average basic salary	0,683	0.779
	EV2	An attractive overall compensation package	0,668	
	EV3	Good promotion opportunities within the organization	0,541	
	EV4	A springboard for future employment	0,416	
	EV5	Job security within the organization	0,345	

* Items were derived from Berthon, Ewing and Hah's EmpAt scale (Berthon et al., 2005)
Source: Authors' own elaboration

Figure 2. The AirEmpAt measurement model
Source: Authors' own elaboration

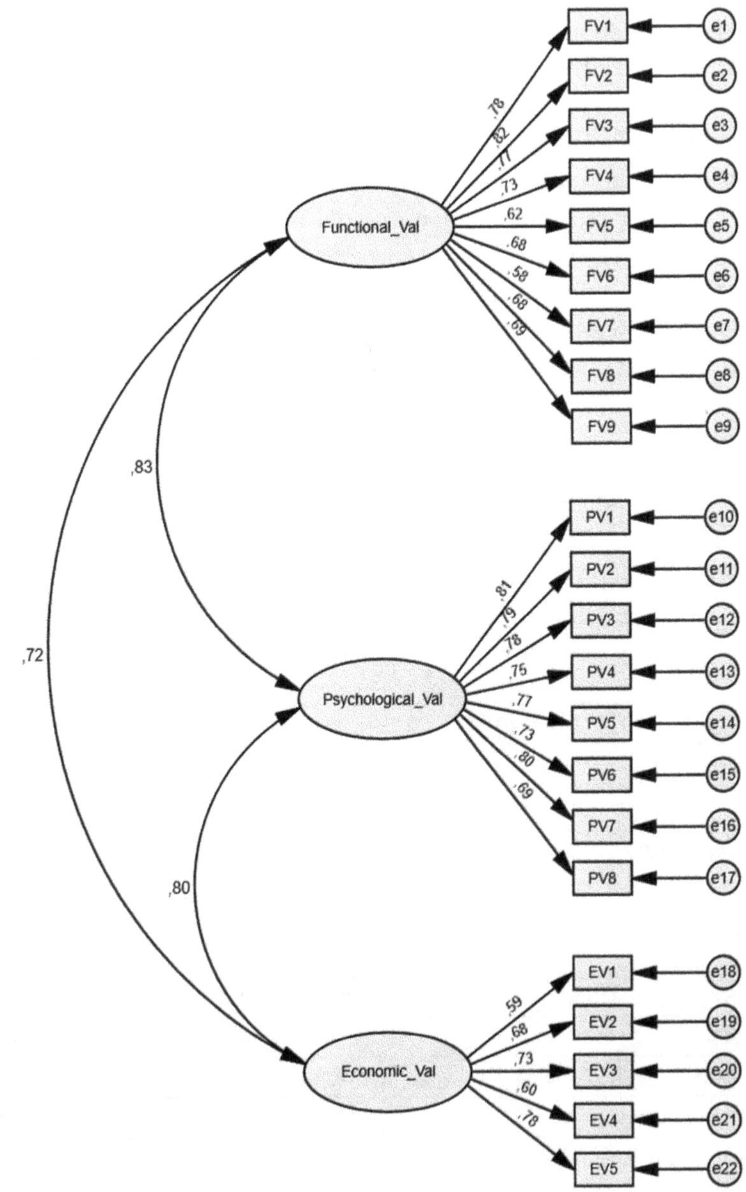

Then, the convergent and discriminant validities of the model were assessed. The composite reliability (CR), average extracted variance (AVE), and heterotrait-monotrait ratio analysis of correlations (HTMT) values of the measurement model are presented in Table 3. To ensure reliability and convergent validity of the measurement model, CR should be higher than 0.7, and AVE should be higher than 0.5 (Hair, Black, et al., 2019). However, when CR is higher than 0.6, AVE values higher than 0.4 can be accepted, and the convergent validity of the construct can still be assessed as appropriate (Fornell

& Larcker, 1981). Therefore, the convergent validity of the construct economic value (AVE = 0.463) was considered adequate. In recent years, researchers have widely used HTMT analysis to determine discriminant validity in both covariance-based and variance-based structural equation modeling studies (Hosen et al., 2021). If the correlations between the construct indicators in the HTMT matrix are less than 0.85, it indicates that the structures have different characteristics and provides discriminant validity (Gaskin, 2021c; Hair, Risher, et al., 2019).

In assessing the reliability and validity of the measurement model, it was found that all constructs were reliable, items converged within structures, and structures were separated from each other. In addition, the goodness of fit of the first-order measurement model and the alternative models were compared using the procedure recommended by Gilal, Zhang and Gilal (2018) to further test discriminant validity. The comparisons of the alternative models are given in Table 4.

As indicated in Table 4, the three-factor AirEmpAt measurement model fitted data better than alternatives, and the inference was that constructs in the proposed model significantly differed compared to other alternative models.

Table 3. Reliability, convergent validity, and discriminant validity

Construct	CR	AVE	HTMT Results		
			Functional Value	Psychological Value	Economic Value
Functional Value	0.900	0.504			
Psychological Value	0.919	0.588	0.846		
Economic Value	0.810	0.463	0.729	0.780	

Source: Authors' own elaboration

Table 4. Alternative model comparisons

Model	Factor Loaded	χ2/df	RMSEA	TLI	CFI	SRMR
1	3 factors: FV, PV, EV	2.705	0.065	0.921	0.930	0.0455
2	2 factors: FV/PV*, EV	3.893	0.085	0.866	0.880	0.0539
3	2 factors: FV, PV/EV*	3.356	0.077	0.891	0.902	0.0507
4	2 factors: FV/EV*, PV	3.702	0.082	0.875	0.888	0.0547
5	1 factor: FV/PV/EV* (All combined)	4.572	0.095	0.835	0.851	0.0592

* Slash symbol refers to combining related constructs.
Source: Authors' own elaboration

Nomological Validity

The literature emphasizes that there is a positive relationship between organizational familiarity and employer attractiveness. In order to test nomological validity, five items were included in the survey, based on the literature (Berthon et al., 2005; Cable & Turban, 2001; Saini et al., 2014, 2015; Turban, 2001). These items measured the organizational familiarity of potential employees through interest, social, economic, development, and application opportunities of employers. The arithmetic mean of these five items was used to create a single variable for organizational familiarity, which was then used in the correlation analysis. Results of the correlation analysis indicated that functional value ($r = 0.359$, $p < 0.01$), psychological value ($r = 0.210$, $p < 0.01$), and economic value ($r = 0.189$, $p < 0.01$) were all positively correlated with organizational familiarity. These results support the nomological validity of the AirEmpAt scale through significant correlations with organizational familiarity.

SOLUTIONS AND RECOMMENDATIONS

In this chapter, the employer attractiveness scale for the air transport industry, namely the AirEmpAt, was presented by adapting the EmpAt scale to the air transport industry domain. The analyses showed that the three-factor model met the requirements for validity and reliability. Also, in the context of nomological validity, it can be expressed that the proposed instrument is adequate to measure the expected values related to employer attractiveness in the air transport industry. The EFA revealed that the original items of the EmpAt scale were distributed among three factors. Therefore, the revealed three-factor structure was accepted and tested by the CFA for adapting the scale to a different business-domain. Moreover, alternative model comparisons showed that the proposed three-factor AirEmpAt model fitted the data best ($\chi 2/df = 2.705$, $p < 0.001$, RMSEA = 0.065, TLI = 0.921, CFI = 0.930, SRMR = 0.0455).

Based on scale adjustments and validation analyses in the final structure, three main aspects that determine the employer attractiveness of air transport companies were identified as functional, psychological, and economic values, respectively. Three primary components of the employer brand concept proposed by Ambler and Barrow (1996) are identical to the three primary values examined in this chapter. The operational definitions of the AirEmpAt dimensions were made from the perspective of value expectations of potential employees and were presented in the following sentences. Functional value is the sum of useful features of jobs offered by the employer that attracts potential employees, such as innovation, gaining work experience, and applying knowledge. The functional value dimension of the AirEmpAt is also supported by several previous studies in the literature (Agrawal & Swaroop, 2009; Berthon et al., 2005). Psychological value is the sum of the social and emotional aspects of the work environment offered by the employer that attracts potential employees, such as relationships with colleagues, acceptance, affiliation, and feelings. The fact that many studies in the literature, which addressed social and psychological value concerning employer attractiveness, support the psychological value dimension of the AirEmpAt (Berthon et al., 2005; Roy, 2008; Schlager et al., 2011). Economic value is the sum of monetary aspects and future career benefits of a job offered by an employer that attracts potential employees, such as salary, compensation, job security, and future employment benefits. Almost all of the previous research on employer attractiveness has focused on the economic value component. So, the economic value dimension of the AirEmpAt is also supported by earlier studies.

Considering the composition of the products produced by companies in the air transport industry, it is clear that the level of services is higher than that of tangible goods. Consequently, the characteristics and skills of employees are quite effective in ensuring that service production meets the quality expected by customers (Brochado et al., 2019). To be preferred by the employees who take air transport companies a step forward, companies need to increase their employer attractiveness and improve their employer brands. The AirEmpAt scale is expected to provide more accurate results in measuring the expectations of potential employees in the air transport industry due to its particular adaptation to this industry.

The AirEmpAt scale, which was adapted to the air transport industry in this chapter, provides valuable insight into how air transport companies will attract potential employees. Particularly given the critical role HR management plays in the air transport industry (Harvey & Turnbull, 2020), employer attractiveness studies will enable HR marketing policies to be more coherent and precise with the HR strategies of air transport companies. Today, it is especially important for HR management and HR marketing to collaborate. The border between HR management and HR marketing is gradually fading, and both have mutual objectives. The AirEmpAt will enable air transport companies to accomplish HR management and HR marketing activities in harmony and to find out how potential employees perceive themselves in the environment of the labor market. In addition, these companies will find out what they have to offer for future employees to recruit them.

As mentioned at the beginning of this chapter, the air transportation industry has started to recover rapidly with the end of the COVID-19 pandemic and its applications, which have adversely affected the air transportation industry, such as closed borders and social distancing in the airliner cabin. When passenger traffic statistics (ICAO, 2022) and scientific studies (Gudmundsson et al., 2021) are examined, it is predicted that the air transport industry will recover to a large extent by the end of 2022. This recovery process will lead to an increase in the human resource needs of the air transport industry. In this period, hiring the best employees will provide a competitive advantage for the air transport companies. Therefore, air transport companies need to understand the expectations of their potential employees in the most accurate way and prepare their value proposition packages that is the most valuable for their employees. In the light of the results discussed in this chapter and the AirEmpAt scale presented, the recommendations developed for air transport companies are summarized in the following items:

- Psychological value offers the greatest contribution to the perception of employer attractiveness of potential employees. In this context, to build a strong employer brand, it is recommended that air transport companies should provide their employees with a happier work environment, implement practices that will improve employee belonging, and create a supportive and encouraging work environment.
- In the process of building a strong employer brand, air transport companies should increase the functional value offering within the scope of employer attractiveness perception of potential employees. For this, it is recommended that employers should develop applications to increase the level of innovation in their business, implement a structure that contributes to the development of their employees on their own expertise, and develop a flexible work environment where employees can use their creativity in the business processes.
- In order to increase the economic value in improving the perception of employer attractiveness of potential employees, air transport companies should meet employee expectations on salaries and compensations, be transparent in employee promotion mechanism and offer job security to their employees. It also can be seen that layoffs in the air transport industry during the period of COVID-19 increased the sensitivity of potential employees about job security.
- In the post-COVID-19 period, air transport companies will compete with each other to bring the best employees to their organizations to fill the new staff requirements that will arise during the recovery process of the air transport industry. In this process, it is recommended that air transport companies should support their HR management processes with successful HR marketing strategies. So, it will be possible for them to establish a stronger employee team in the post-COVID-19 period. In this direction, applications such as determining the expectation levels of values that will

attract potential employees, determining the level of company capabilities to respond to expectations of potential employees, and the management of issues to be developed to meet employee expectations are areas where the use of the AirEmpat scale is highlighted.

- Finally, it is suggested that the use of the AirEmpAt scale, which is adapted directly to the expectations of future employees of the air transport industry, as an instrument to manage the employer brand will be beneficial for air transport companies to employ effective and qualified human resources in the recovery process. Lastly, it is proposed that using the AirEmpAt scale as a tool to manage the employer brand will help air transport companies hire qualified and efficient human resources during the recovery process. This is because the AirEmpAt scale is directly tailored to the expectations of future workers in the air transport industry.

In this chapter, a reliable and valid scale for measuring the employer attractiveness perceived by potential employees was adopted for the air transport industry. The proposed AirEmpAt scale can be both useful to determine the expectation level of a specific potential employee group or to measure the level of fulfillment of current employees. Also, an effective tool has been presented to develop an employer brand strategy for companies such as airlines, airports, and ground handling services in the growing air transport industry, which is recovering itself expeditiously in the post-COVID era. Finally, this chapter contributes to the existing literature as it is the first to examine employer attractiveness exclusively in the air transport industry.

FUTURE RESEARCH DIRECTIONS

In this chapter, the AirEmpAt scale for measuring employer attractiveness was presented in line with the responses of future employees of the air transport industry. When the age range of the participants (between 15-29) is examined, it is clear that the sample included members of Generation Y and Generation Z. In the context of generational cohort theory, generations have been assumed to share common beliefs, values, and behaviors (Mannheim, 1952). Several studies have shown that the concept of employer attractiveness indicates generational changes (Forgas et al., 2010; Ng et al., 2010). Also, these two generations have significant differences in terms of their behavior in their work lives and career expectations (Maloni et al., 2019). In the air transport industry, as in other industries, Generation Y currently constitutes a large segment of the professional workforce. Generation Z will replace the Generation Y in the workforce market, but there is no study on the employer attractiveness that specifically addresses Generation Z. Since this chapter focuses on the understanding of the expectations of the employees without generational differentiating, the AirEmpAt scale was presented for the whole air transport industry. However, determining the expectations of different generations in the air transport industry will also offer advantages in terms of HR management and marketing. It is recommended that future studies may examine the generational differences on the employer attractiveness and employee expectations in the air transport industry.

Since the AirEmpAt scale was only applied to a sample of potential employees of the air transport industry, it is more appropriate to use it in the future studies to measure expected employer attractiveness (Schlager et al., 2011). Furthermore, it is recommended that future studies may test construct validity toward perceived employer attractiveness by including current employees of the air transport industry.

CONCLUSION

With emergence of the employer brand concept (Ambler & Barrow, 1996) and the growing importance of internal marketing (George, 1990), the field of HR marketing, which includes the activities of retaining, motivating and attracting the best employees, has developed and became prominent (Urbancová & Hudáková, 2017). Nowadays, it is seen that HR marketing applications are gaining more and more importance for businesses to gain competitive advantage. Although there are slight differences according to the business-domains of the enterprises, the share of the employees in the success of the enterprises is undeniable. For this reason, businesses seek an answer to the question of how they can offer more value to their employees, which is one of their most important stakeholders. When the characteristics of the air transport industry are considered, it is obvious that the importance of employees is one-fold higher. Due to the main characteristics of the service industry, employees are the most important production source of air transport companies and the most important part of the service delivery process (Alzaydi et al., 2018). For this reason, air transport companies should benefit from HR marketing applications to attract, retain and motivate the best employees. Herein, starting with attracting the best potential employees could be considered primary step in HR marketing. Because businesses should know both themselves and their potential employees accurately in order to implement a successful employer brand strategy. Only in this way the expectations of potential employees could be determined accurately and the value proposition of the company could be developed successfully.

In this chapter, the three-factor AirEmpAt scale was presented for the air transport industry. As a result of the analyses conducted, proposed AirEmpAt scale was ensured to be valid and reliable for the air transport industry. In conclusion, three dimensions of the employer attractiveness were determined as functional, psychological, and economic value, respectively. This dimensions identify the main subjects that are effective to attract potential employers of air transport companies. Therefore, using the AirEmpAt instrument as a tool in HR management and HR marketing will provide an opportunity to develop value packages to attract and hire qualified employees. In this way, air transport companies can acquire a competitive edge, increase employee satisfaction, and build stronger employer brand concurrently.

The main limitation of this scale adaptation research is that it focuses exclusively on the air transport industry. Therefore, the AirEmpAt scale can be used within this limitation. Like other scale adaptation studies, the research contains cultural limitations based on business-domain (Chatman & Jehn, 1994; Gjersing et al., 2010). For another industry or business-domain, the AirEmpAt scale may need to be revalidated.

The data of the research were collected during COVID-19, and the answers of the participants were given by taking into account the negativities experienced by the air transport industry during the COVID-19 period. In this respect, the research highlights the expectations of potential employees from the air transport industry regarding the COVID-19 period and beyond. In line with the findings obtained in this section, we can express the following issues about potential employees in the recovery process of air transportation industry:

- According to the variances of employer attractiveness dimensions, psychological value comes first among the expectations of potential employees from air transport companies. The psychological value dimension includes items such as acceptance and belonging, happy work environment, feeling more self-confident. During the COVID-19 era, adversities such as pay cuts and layoffs faced by air transport industry employees are estimated to increase the importance of psychological value offerings like acceptance and belonging.

- In the ranking of employer attractiveness dimensions according to their variances, the second dimension that explains the expectations of potential employees the most is functional value, and the third dimension is economic value. Potential employees are more interested in the developmental and applicational utilities of air transport companies than in materials and monetary rewards.

Finally, in the post-COVID-19 period, it is predicted that AirEmpAt scale will provide an advantage to companies in attracting the most effective and productive human resources during the recovery of the air transport industry. This study will also help air transport companies to build a stronger employer brand in line with the expectations of their potential employees. As a result, this chapter is expected to provide an administrative application advantage for the air transportation companies in order to be preferred by the best employees.

ETHICAL CONSIDERATIONS

All authors declare that they have no conflict of interest. This research was approved by the decision of "The Social and Human Sciences Scientific Research and Publication Ethics Committee of Eskisehir Technical University", on 06/30/2020 with number 4/10. Permission to use the Employer Attractiveness Scale (Berthon et al., 2005) was obtained from Pierre Berthon via email. Informed consents were given by each participant of this research.

ACKNOWLEDGMENT

This research was supported by the Eskisehir Technical University [grant number 20ADP168].

REFERENCES

Aaker, D. A. (1996). Measuring Brand Equity Across Products and Markets. *California Management Review, 38*(3), 102–120. Advance online publication. doi:10.2307/41165845

Agrawal, R. K., & Swaroop, P. (2009). Effect of Employer Brand Image on Application Intentions of B-School Undergraduates. *Vision: The Journal of Business Perspective, 13*(3), 41–49. doi:10.1177/097226290901300304

Alzaydi, Z. M., Al-Hajla, A., Nguyen, B., & Jayawardhena, C. (2018). A review of service quality and service delivery. *Business Process Management Journal, 24*(1), 295–328. doi:10.1108/BPMJ-09-2016-0185

Ambler, T., & Barrow, S. (1996). The employer brand. *Journal of Brand Management, 4*(3), 185–206. doi:10.1057/bm.1996.42

Arachchige, B. J. H., & Robertson, A. (2011). Business Student Perceptions of a Preferred Employer: A Study Identifying Determinants of Employer Branding. *The IUP Journal of Brand Management, 8*(3), 25–46.

ATAG. (2020). *Aviation: Benefits Beyond Borders*. Air Transport Action Group.

Backhaus, K., & Tikoo, S. (2004). Conceptualizing and researching employer branding. *Career Development International, 9*(5), 501–517. doi:10.1108/13620430410550754

Benraïss-Noailles, L., & Viot, C. (2021). Employer brand equity effects on employees well-being and loyalty. *Journal of Business Research, 126*, 605–613. doi:10.1016/j.jbusres.2020.02.002

Berry, L. L., Hensel, J. S., & Burke, M. C. (1976). Improving retailer capability for effective consumerism response. *Journal of Retailing, 52*(3), 3–14.

Berthon, P., Ewing, M., & Hah, L. L. (2005). Captivating company: Dimensions of attractiveness in employer branding. *International Journal of Advertising, 24*(2), 151–172. doi:10.1080/02650487.2005.11072912

Biswas, M. K., & Suar, D. (2016). Antecedents and Consequences of Employer Branding. *Journal of Business Ethics, 136*(1), 57–72. doi:10.100710551-014-2502-3

Bouranta, N., Chitiris, L., & Paravantis, J. (2009). The relationship between internal and external service quality. *International Journal of Contemporary Hospitality Management, 21*(3), 275–293. doi:10.1108/09596110910948297

Brochado, A., Rita, P., Oliveira, C., & Oliveira, F. (2019). Airline passengers' perceptions of service quality: Themes in online reviews. *International Journal of Contemporary Hospitality Management, 31*(2), 855–873. doi:10.1108/IJCHM-09-2017-0572

Brown, S. P., & Lam, S. K. (2008). A Meta-Analysis of Relationships Linking Employee Satisfaction to Customer Responses. *Journal of Retailing, 84*(3), 243–255. doi:10.1016/j.jretai.2008.06.001

Brown, T. A. (2015). *Confirmatory factor analysis for applied research*. The Guilford Press.

Cable, D. M., & Turban, D. B. (2001). Establishing the dimensions, sources and value of job seekers' employer knowledge during recruitment. *Research in Personnel and Human Resources Management, 20*, 115–163. doi:10.1016/S0742-7301(01)20002-4

CDC. (2022). *CDC Museum COVID-19 Timeline*. David J. Sencer CDC Museum. https://www.cdc.gov/museum/timeline/covid19.html# Charumilind,

Charumilind, S., Craven, M., Lamb, J., Sabow, A., Singhal, S., & Wilson, M. (2022). *When will the COVID-19 pandemic end?* Retrieved November 20, 2022, from https://www.mckinsey.com/industries/healthcare-systems-and-services/our-insights/when-will-the-covid-19-pandemic-end

Chatman, J. A., & Jehn, K. A. (1994). Assessing the Relationship between Industry Characteristics and Organizational Culture: How Different Can You Be? *Academy of Management Journal*, *37*(3), 522–553. doi:10.2307/256699

Chen, C.-F. (2006). Job satisfaction, organizational commitment, and flight attendants' turnover intentions: A note. *Journal of Air Transport Management*, *12*(5), 274–276. doi:10.1016/j.jairtraman.2006.05.001

Dabirian, A., Berthon, P., & Kietzmann, J. (2019). Enticing the IT crowd: Employer branding in the information economy. *Journal of Business and Industrial Marketing*, *34*(7), 1403–1409. doi:10.1108/JBIM-11-2018-0333

Dabirian, A., Paschen, J., & Kietzmann, J. (2019). Employer Branding: Understanding Employer Attractiveness of IT Companies. *IT Professional*, *21*(1), 82–89. doi:10.1109/MITP.2018.2876980

Devanna, M. A., Fombrun, C., & Tichy, N. (1981). Human resources management: A strategic perspective. *Organizational Dynamics*, *9*(3), 51–67. doi:10.1016/0090-2616(81)90038-3

Deveci, M., Çiftçi, M. E., Akyurt, İ. Z., & Gonzalez, E. D. R. S. (2022). Impact of COVID-19 pandemic on the Turkish civil aviation industry. *Sustainable Operations and Computers*, *3*, 93–102. doi:10.1016/j.susoc.2021.11.002

Dogan, V. (2018). The Use and Misuse of Structural Equation Modeling in Turkish Marketing Academia: Problems and Solutions. *The Journal of American Science*, *16*(32), 201–230.

Douglas, S. P., & Craig, C. S. (2007). Collaborative and iterative translation: An alternative approach to back translation. *Journal of International Marketing*, *15*(1), 30–43. doi:10.1509/jimk.15.1.030

Eichhorst, W., Marx, P., Rinne, U., & Brunner, J. (2022). *Job retention schemes during COVID-19: A review of policy responses*. International Labour Organization publications.

Farooq, M. S., Salam, M., Fayolle, A., Jaafar, N., & Ayupp, K. (2018). Impact of service quality on customer satisfaction in Malaysia airlines: A PLS-SEM approach. *Journal of Air Transport Management*, *67*, 169–180. doi:10.1016/j.jairtraman.2017.12.008

Forgas, S., Moliner, M. A., Sánchez, J., & Palau, R. (2010). Antecedents of airline passenger loyalty: Low-cost versus traditional airlines. *Journal of Air Transport Management*, *16*(4), 229–233. doi:10.1016/j.jairtraman.2010.01.001

Fornell, C., & Larcker, D. F. (1981). Evaluating Structural Equation Models with Unobservable Variables and Measurement Error. *JMR, Journal of Marketing Research*, *18*(1), 39–50. doi:10.1177/002224378101800104

Frow, P., & Payne, A. (2011). A stakeholder perspective of the value proposition concept. *European Journal of Marketing*, *45*(1/2), 223–240. doi:10.1108/03090561111095676

Gaskin, J. (2021a). "CFA". *Gaskination's StatWiki*. http://statwiki.gaskination.com/

Gaskin, J. (2021b). "EFA". *Gaskination's StatWiki*. http://statwiki.gaskination.com/

Gaskin, J. (2021c). "Plugins". *Gaskination's StatWiki*. http://statwiki.gaskination.com

George, W. R. (1990). Internal marketing and organizational behavior: A partnership in developing customer-conscious employees at every level. *Journal of Business Research*, *20*(1), 63–70. doi:10.1016/0148-2963(90)90043-D

Gilal, N. G., Zhang, J., & Gilal, F. G. (2018). The four-factor model of product design: Scale development and validation. *Journal of Product and Brand Management*, *27*(6), 684–700. doi:10.1108/JPBM-11-2017-1659

Gjersing, L., Caplehorn, J. R., & Clausen, T. (2010). Cross-cultural adaptation of research instruments: Language, setting, time and statistical considerations. *BMC Medical Research Methodology*, *10*(1), 13. doi:10.1186/1471-2288-10-13 PMID:20144247

Gounaris, S. (2008). The notion of internal market orientation and employee job satisfaction: Some preliminary evidence. *Journal of Services Marketing*, *22*(1), 68–90. doi:10.1108/08876040810851978

Greene, W. E., Walls, G. D., & Schrest, L. J. (1994). Internal Marketing: The Key to External Marketing Success. *Journal of Services Marketing*, *8*(4), 5–13. doi:10.1108/08876049410070682

Gudmundsson, S. V., Cattaneo, M., & Redondi, R. (2021). Forecasting temporal world recovery in air transport markets in the presence of large economic shocks: The case of COVID-19. *Journal of Air Transport Management*, *91*, 102007. doi:10.1016/j.jairtraman.2020.102007 PMID:36568736

Gummesson, E. (2007). Exit services marketing - enter service marketing. *Journal of Customer Behaviour*, *6*(2), 113–141. doi:10.1362/147539207X223357

Gupta, B. (2011). A comparative study of organizational strategy and culture across industry. *Benchmarking*, *18*(4), 510–528. doi:10.1108/14635771111147614

Gürbüz, S. (2019). *Structural Equation Modelling with AMOS*. Seçkin Yayıncılık.

Hair, J. F., Black, W. C., Babin, B. J., & Anderson, R. E. (2019). *Multivariate Data Analysis* (8th ed.). Cengage Learning.

Hair, J. F., Risher, J. J., Sarstedt, M., & Ringle, C. M. (2019). When to use and how to report the results of PLS-SEM. *European Business Review*, *31*(1), 2–24. doi:10.1108/EBR-11-2018-0203

Harvey, G., & Turnbull, P. (2020). Ricardo flies Ryanair: Strategic human resource management and competitive advantage in a Single European Aviation Market. *Human Resource Management Journal*, *30*(4), 553–565. doi:10.1111/1748-8583.12315

Hosen, M., Ogbeibu, S., Giridharan, B., Cham, T.-H., Lim, W. M., & Paul, J. (2021). Individual motivation and social media influence on student knowledge sharing and learning performance: Evidence from an emerging economy. *Computers & Education*, *172*, 104262. doi:10.1016/j.compedu.2021.104262

Hu, L., & Bentler, P. M. (1999). Cutoff criteria for fit indexes in covariance structure analysis: Conventional criteria versus new alternatives. *Structural Equation Modeling*, *6*(1), 1–55. doi:10.1080/10705519909540118

ICAO. (2021). *Presentation of 2020 Air Transport Statistical Results*. ICAO. https://www.icao.int/annual-report-2020/Documents/ARC_2020_Air Transport Statistics_final_sched.pdf

ICAO. (2022). *2021 global air passenger totals show improvement from 2020, but still only half pre-pandemic levels*. ICAO. https://www.icao.int/Newsroom/Pages/2021-global-air-passenger-totals-show-improvement.aspx

Jiang, T., & Iles, P. (2011). Employer-brand equity, organizational attractiveness and talent management in the Zhejiang private sector, China. *Journal of Technology Management in China*, *6*(1), 97–110. doi:10.1108/17468771111105686

Kissel, P., & Büttgen, M. (2015). Using social media to communicate employer brand identity: The impact on corporate image and employer attractiveness. *Journal of Brand Management*, *22*(9), 755–777. doi:10.1057/bm.2015.42

Leekha Chhabra, N., & Sharma, S. (2014). Employer branding: Strategy for improving employer attractiveness. *The International Journal of Organizational Analysis*, *22*(1), 48–60. doi:10.1108/IJOA-09-2011-0513

Lloyd, S. (2002). Branding from the inside out. *Business Review Weekly - BRW, 24*(10), 64-66.

MacCallum, R. C., Widaman, K. F., Zhang, S., & Hong, S. (1999). Sample size in factor analysis. *Psychological Methods*, *4*(1), 84–99. doi:10.1037/1082-989X.4.1.84

Maloni, M., Hiatt, M. S., & Campbell, S. (2019). Understanding the work values of Gen Z business students. *International Journal of Management Education*, *17*(3), 100320. doi:10.1016/j.ijme.2019.100320

Mannheim, K. (1952). The Problem of Generations. In P. Kecskemeti (Ed.), *Essays on the Sociology of Knowledge* (pp. 276–320). Routledge and Kegan Paul.

Maxwell, R., & Knox, S. (2009). Motivating employees to "live the brand": A comparative case study of employer brand attractiveness within the firm. *Journal of Marketing Management*, *25*(9–10), 893–907. doi:10.1362/026725709X479282

Mundfrom, D. J., Shaw, D. G., & Ke, T. L. (2005). Minimum Sample Size Recommendations for Conducting Factor Analyses. *International Journal of Testing*, *5*(2), 159–168. doi:10.120715327574ijt0502_4

Ng, E. S. W., Schweitzer, L., & Lyons, S. T. (2010). New Generation, Great Expectations: A Field Study of the Millennial Generation. *Journal of Business and Psychology*, *25*(2), 281–292. doi:10.100710869-010-9159-4

Nižetić, S. (2020). Impact of coronavirus (COVID-19) pandemic on air transport mobility, energy, and environment: A case study. *International Journal of Energy Research*, *44*(13), 10953–10961. doi:10.1002/er.5706

Nunnally, J. (1978). *Psychometric Methods* (2nd ed.). McGraw-Hill.

Payne, A., Frow, P., & Eggert, A. (2017). The customer value proposition: Evolution, development, and application in marketing. *Journal of the Academy of Marketing Science, 45*(4), 467–489. doi:10.100711747-017-0523-z

Ramayah, T., Samat, N., & Lo, M. (2011). Market orientation, service quality and organizational performance in service organizations in Malaysia. *Asia-Pacific Journal of Business Administration, 3*(1), 8–27. doi:10.1108/17574321111116379

Reis, G. G., & Braga, B. M. (2016). Employer attractiveness from a generation perspective: Implications for employer branding. *Revista ADM, 51*(1), 103–116. doi:10.5700/rausp1226

Revilla, M. A., Saris, W. E., & Krosnick, J. A. (2014). Choosing the Number of Categories in Agree–Disagree Scales. *Sociological Methods & Research, 43*(1), 73–97. doi:10.1177/0049124113509605

Ritz, A., & Waldner, C. (2011). Competing for Future Leaders. *Review of Public Personnel Administration, 31*(3), 291–316. doi:10.1177/0734371X11408703

Roy, S. K. (2008). Identifying the dimensions of attractiveness of an employer brand in the Indian context. *South Asian Journal of Management, 15*(4), 110–130.

Saini, G. K., Gopal, A., & Kumari, N. (2015). Employer Brand and Job Application Decisions: Insights from the Best Employers. *Management and Labour Studies, 40*(1–2), 34–51. doi:10.1177/0258042X15601532

Saini, G. K., Rai, P., & Chaudhary, M. K. (2014). What do best employer surveys reveal about employer branding and intention to apply? *Journal of Brand Management, 21*(2), 95–111. doi:10.1057/bm.2013.10

Schlager, T., Bodderas, M., Maas, P., & Luc Cachelin, J. (2011). The influence of the employer brand on employee attitudes relevant for service branding: An empirical investigation. *Journal of Services Marketing, 25*(7), 497–508. doi:10.1108/08876041111173624

Schmitt, D., & Gollnick, V. (2016). The Air Transport System. In *Air Transport System* (pp. 1–17). Springer Vienna. doi:10.1007/978-3-7091-1880-1_1

Shah, D., Rust, R. T., Parasuraman, A., Staelin, R., & Day, G. S. (2006). The Path to Customer Centricity. *Journal of Service Research, 9*(2), 113–124. doi:10.1177/1094670506294666

Shah, F. T., Syed, Z., Imam, A., & Raza, A. (2020). The impact of airline service quality on passengers' behavioral intentions using passenger satisfaction as a mediator. *Journal of Air Transport Management, 85*, 101815. doi:10.1016/j.jairtraman.2020.101815

Shrivastava, N. K., & Shukla, A. V. (2021). Measuring an employer brand: a study towards valid scale development (as a second-order factor of a structural model). *International Journal of Organizational Analysis*. doi:10.1108/IJOA-02-2021-2615

Sivertzen, A.-M., Nilsen, E. R., & Olafsen, A. H. (2013). Employer branding: Employer attractiveness and the use of social media. *Journal of Product and Brand Management, 22*(7), 473–483. doi:10.1108/JPBM-09-2013-0393

Slåtten, T., Lien, G., & Svenkerud, P. J. (2019). The role of organizational attractiveness in an internal market-oriented culture (IMOC): A study of hospital frontline employees. *BMC Health Services Research, 19*(1), 307. doi:10.118612913-019-4144-8 PMID:31088463

Steenkamp, J.-B. E. M., & Maydeu-Olivares, A. (2021). An updated paradigm for evaluating measurement invariance incorporating common method variance and its assessment. *Journal of the Academy of Marketing Science, 49*(1), 5–29. doi:10.100711747-020-00745-z

Tabares, D. A., & Mora-Camino, F. (2017). Aircraft ground handling: Analysis for automation. *17th AIAA Aviation Technology, Integration, and Operations Conference*, 1–16. 10.2514/6.2017-3425

Tanrıverdi, G., Bakır, M., & Merkert, R. (2020). What can we learn from the JATM literature for the future of aviation post Covid-19? - A bibliometric and visualization analysis. *Journal of Air Transport Management, 89*, 101916. doi:10.1016/j.jairtraman.2020.101916 PMID:32901183

Tanwar, K., & Prasad, A. (2017). Employer brand scale development and validation: A second-order factor approach. *Personnel Review, 46*(2), 389–409. doi:10.1108/PR-03-2015-0065

Tkalac Verčič, A., Galić, Z., & Žnidar, K. (2021). The Relationship of Internal Communication Satisfaction With Employee Engagement and Employer Attractiveness: Testing the Joint Mediating Effect of the Social Exchange Quality Indicators. *International Journal of Business Communication, 232948842110538*. doi:10.1177/23294884211053839

Turban, D. B. (2001). Organizational Attractiveness as an Employer on College Campuses: An Examination of the Applicant Population. *Journal of Vocational Behavior, 58*(2), 293–312. doi:10.1006/jvbe.2000.1765

Urbancová, H., & Hudáková, M. (2017). Benefits of Employer Brand and the Supporting Trends. *Economia e Sociologia, 10*(4), 41–50. doi:10.14254/2071-789X.2017/10-4/4

Vidovic, L. (2022). Industries Most and Least Impacted by COVID-19 from a Probability of Default Perspective - January 2022 Update. *S&P Global Market Intelligence*. https://www.spglobal.com/marketintelligence/en/news-insights/blog/industries-most-and-least-impacted-by-covid-19-from-a-probability-of-default-perspective-january-2022-update

Wang, R., Shu-Li, Hsu, Lin, Y. H., & Tseng, M.-L. (2011). Evaluation of customer perceptions on airline service quality in uncertainty. *Procedia: Social and Behavioral Sciences, 25*, 419–437. doi:10.1016/j.sbspro.2012.02.054

Yaşlıoğlu, M. M. (2017). Sosyal bilimlerde faktör analizi ve geçerlilik : Keşfedici ve doğrulayıcı faktör analizlerinin kullanılması [Factor analysis and validity in social sciences: using exploratory and confirmatory factor analysis]. *Istanbul University Journal of the School of Business, 46* (Özel Sayı 2017), 74–85.

ADDITIONAL READING

Eriksson, T., Näppä, A., & Robertson, J. (2022). All for one and one for all: Encouraging ecosystem citizenship behaviour to strengthen employer branding. *Scandinavian Journal of Management*, *38*(2), 101211. doi:10.1016/j.scaman.2022.101211

Hong, S. J., Savoie, M., Joiner, S., & Kincaid, T. (2022). Analysis of airline employees' perceptions of corporate preparedness for COVID-19 disruptions to airline operations. *Transport Policy*, *119*, 45–55. doi:10.1016/j.tranpol.2022.02.008 PMID:35221549

Kucherov, D. G., Tsybova, V. S., Lisovskaia, A. Y., & Alkanova, O. N. (2022). Brand orientation, employer branding and internal branding: Do they effect on recruitment during the COVID-19 pandemic? *Journal of Business Research*, *151*, 126–137. doi:10.1016/j.jbusres.2022.06.053 PMID:35791374

Mostafa, B. A. (2022). Leveraging workforce insights to enhance employer attractiveness for young job seekers during pandemic era. *Heliyon*, *8*(6), e09699. doi:10.1016/j.heliyon.2022.e09699 PMID:35756108

Saruhan, N., Yıldız, E., Anuk, D., & Ünsal, P. (2022). Exploring the Relationship between the Fear of Covid-19, Job Insecurity, Employee Well-Being, and Job Involvement in Flight Personnel. *Transportation Research Procedia*, *66*, 167–178. doi:10.1016/j.trpro.2022.12.018

Waramontri, R., Charoenit, K., & Guzikova, L. (2022). Tourism Strategy responses to Covid-19 Pandemic: Case Study in Air Transport. *Transportation Research Procedia*, *63*, 2500–2505. doi:10.1016/j.trpro.2022.06.287

Zhu, C., Wu, J., Liu, M., Wang, L., Li, D., & Kouvelas, A. (2021). Recovery preparedness of global air transport influenced by COVID-19 pandemic: Policy intervention analysis. *Transport Policy*, *106*, 54–63. doi:10.1016/j.tranpol.2021.03.009 PMID:33785994

KEY TERMS AND DEFINITIONS

Business-Domain: The overall area of activity or expertise of a company.

Employer Attractiveness: The total benefits that an employee anticipates for working a particular organization.

Employer Brand: Branding and marketing the entire employment experience.

Employee Satisfaction: It is a term that describes how satisfied employees are with their jobs, workplace experiences and the organization they work for.

HR Marketing: HR marketing is a broad scale concept which involves the activities used by a company to attract, motivate, and retain the best employees.

Internal Marketing: It is a marketing concept that considers not only the buyers of the product but also the employees as customers.

Resilience: The ability of a company to respond rapidly and adapt effectively to disruptions or unforeseen changes that could harm its operations, personnel, assets, brand, or reputation.

Value Proposition: It is the set of values that a company or organization promises to offer to its customers.

Chapter 2
Analysis of Airport Stakeholder Collaboration With AHP Method to Provide Safe Air Travel Adventure Through Digital Means to Disabled Passengers

Vildan Durmaz
Eskisehir Technical University, Turkey

Kübra Nur Cingöz
Gaziantep University, Turkey

ABSTRACT

Air transportation demand increases all over the world every day, and airports have been serving more passengers. Furthermore, the number of passengers with disabilities who get service at the airport has been increasing recently. But the service which disabled passengers get is not enough, and these passengers are confronted with difficulties while using air transportation. The biggest issue these passengers face is not getting efficient guidance service and not benefiting from some right which other passengers do. Hence, this study aims to analyse collaboration of airport stakeholders to provide safe air transportation journey and reduce or remove difficulties for disabled passengers. The research is based on literature review, interviews with airport stakeholders, and analysis of these with AHP method.

INTRODUCTION

Airports are used by people from all groups. Recently, airports have been serving a bigger variety and increasing number of people with disabilities and mobility issues, especially those related to aging, obesity, and medical needs (Ancell & Graham, 2015). Over 650 million people in the whole world (including 50 million in Europe and 13 million in the United Kingdom) have some sort of disability and individuals with

DOI: 10.4018/978-1-6684-6835-7.ch002

disabilities represent largest minority of the world (Budd & Ison, 2020) And at some airports, the annual growth rate passenger with mobility issues of traffic is six times the overall passenger growth (Budd & Ison, 2020). As the population of disabled passengers increases, aviation sector which has already used many new technological products such as mobile applications, biometric systems, blockchain, robot and autonomous vehicles, artificial intelligence, virtual reality, sound technology, cyber security, in-flight internet, assistive technology (for disabled individuals) needs new technologies for growing demand of disabled passengers (Güreş et al., 2019).

In other respects, the aviation industry has confronted with destructive crisis, COVID-19, in 2019. While previous health crises such as SARS have affected the transportation sector but the COVID-19 pandemic is unprecedented and has had a particular impact on the sector (Benitez, 2021; Gössling, 2020). It has commenced a wave of recessions all around the world due to supply and demand chain disruptions in travel market (Dube et al., 2021). Mass transportation has limited because of government restrictions and fears of contracting the virus (Rayash & Dincer, 2020) and reduced passenger demand (combined with flight bans at the national level) has resulted in most airlines ceasing; many have had to stop nearly all operations and ground their entire fleets (Sun et al, 2021), majority of the flights have been called off by airlines or due to the closure of airports and routes by the authorities, many airlines went bankrupt (Heiets & Xie, 2021).

Despite the fact that Covid-19 pandemic have caused bankruptcy, airport's closure, flight cancellation, it also has ushered in a new era of digitalization such as distance education, working from home, contact- free mobile payment, contact free delivery, increasing rate of online transaction (Elfirdoussi et al, 2020; Zhao & Bacao, 2021; Nguyen & Vu, 2021; Bolisani et al., 2021; Dockery & Bawa, 2020; Bick et al., 2020). With the removal of travel bans, flights have started in June 2020 and air traffic demand which fell to 10-15% in in April and May 2020 compared to 2019 (Gössling, 2020) have increased as it is compared to 2020. Latest traffic data to show how airports, airlines and States are recovering from the COVID-19 pandemic with further analysis on traffic flows, market segments, oil prices, aviation sustainability and delays (ICAO, 2022). With the increasing air traffic demand, disabled passengers whose purchasing behaviors have been changed because of the Covid-19 pandemic will demand less contact with stuff and items and the airport in post-Covid-19 period. To provide quality service for disabled passengers, who faces the difficulty more than other passengers and facilitate the transaction and services through digital means in post-Covid-19 period, it is crucial to analyze and prioritize the collaboration of airport stakeholders which existed in the system that one's action effects another's. Consequently, the necessity of this chapter has been emphasized.

In this context the aim of this chapter is to analyze and prioritize the collaboration of airport stakeholders to provide safe air transportation journey to disabled passengers with advantage of digitalization. To analyze and prioritize the collaboration of airport stakeholders, data which obtained through the interviews with airport stakeholders has been examined. Analytic Hierarchy Process (AHP) method, which is one of the multi-criteria decision-making methods and can be used for the purpose of classification and prioritization has been used for the examination. By applying the AHP, which collaboration is more important to provide services to disabled passengers through digital means can be determined. To serve this objective 3 main criteria, 9 sub-criteria and 3 alternatives has been determined and analysis has been performed on Excel.

As a result of analysis, it has been concluded that airport administration-ground handling companies' collaboration is the most important collaboration to provide service for the disbaled passenger through digital means. In the lights of the results obtained it has been foreseen that with the given importance to the airport administration-ground handling companies' collaboration, both the service quality and demand for the air travel will increase in the post-Covid-19.

BACKGROUND

The number of studies concerning digitalization has increased recently. Digitalization is a considerable factor which have created new opportunities for companies and most important technology trend facing the world, and it has an impact on the transportation sector and among others. It can lead to efficiency gains through lower transaction costs and reduced information asymmetries supported by rating systems, lower consumer prices, increased market access, more competition, better use of underutilized resources and increased flexibility for the providers of services (Molchanova et al., 2020).

Bloomberg (2018) defines digitization as taking analog information and encoding it into zeroes and ones so that computers can store, process, and transmit such information (Bloomberg, 2018). Parviainen (2017) identified digitization as one of the major trends transforming society and business (Parviainen, 2017).

On the other hand, since the Covid-19 pandemic, the way of businesses and consumers' behaviors have changed (Eger et al., 2021) and power of digitalization has been noticed in this process. Services like distance education, online shopping, distance working, cashless payment or contactless mobile payment, contactless delivery, increased rate of online transactions have become possible (Elfirdoussi et al, 2020; Zhao & Bacao, 2021; Nguyen & Vu, 2021; Bolisani et al., 2021; Dockery & Bawa, 2020; Bick et al., 2020; Eger et al., 2021).

With the digitalization which has also great impact on aviation industry, disabled passengers who were identified as passengers with reduced mobility (PRM) because of physical incapacity (sensory or locomotor), illness, age, intellectual deficiency or any other disability while using transport and whose condition requires special attention and adapting services offered to all passengers to the individual's needs (ICAO, 2022) and faces more challenges and difficulties while travelling and using the public transport (Soltanii, 2012) may have opportunity to make transaction through digital platforms and access certain services which non-disabled passengers have while the disabled passengers don't. While performing the passenger's transportation service, the service process requires the cooperation of the three airport stakeholders: terminal administration, ground handling companies and airlines. In this context to be able to offer quality service with the power of the digitalization, collaboration of three airport stakeholders which passengers take services needs to be prioritized.

On the other hand, when the literature has been review studies about disabled passengers (Richards, 1968: Lauder, 1969), problems which they have faced during air travel (Muller, 1977; Reida, 1984; Speidel, 1984, Matson, 1985; Abeyratne, 1995), regulations, laws and rights of them (Boyer, 2002; Frye, 2010; Prager, 2011), the cost of their travel (Ancell & Graham, 2016), digitalization concept and its impact on commercial aviation (Kuisma, 2018; Chang et al., 2019) and change brought about by digitization (Kurtcebe, 2020) has been discussed.

The most basic form of the digital term, often synonymous, is digitization and digitization. Digitization refers to a technological process that "incorporates digital technologies into everyday life" (Bican & Brem, 2020). This technical process of digitization as taking analog information and encoding it into zeroes and ones so that computers can store, process, and transmit such information (Bloomberg, 2018).

The broad impact of digitalization affects everything from social media and the enhanced personal relationships of its services to other relationships such as how citizens interact with support services in e-government and business models. Many industrial companies use digital technologies to innovate their business models and explore new innovative products. These digital business models often add service elements to physical products so that companies can offer customized solutions (Gray & Rumpe, 2015; Linde et al., 2021). And herein, the concept of digitalization has been identified as the use of digital technologies to change a business model and provide new revenue and value-producing opportunities; it is the process of moving to a digital business (Gray & Rumpe, 2015).

Besides, concept of digitalization has also become important concept and been used in literature Sanjog et al. (2013) reviewed the literature highlighting state-of-the-art applications of digital human modeling (DHM) in the aerospace industry and sparking research interest and promoting the application of digital human modeling technologies. They pointed out that the aerospace industry was the first to adopt DHM technology, and most of the current users are in the automotive and aerospace engineering fields (Sanjog et al., 2013). Molchanova et al. (2020) conducted study to examine the concept of digital platforms and their place in the modern digital world. As the light of the results obtained, they found that finding customers, reducing transaction costs and monetizing underutilized assets are easier through digital platforms (Molchanova et al., 2020). Molchanova et al. (2020) examine the development of the definition of digital twin (DT) and compare its concept with Internet of Things, cyber-physical systems, and digital threads, and present a research timeline for aviation DT. Authorities such as the U.S. Air Force Research Laboratory and NASA over the past 10 years. They concluded that the aviation DT is the most widely used in manufacturing and maintenance (Molchanova et al., 2020).

Besides, the studies about disabled people have been review in the literature.

Soltani et al. (2012) conducted study to demonstrate the impact of current disability-specific legislation and standard design on accessibility at public transport terminals. They pointed out that people with disabilities have fewer opportunities and a lower quality of life than non-disabled people.

Chiscano and Darcy (2022) conducted this study in response to a call for further research resources to restore public transport accessibility during the ongoing COVID19 pandemic and any future pandemics to ensure equitable access to public transport. They explore how a combination of digital and non-digital resources provided by transport operators can create the greatest value for people with disabilities. The results suggest that transport operators can better manage the improvement of users' subway experience by taking a more targeted approach.

Bigby et al., (2019) investigated the experiences of people with communication disabilities on the rail network in Victoria, Australia, to identify the barriers they encountered. The results shows that difficulties identified during travel were grouped into three themes: (a) variable availability of information, (b) negative impact of a large and complex service system, and (c) culture of uncertainty in asking for and giving.

Wu et al., (2020) points out the concerns of persons with disabilities about the difficulty of booking accessible vehicles, the lack of a unified vehicle booking system, the fact that vehicle booking systems are not mobile-based, the inflexibility of such systems, restrictions on inter-regional travel, and the lack of an evaluation system and they propose a single mobile-based platform, will integrate government resources to help the disabled connect with drivers and improve the satisfaction of both parties.

Since the Covid-19 pandemic has changed the way of businesses and consumers' behavior (Eger et al., 2021) and digitalization has come into prominence Polishchuk et al. (2022) disabled passengers got the opportunity to get quality service with the digital tools.

A review of the literature it has been seen that there are studies about the difficulties disabled passengers faced and few studies have attempted to introduce mobile service systems suitable for people with disabilities in social, transportation, or travel (Wu et al., 2020). Emphasis on digital services which disabled passengers can get in aviation industry, and importance of collaboration while serving the disabled passengers is absent.

MAIN FOCUS OF THE CHAPTER

One of the results of the deregulation policies of air transport around the world is the increasing number of passengers, involving people with disabilities. Because social equality is considered a worthy goal, providing an environment in airport terminals and on-board aircraft to facilitate the travel of people with different disabilities is a goal that airlines and airport authorities are increasingly focusing on and monitoring (Chang & Chen, 2011).

Accessibility is important in daily life especially when dealing with external and internal environment. Realizing that most public transport terminals still lack in terms of good design and facilities thus a serious concern for the matters is needed to ensure the convenient for all. It has been widely accepted that disabled people, have fewer opportunities and lower quality of life than non-disabled. Added with poor accessibility, the disabled people face more challenges and difficulties while travelling and using the public transport. Therefore, it becomes increasingly difficult to ignore the disabled issues while using the facilities of the public transport (Soltani et al., 2012).

Even they have focused on passengers with disabilities, studies shows that passenger with disabilities have still been facing with difficulties (Soltani et al., 2012).

The findings of the study which conducted in 2011 with 3 level of disabilities, severe, moderate and minor shows that toilet accessibility and seat arrangement are the least satisfied attributes for people with disabilities. When three level of impairment have been compared in terms of satisfaction, it has been found that satisfaction level for both moderate and severe impairment groups is lower than minor group (Chang & Chen, 2011).

However, with the Covid-19 pandemic, disabled passengers whose purchasing behaviour has been changed and they became inclined to demand less contact with stuff and items and the airport in post-Covid-19 period. To provide quality service for disabled passengers, who faces the difficulty more than other passsegers and facilitate the transaction and services through digital means in post-Covid-19 period, it is crucial to analyze and prioritize the collaboration of airport stakeholders which existed in the system that one's action effects another's. Consequently, the necessity of this chapter has been emphasized.

In this context the aim of this chapter is to analyze and prioritize the collaboration of airport stakeholders to provide safe air transportation journey to disabled passengers with advantage of digitalization.

To be understood, some key terms and connections about the topic have been given below.

1. Disabled Passenger

According to ICAO (International Civil Aviation Organization), disabled passenger is a passenger with reduced mobility (PRM) because of physical incapacity (sensory or locomotor), illness, age, intellectual deficiency or any other disability while using transport and whose condition requires special attention and adapting services offered to all passengers to the individual's needs (ICAO, 2022).

Directorate General of Civil Aviation (DGCA) defines a disabled or mobility-impaired person as a person who requires special care and adaptation of the services provided to other passengers due to any physical, mental, old age or any other disability related to permanent or temporary, sensory or locomotor skills, while using transportation vehicles (DGCA, 2022). The United Nations defines person with disabilities as a person with long-run (long-dated) intellectual, mental, physical, or sensory impairments that in interaction with a variety of impediment may block them from participating fully and effectively in society on an equal basis with others (UN, 2022).

As to World Health Organization (WHO), there are three dimensions of disabilities. These are activity restraint (limitation), like hardship problem solving, walking, seeing or hearing; impairment of a person's anatomy (body structure) or function, or mental functioning; participation limitation in everyday life activities like social activities, work issues, entertainment and having healthcare and preventive services (CDC, 2022). Acar and Nur identify disabled like being unable to comply with the requirements of normal life due to the loss of physical, mental, spiritual, emotional and social abilities in various degrees as a result of congenital or subsequent illness or accident; It is defined as the person who needs protection, care, rehabilitation, counseling and support services (Acar and Nur, 2018).

In this chapter disabled word defines a person who has impairment of body structure or function and has difficulty walking, seeing and hearing. Because the chapter aims to provide safe air transportation journey with the power of the digitalization, yet mentally disabled passengers are not able to make their transactions on their own and need assistance. However, their companion can take advantage of the digitalization.

1. Airport Stakeholders Collaboration

Airports provides access to air transportation services and collaborations of stakeholders are extremely important. Airport stakeholders may be external and internal and can include tenants, passengers, government, airlines, national and international authorities, local communities, unions, suppliers, media, mil/defence agencies, terminal administration, ICAO Regional Offices, construction companies, oil companies, catering companies, ground handling companies (Long, 2017; ICAO, 2011).In this chapter, however, most needy stakeholders in providing services to passengers with disabilities through digital means will be examined.

The disabled passengers whose purchasing behavior changed got the opportunity to get their desired digital services which they can't before the pandemic.

To provide quality service for disabled passengers, who faces the difficulty more than other passengers and facilitate the transaction and services through digital means in post-Covid-19 period, it is crucial to analyze and prioritize the collaboration of airport stakeholders which existed in the system that one's action effects another's. Consequently, the necessity of this chapter has been emphasized.

To be able to provide safe and digital air journey to disabled passengers, these problems are considered and stakeholders' collaboration has been prioritized with AHP model to serve this object.

METHODOLOGY

One of the most preferred multi-criteria decision-making models; Analytical Hierarchy Process (AHP) which was developed by Saaty (Al-Harbi, 2001) and most suitable method for prioritization has been used.

It is a multi-criteria decision-making tool. This is an Eigen Value approach for pairwise comparisons. It also provides a way to calibrate numerical scales to measure quantitative and qualitative performances (Vaidya and Kumar, 2006). It is nine-point scale which was developed by Saaty and given in Table 1.

Some key and basic steps involved in this methodology has (Russoa and Camanho, 2015; Al-Harbi, 2001; Saaty, 2008; Vaidya and Kumar, 2006) been given below.

1. Defining the problem and identifying the goal.
2. Building hierarchies from the top level (the goal of the decision maker's point of view) through middle level (the criteria on which subsequent layers depend) to the lowest level, which usually contains a list of alternatives.
3. Constructing a set of pair-wise comparison matrices for each of the lower levels with one matrix at a time for each element in the level by using the relative scale measurement shown in Table 1. The pair-wise comparisons are done in terms of which element dominates the other.
4. n(n-1)/ judgments required to develop the matrix set. Automatically assign reciprocals in each pairwise comparison
5. Hierarchical synthesis is now used to weight the eigenvectors by the weights of the criteria and the sum is taken over all weighted eigenvector entries corresponding to those in the next lower level of the hierarchy.
6. To perform some calculation to find eigen value (λmax), consistency index (CI) (CI=(λmax-n)(n-1). Judgment consistency can be checked by taking the consistency ratio (CR) of CI with the appropriate value in Table 7. The CR is acceptable, if it does not exceed 0.10. If it is more, the judgment matrix is inconsistent. To obtain a consistent matrix, judgments should be reviewed and improved.

These actions have been applied as the following.

Step 1

At this step, the purpose of decision problem and hierarchical structure has been determined. To be able to create hierarchical structure, the aim, main criterion, sub-criterion and alternatives has been specified according to the most important criteria which the companies are concerned with when servicing passengers with disabilities and alternatives which were most needed collaboration while servicing disabled passenger.

Figure 1 identifies the Hierarchical Structure. The aim of the study is to prioritize the collaboration of airport stakeholders to provide safe air travel journey to disabled passenger. As it has been seen in Figure 1, main criteria are Quality (C1), Operation (C2) and Cost (C3) while the sub-criteria are Employee Behavior (SC1), Flexibility (SC2), Accessibility (SC3), Safety (S4), Efficiency (SC5), Innovativeness (SC6), Operating Cost (SC7), Labor (SC8) and Technological Competence (SC9). Alternatives are Airport Administration- Terminal Administration Collaboration (A1), Airport Administration-Ground Handling Companies' Collaboration (A2) and Airport Administration-Airlines Collaboration (A3).

Figure 1. Hierarchial structure
Source: *Developed by authors*

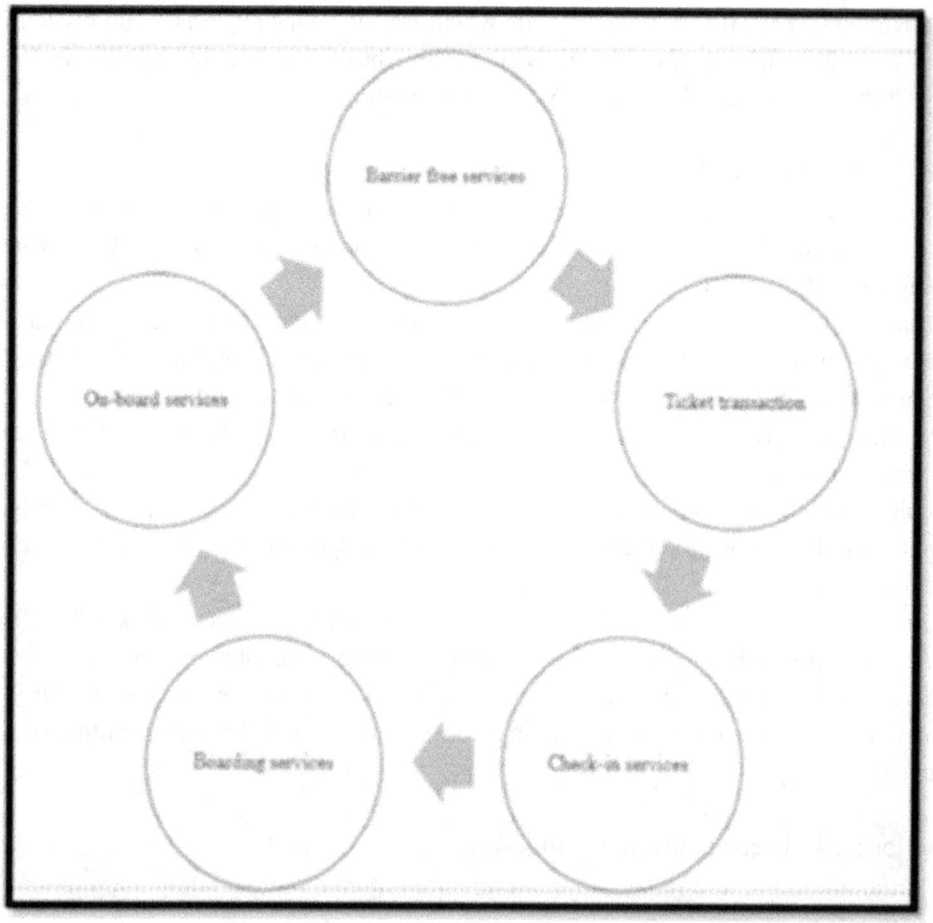

Step 2

This step of AHP, paired comparison matrix between main criteria and sub-criteria on the same level has been created.

To serve this objective Saaty's nine-point scale which was given in Table 1 has been utilized and comparison matrix for the main criteria has been created.

Step 3

Thereafter the creation of comparison matrix, normalized matrix has been calculated. To calculate normalized each column values are summed up, and then each element in the column divided by sum of the column to calculate normalized matrix.

Local priorities are calculated by taking average of each line in the matrix.

Table 1. Saaty nine-point importance scale

Intensity of Importance	Definition
1	Equal
2	Weak or slight
3	Modarate
4	Moderate plus
5	Strong
6	Strong plus
7	Very strong or demonstared
8	Very, very strong
9	Extreme importance

Source: Akdeniz (2018)

Table 2. Pair-wise comparison matrix for the main criteria

Main Criteria	C1	C2	C3
C1	1,00	1/5	1,00
C2	5,00	1,00	6,00
C3	1,00	1/6	1,00
Total	7,000	1,366	8,000

Step 4

All priorities matrix has been calculated. To calculate all priorities matrix the local priorities are multiplied by comparison matrix given in the beginning.

Table 3. Normalized pair-wise matrix and local priorities for the main criterion

	C1	C2	C3	Local Priorities
C1	0,143	0,146	0,125	0,138
C2	0,714	0,732	0,750	0,732
C3	0,143	0,122	0,125	0,130

Table 4. All priorities of main criteria

	C1	C2	C3	All Priorities
C1	0,138	0,146	0,130	0,414
C2	0,690	0,732	0,780	2,202
C3	0,138	0,122	0,130	0,390

Step 5

Subsequent to the completion of matrix calculation, the consistency index calculated through usage of Equation 1.

$$CR= \frac{CI}{RI} \quad (1)$$

CI: Consistency Index
RI: Random Index
To be able to calculate consistency index Equation 2 has been used.

$$CI= \frac{(\lambda max - n)}{n-1} \quad (2)$$

λmax refers to eigen value and is calculated by the division of all priorities to local priorities and the average of these values. Calculation of for λmax sub-criterion has been shown in Table 5.

Table 5. Eigen value calculation for main criterion

All Priorities (T)	Local priorities (L)	T/L	λmax
0,414	0,138	3,001	3,004
2,202	0,732	3,008	
0,390	0,130	3,001	

To calculate Consistency Index Equation 2 has been used;

$$CI= \frac{(\lambda max - n)}{n-1}$$

For main criterion;

$$CI= \frac{(3{,}004 - 3)}{3-1}$$

=0,002

To calculate Consistency Ratio, Random Consistency Index Tabled which was created by Saaty (1980) has been used and it has been shown in Table 6.

Table 6. Random consistency index (RI) table

n	2	3	4	5	6	7	8
RI	0,0	0,58	0,90	1,12	1,24	1,32	1,41
n	9	10	11	12	13	14	15
RI	1,45	1,49	1,51	1,48	1,56	1,57	1,59

Source: Saaty (1980)

Consistency Ratio for main criterion has been calculated according to Equation 1:

$$CR=\frac{CI}{RI}$$

$$=\frac{0,002}{0,58}$$

$$=0,003$$

The same process has been repeated for sub-criteria and alternatives and results have been shown below Tables 8-19 in Appendix.

The consistency index should be between 0,001≤ 0.1 to get consistent results (Saaty, 1980; Şahin and Yurdagül, 2018). According to results all comparisons are consistent.

Alternatives have been compared pair-wise with respect to each sub-criterion. Tables 11-19 shows experts' comparison of alternatives against each sub-criterion. The weight of each sub-criterion from nine matrices (Tables 11-19) and the weights of alternatives with reference to the sub-criteria are multiplied to acquire the priorities of alternatives on the basis of main criteria. For instance, the rating of Alternative 1 with regard to Quality main criterion has been calculated as follows:

$$(0,084\times0,703)+(0,444\times0,109)+(0,472\times0,084)=0,147$$

The outcome has been shown in Table 7.

Table 7. Final Priorities of the alternatives

	Local prioties of main criteria			AHP Results
	C1 0,138	C2 0,732	C3 0,130	
A1	0,147	0,155	0,175	0,156
A2	0,468	0,542	0,344	0,506
A3	0,385	0,303	0,381	0,325

The overall weights of the three alternatives has been obtained by multiplying the priority of each main criterion in Table 7 by the priority of each alternative. The overall ranking of alternatives have been calculated as follows:

(0,138 x 0,147)+(0,732 x 0,155)+(0,130 x 0,175)= 0,156 (Alternative 1)

(0,138 x 0,468)+(0,732 x 542)+(0,130 x 0,344)= 0,506 (Alternative 2)

(0,138 x 0,385)+(0,732 x 0,303)+(0,130 x 0,381)= 0,325 (Alternative 3)

Data Analysis and Tools

The data were collected through structured interviews with three airport stakeholders; terminal administration, ground handling companies and airlines. By reviewing the literature, the criteria for providing quality service to the passengers were determined and the experts were asked to point out the degree of importance these criteria. The analysis was performed on Microsoft Excel.

FINDINGS

As it has been seen in Table 7, it is apparent that the operation main criterion (C2) is the most important factor with the 0,732 priority value, followed by quality (C1) with 0,138 priority value and cost (C3) with 0,130 priority value.

Tables 8-10 indicates the priority values of each sub-criterion. According to result obtained from Tables 8-10 *Accessibility (SC3)* is the most important sub-criterion with a score of 0,472 in *Quality* main criterion, *Safety (SC4)* is the most important sub-criterion with a score of 0,703 in *Operation* main criterion and *Operating Cost (SC7)* is the most important sub-criterion with a score of 0,703 in *Cost* main criterion.

It can be concluded from Table 19 that Alternative 2 (airport administration-ground handling companies' collaboration) with a rating of 0,506 is most important collaboration, followed by Alterative 3 (airport administration-airlines' collaboration) and Alternative 1 (airport administration-terminal administration collaboration).

CONCLUSION

With the Covid-19 pandemic, digitalization has been put forward. The pandemic has furnished an occasion to the passengers who can't get enough services in terms of transaction, online services.

In this chapter, airport stakeholders' collaboration has been prioritized in order to give quality service for disabled passengers through the advantage of digitalization Covid-19 pandemic drove forward.

In conclusion, it has been seen that airport administration-ground handling companies' collaboration is the most important collaboration to provide quality service for disabled passengers, followed by airport administration-airlines' collaboration and airport administration-terminal administration collaboration.

The collaboration of airport administration-ground handling companies can be opportune for quality service in post-Covid-19 era. This chapter will conduce to getting quality service through the power of digitalization and make some transaction which they can't make while non-disabled can.

SOLUTION AND RECOMMENDATION

In this chapter analysis of airport stakeholder collaboration has been forwarded. And results have illustrated that collaboration of ground handling companies and airport administration is the most important collaboration in order to offer a quality service through digital means for disabled passengers.

It also has revealed that *Accessibility (SC3)* is the most important sub-criterion with a score of 0,472 in *Quality* main criterion, *Safety (SC4)* is the most important sub-criterion with a score of 0,703 in *Operation* main criterion and *Operating Cost (SC7)* is the most important sub-criterion with a score of 0,703 in *Cost* main criterion.

Paying regard to challenges of disabled passenger through their air travel journey, it is important to offer them a quality service. In this context, collaboration of airport stakeholders and prioritization is crucial.

The pandemic leading to bankruptcies, airport closures, flight annulment has brought many innovations to the aviation industry. For instance, distance education/training, working distance, contactless payments and deliveries, increased rates of online transactions (Elfirdoussi et al, 2020; Zhao and Bacao, 2021; Nguyen and Vu, 2021; Bolisani et al, 2021; Dockery and Bawa, 2020; Bick et al., 2020). With the lifted travel bans and flights commencing in June 2020, air traffic demand which dropped to 10-15% in April and May 2020 compared to same period in 2019 (Gössling, 2020), has scaled up compared to 2020 increased (Gössling, 2020). The latest air traffic data demonstrate how states and aviation industry are recovering from the COVID-19 pandemic (ICAO, 2022). Along with the rise of air traffic demand, passengers with disabilities, whose purchasing behavior has been changed because of the Covid-19 pandemic, will request less contact with stuff and items and the airport in post-Covid-19 period.

In this chapter the prioritization of airport stakeholders gives opportunity to disabled passenger not only to get quality service but also to make transaction with digitalization empowerment. With that prioritization it is expected that disabled passengers get more quality and digital service than before in post-Covid-19 period.

FUTURE RESEARCH DIRECTION

The chapter has put forward the airport stakeholder's collaboration to provide quality service with the advantage of digitalization which Covid-19 pandemic brought out for disabled passenger.

When to be considered the challenges disabled passenger encounter, authors are conducting study to take advantage of digitalization and develop a software which can be share with passenger for per flight and uploaded cellphone or digital watch. Through that program disabled passenger;

- Will be able to make ticket transaction through that program (changes and cancellation, online check-in)
- Disabled passenger can deliver their request through their phone or watch (assistance, wheelchair etc.)

- Airport location, terminal map, information about park position, gate, lounge, duty free etc. will be uploaded, so extra personnel for guiding won't be necessary
- Read-aloud option for passenger with visually impairment

Through the power of technology and digitalization, thanks to the one-time (per flight) application that can be downloaded to the watch or phone, both the problems faced by disabled passengers will be solved and an equality will be provided in terms of some rights that other passengers can benefit from while disabled passengers cannot. In that scope, interviews with airport stakeholders have been executed and responsibility of them have been analyzed with AHP Method.

REFERENCES

Abeyratne, R. I. R. (1995). Proposals and Guidelines for the Carriage of Elderly and Disabled Persons by Air. *Journal of Travel Research, 33*(3), 52–59. doi:10.1177/004728759503300310

Abu-Rayash, A., & Dincer, I. (2020). Analysis of mobility trends during the COVID-19 coronavirus pandemic: Exploring the impacts on global aviation and travel in selected cities. *Energy Research & Social Science, 68*, 101693. doi:10.1016/j.erss.2020.101693 PMID:32839706

Acar, N., & Nur, R. (2018). *Evaluation of the Airport Services According to Handicap of Passengers.* Government Accountability Office.

Al-Harbi, K. M. A. S. (2001). Application of the AHP in project management. *International Journal of Project Management, 19*(1), 19–27. doi:10.1016/S0263-7863(99)00038-1

Ancell, D., & Graham, A. (2016). A Framework for Evaluating the European Airline Costs of Disabled Persons and Persons with Reduced Mobility. *Journal of Air Transport Management, 50*, 41–44. doi:10.1016/j.jairtraman.2015.10.001

Bican, P. M., & Brem, A. (2020). Digital business model, digital transformation, digital entrepreneurship: Is there a sustainable "digital"? *Sustainability, 12*(13), 5239. doi:10.3390u12135239

Bick, A., Blandin, A., & Mertens, K. (2020). *Work from home after the COVID-19 Outbreak.* Federal Reserve Bank of Dallas.

Bigby, C., Johnson, H., O'Halloran, R., Douglas, J., West, D., & Bould, E. (2019). Communication access on trains: A qualitative exploration of the perspectives of passengers with communication disabilities. *Disability and Rehabilitation, 41*(2), 125–132. doi:10.1080/09638288.2017.1380721 PMID:28949244

Bloomberg, J. (2018). Digitization, digitalization, and digital transformation: confuse them at your peril. *Forbes.*

Bolisani, E., Scarso, E., Ipsen, C., Kirchner, K., & Hansen, J. P. (2020). Working from home during COVID-19 pandemic: Lessons learned and issues. *Management & Marketing. Challenges for the Knowledge Society, 15*(1), 458–476. doi:10.2478/mmcks-2020-0027

Boyer, G. (2002). English Poor Laws. *Eh.Net Encyclopedia.* https://eh.net/encyclopedia/english-poor-laws/

Budd, L., & Ison, S. (2020). Supporting the Needs of Special Assistance (Including PRM) Passengers: An International Survey of Disabled Air Passenger Rights Legislation. *Journal of Air Transport Management, 87*, 101851. doi:10.1016/j.jairtraman.2020.101851

Çakır, E., & Ulukan, Z. (2022). Digitalization on Aviation 4.0: Designing A Scikit-Fuzzy Control System for In-Flight Catering Customer Satisfaction. In *Intelligent and Fuzzy Techniques in Aviation 4.0* (pp. 123–146). Springer. doi:10.1007/978-3-030-75067-1_6

Chang, Y. C., & Chen, C. F. (2011). Identifying Mobility Service Needs for Disabled Air Passengers. *Tourism Management, 32*(5), 1214–1217. doi:10.1016/j.tourman.2010.11.001

Chiscano, M. C., & Darcy, S. (2022). An accessible and inclusive public transportation management response to COVID-19 through a co-creation process with people with disability. The case of Metro Barcelona. *Research in Transportation Business & Management, 45*, 100880. doi:10.1016/j.rtbm.2022.100880

DGCA. (2015), Instruction for Airline Passengers with Disabilities or Restricted Mobility. *SHT Barrier Free.* https://web.shgm.gov.tr/documents/sivilhavacilik/files/mevzuat/sektorel/talimatlar/sht-engelsız.pdf

Dockery, M., & Bawa, S. (2020). Working from Home in the COVID-19 Lockdown. *BCEC, 19*, 1–5.

Dube, K., Nhamo, G., & Chikodzi, D. (2021). COVID-19 pandemic and prospects for recovery of the global aviation industry. *Journal of Air Transport Management, 92*, 102022. doi:10.1016/j.jairtraman.2021.102022 PMID:36567961

Eger, L., Komárková, L., Egerová, D., & Mičík, M. (2021). The effect of COVID-19 on consumer shopping behaviour: Generational cohort perspective. *Journal of Retailing and Consumer Services, 61*, 102542. doi:10.1016/j.jretconser.2021.102542

Elfirdoussi, S., Lachgar, M., Kabaili, H., Rochdi, A., Goujdami, D., & El Firdoussi, L. (2020). Assessing distance learning in higher education during the COVID-19 pandemic. *Education Research International, 2020*, 2020. doi:10.1155/2020/8890633

Frye, A. (2010). *Air Passenger Rights for Disabled People.* Research Gate.

Gössling, S. (2020). Risks, resilience, and pathways to sustainable aviation: A COVID-19 perspective. *Journal of Air Transport Management, 89*, 101933. doi:10.1016/j.jairtraman.2020.101933 PMID:32952322

Gray, J., & Rumpe, B. (2015). Models for digitalization. *Software & Systems Modeling, 14*(4), 1319–1320. doi:10.100710270-015-0494-9

Güreş, N., Arslan, S., Yüksel, C., & Yılmaz, H. (2019). Mobil Uygulamaya Sahip Havayolu İşletmelerinin Yolculara Yönelik Hizmetlerinin Araştırılması [Investigation of passenger services of airlines with mobile applications]. *Çukurova Üniversitesi Sosyal Bilimler Enstitüsü Dergisi, 28*(3), 252-265.

Heiets, I., & Yibing, X. İ. E. (2021). The Impact of the COVID-19 Pandemic on the Aviation Industry. *Journal of Aviation, 5*(2), 111–126. doi:10.30518/jav.933296

ICAO. (2011). Identify stakeholders map. ICAO. https://www.icao.int/apac/meetings/2011_fpl_am_tf4_seminar/ıdentify%20stakeholders.pdf

ICAO. (2022). Eurocontrol Comprehensive Aviation Assessment. ICAO. https://www.eurocontrol.int/publication/eurocontrol-comprehensive-aviation-assessment

ICAO. (2022). Passenger with disablities. ICAO. https://store.icao.int/en/manual-on-access-to-air-transport-by-persons-with-disabilities-doc-9984

James, L. (2017), *Importance of Airport Stakeholder Outreach.* Aviation Pros. https://www.aviationpros.com/airports/blog/12388063/importance-of-airport-stakeholder-outreach

Kuisma, N. (2018). *Digitalization and Its Impact on Commercial Aviation.* MDPI.

Kurtcebe, M. (2020). Tıme And Money in Dıgıtalızatıon of Industry: A Socıologıcal Research on Aviation in Turkey. *And Differentiation II*, 24.

Lauder, R. (1969). The Goal Is: Mobility! Environmental and Transportation Barriers Encountered by The Disabled. *National Citizens Conference on Rehabilitation of the Disabled and Disadvantaged*. OhioLINK.

Linde, L., Sjödin, D., Parida, V., & Gebauer, H. (2020). Evaluation of digital business model opportunities: A framework for avoiding digitalization traps. *Research Technology Management, 64*(1), 43–53. doi:10.1080/08956308.2021.1842664

Matson, A. (1985). Service for the Disabled at United Kingdom Airports. *Nınnescah Internatıonal Servıce, 3*(4). *Upational Therapy in Health Care, 1*(4), 55–67.

Mhalla, M. (2020). The impact of novel coronavirus (COVID-19) on the global oil and aviation markets. *Journal of Asian Scientific Research, 10*(2), 96–104. doi:10.18488/journal.2.2020.102.96.104

Molchanova, K. M., Trushkina, N. V., & Katerna, O. K. (2020). Digital platforms and their application in the aviation industry. *Intellectualization of logistics and Supply Chain Management*, (3), 83-98.

Muller, H., & Rolen, G. (1977). *Airlines And Disabled Travellers*. Department of Transportation.

Nguyen, T. H., & Vu, D. C. (2020). Food delivery service during social distancing: Proactively preventing or potentially spreading coronavirus disease–2019? *Disaster Medicine and Public Health Preparedness, 14*(3), e9–e10. doi:10.1017/dmp.2020.135 PMID:32367789

Parviainen, P., Tihinen, M., Kääriäinen, J., & Teppola, S. (2017). Tackling the digitalization challenge: how to benefit from digitalization in practice. *International journal of information systems and project management, 5*(1), 63-77.

Polishchuk, O., Kulinich, T., Martynovych, N., & Popova, Y. (2022). Digitalization and Sustainable Development: The New COVID-19 Challenge Requires Non-Standard Solutions. *Problemy Ekorozwoju, 17*(2).

Prager, S. (2011, July). Pioneering Passengers' Rights: Legislation and Jurisprudence from the Aviation Sector. In *Era Forum* (*Vol. 12,* No. 2, Pp. 301-320). Springer-Verlag.

Reida, E. (1984, October). Available Research on Disabled and Elderly Air Passengers. In *Thırd Internatıonal Conference on Mobılıty and Transport of Elderly and Handıcapped Personsflorida State University.* Tallahasseeloughborough University of Technology.

Richards, P. R. (1968). Modern Air Transport and Invalids. *The Journal of the Royal College of General Practitioners, 15*(5), 336–343. PMID:5689750

Russo, R. F. S. M., & Camanho, R.de Russo, F.S.M. (2015). Criteria in AHP: A systematic review of literature. *Procedia Computer Science, 55*, 1123–1132. doi:10.1016/j.procs.2015.07.081

Saaty, T. L. (2008). Decision making with the analytic hierarchy process. *International Journal of Services Sciences, 1*(1), 83–98. doi:10.1504/IJSSCI.2008.017590

Sanjog, J., Karmakar, S., Patel, T., & Chowdhury, A. (2015). Towards virtual ergonomics: aviation and aerospace. *Aircraft Engineering and Aerospace Technology: An International Journal.*

Soltani, S. H. K., Sham, M., Awang, M., & Yaman, R. (2012). Accessibility for disabled in public transportation terminal. *Procedia: Social and Behavioral Sciences, 35*, 89–96. doi:10.1016/j.sbspro.2012.02.066

Speidel, L. K. (1984). Air Transporation and the Disabled: Current and Future Trends. *Occupational Therapy in Health Care, 1*(4), 55-67.

Sun, X., Wandelt, S., Zheng, C., & Zhang, A. (2021). COVID-19 pandemic and air transportation: Successfully navigating the paper hurricane. *Journal of Air Transport Management, 94*, 102062. doi:10.1016/j.jairtraman.2021.102062 PMID:33875908

Vaidya, O. S., & Kumar, S. (2006). Analytic hierarchy process: An overview of applications. *European Journal of Operational Research, 169*(1), 1–29. doi:10.1016/j.ejor.2004.04.028

Xiong, M., & Wang, H. (2022). Digital twin applications in aviation industry: A review. *International Journal of Advanced Manufacturing Technology, 121*(9-10), 1–16. doi:10.100700170-022-09717-9

Zhao, Y., & Bacao, F. (2021). How does the pandemic facilitate mobile payment? An investigation on users' perspective under the COVID-19 pandemic. *International Journal of Environmental Research and Public Health, 18*(3), 1016. doi:10.3390/ijerph18031016 PMID:33498863

APPENDIX

Table 8. Comparisons of sub-criteria with respect to "quality" criteria

	SC1	SC2	SC3	Local Priorities
SC1	1,00	1/5	1/6	0,084
SC2	5,00	1,00	1,00	0,444
SC3	6,00	1,00	1,00	0,472
				λmax=3,004 CI=0,002 CR=0,003

Table 9. Comparisons of sub-criteria with respect to "operation" criteria

	SC4	SC5	SC6	Local Priorities
SC4	1	5	5	0,703
SC5	1/5	1	1/2	0,115
SC6	1/5	2	1	0,182
				λmax=3,054 CI=0,027 ≤0,1 CR=0,047

Table 10. Comparisons of sub-criteria with respect to "cost" criteria

	SC7	SC8	SC9	Local Priorities
SC7	1	5	5	0,703
SC8	1/5	1	1/2	0,115
SC9	1/5	2	1	0,182
				λmax=3,054 CI=0,027 ≤0,1 CR=0,047

Table 11. Comparison of the alternatives with respect to "employee behaviour"

	A1	A2	A3	Local Priorities
A1	1	5	5	0,703
A2	1/5	1	1/2	0,115
A3	1/5	2	1	0,182
				λmax=3,054 CI=0,027 ≤0,1 CR=0,047

Table 12. Comparison of the alternatives with respect to "flexibility" sub-criteria

	A1	A2	A3	Local Priorities
A1	1	1/4	1/4	0,109
A2	4	1	1/3	0,297
A3	4	3	1	0,594
				λmax=3,138 CI=0,069 ≤0,1 CR=0,119

Table 13. Comparison of the alternatives with respect to "accessibility" sub-criteria

	A1	A2	A3	Local Priorities
A1	1	1/6	1/4	0,084
A2	6	1	5	0,692
A3	4	1/5	1	0,223
				λmax=3,168 CI=0,084 ≤0,1 CR=0,145

Table 14. Comparison of the alternatives with respect to "safety" sub-criteria

	A1	A2	A3	Local Priorities
A1	1	1/3	1/2	0,611
A2	3	1	2	2,000
A3	2	1/2	1	1,167
				λmax=3,009 CI=0,005 ≤0,1 CR=0,008

Table 15. Comparison of the alternatives with respect to "efficiency" sub-criteria

	A1	A2	A3	Local Priorities
A1	1	1/2	1/3	0,172
A2	2	1	2	0,478
A3	3	1/2	1	0,350
				λmax=3,136 CI=0,068 ≤0,1 CR=0,118

Table 16. Comparison of the alternatives with respect to "innovativeness" sub-criteria

	A1	A2	A3	Local Priorities
A1	1	1/4	1/4	0,109
A2	4	1	3	0,594
A3	4	1/3	1	0,297
Total	9,000	1,583	4,250	
				λmax=3,138 CI=0,069 ≤0,1 CR=0,119

Table 17. Comparison of the alternatives with respect to "operating cost" sub-criteria

	A1	A2	A3	Local Priorities
A1	1	1/2	1/2	0,198
A2	2	1	1/2	0,312
A3	2	2	1	0,490
Total	5,000	3,500	2,000	
				λmax=3,054 CI=0,027 ≤0,1 CR=0,046

Table 18. Comparison of the alternatives with respect to "labor" sub-criteria

	A1	A2	A3	Local Priorities
A1	1	1/2	1/2	0,197619048
A2	2	1	2	0,49047619
A3	2	1/2	1	0,311904762
				λmax=3,054 CI=0,027 ≤0,1 CR=0,046

Table 19. Comparison of the alternatives with respect to "technological competence" sub-criteria

	A1	A2	A3	Local Priorities
A1	1	1/7	1/6	0,072
A2	7	1	1/2	0,373
A3	6	2	1	0,555
Total	14,000	3,143	1,667	
				λmax=3,081 CI=0,040 ≤0,1 CR=0,070

Chapter 3
Impact of Hydrogen Fuel Cell Technology on Aircraft Maintenance:
Reliability, Cost, and Safety

Syed Shaukat
University of New South Wales, Australia

Cheng-Lung Wu
University of New South Wales, Australia

ABSTRACT

The aviation industry is adopting a promising approach to decarbonize the aviation sector dramatically by incorporating hydrogen fuel cell technology and expects to start some commercial services as early as 2035. The authors estimate the aircraft operating cost to become less expensive eventually with this forthcoming fuel cell technology, and the total maintenance cost of hydrogen-powered aircraft to come down in the long run for all segments of the aviation sector, from short haul to long haul services. With the possible incorporation of a new superconducting electrical system, a pneumatic power system, and a fully integrated electrical drive train system, additional maintenance will be required due to their cryogenic factors and associated safety issues. This chapter focuses on the potential impacts of hydrogen fuel cell technology on aircraft maintenance reliability, cost, and safety across the wide range of commuter, regional, short, medium, and long-haul segments of the aviation industry.

INTRODUCTION

It is an undeniable fact that proper execution of extensive and effective aircraft maintenance practices is vital for the safety and quality of aircraft health and airline operation. Negative impacts often manifest in rather unwanted occurrences of unscheduled maintenance stops, prolonged maintenance downtimes, aircraft on ground (AOG) situations, maintenance overloads and over expenditures. Therefore, the

DOI: 10.4018/978-1-6684-6835-7.ch003

maintenance, repair, and overhaul organizations (MROs) must work together with the airline industries to maximize maintenance reliability and minimize total maintenance cost as much as possible by thoroughly incorporating necessary maintenance requirements and optimizing all preventative and non-routine maintenance activities.

While the aviation world is still in some early stage of trials, forecasts, projections, and tentative schedules for the implementation of and the eventual transition to hydrogen fuel cell technology, the impacts on all aspects of aircraft maintenance and reliability must be analysed and estimated now for the successful deployment of this new technology. However, the required estimates and calculations have not yet been carried out thoroughly for the aviation and maintenance industry because some significant developments and design changes are still pending. Therefore, instead of waiting for some bells and whistles, the authors consider all aspects of aircraft maintenance issues resulting from incorporating the hydrogen fuel cell technology in general aviation sectors. For this purpose, this chapter focuses on the impacts of this new technology on aircraft maintenance reliability, cost, and safety, mostly in qualitative terms, with necessary comparisons and indications of the relative magnitudes, intervals, consequences, and challenges of various maintenance issues.

BACKGROUND

The first airship, LZ1, took off on July 2, 1900, near Lake Constance in Germany, with seventeen hydrogen cells and two internal combustion engines, flying for about 17 minutes and reaching a height of 1,300 feet above the ground (Bellis, 2019). This event marked the beginning of the Zeppelins Airship era, which provided one of the most successful air travel services of that time, in contemporary luxury for 30 years, flying more than 2,000 flights over a million miles without a single injury. Then, in a few fiery minutes at the end of the last journey of airship LZ129 on May 6, 1937, in Lakehurst, New Jersey, the golden era of the rigid airships ended abruptly with a shocking disaster. In total, the Hindenburg accident caused 36 souls to depart untimely by killing 13 passengers, 22 crewmen on board, and 1 crew member on the ground, while miraculously sparing the remaining 62 people on board to survive, although some of them suffered serious injuries (The Hindenburg Disaster, 2020).

Scientific tests and research had been carried out for more than 80 years now to determine the most probable cause of the disaster, and the findings coincide with what the world has already gathered from the Hindenburg accident report prepared by the US commerce department in 1937 (Hindenburg Accident Report, 1937). To date, there has been no evidence to date to support a plausible sabotage hypothesis that resulted in such a horrific disaster, and the fabric covering of the zeppelin airship was not blamed much as one would have expected. A static discharge most possibly ignited a mixture of hydrogen and air, which had been built up in considerable quantity in the upper stern part of the ship by a leak at or in the vicinity of cells 4 and 5 (Hindenburg Accident Report, 1937). Simply put, the Hindenburg was destroyed in a sudden fire after a spark most probably ignited the flammable hydrogen gas core and it took just 32 seconds for the hull of the airship to be incinerated. The majestic Hindenburg airship was going to be obsolete in a few years anyway as it experienced significant hydrogen leakages even before the disaster, the exact cause of which is still a mystery (The Hindenburg Disaster, 1997). *"Along with the Titanic sinking in 1912 and the Space Shuttle Challenger exploding in 1986, the Hindenburg burning, and crashing is one of the three most famous passenger craft disasters of the 20th century."* (Frassinelli, 2012).

The cloth bags of those days for Zeppelin's hydrogen storage containment are gone forever and have been replaced by the modern highly insulated, structurally ergonomic, and robust hydrogen storage tanks of our time (Rondinelli et al., 2017). Hydrogen fuel cell technology is slowly making inroads into the commercial aviation sector, but it is nevertheless true that hydrogen is still regarded as a dangerous fuel due to its flammability and cryogenic hazards. Despite the high energy efficiency and green environmental gains, the aviation world must bear in mind that any negative incidence may seriously impact the public's perception of the successful deployment of hydrogen as a clean, safe, and sustainable alternative to fossil fuel.

Some new hydrogen start-up company is already using compressed hydrogen for the proton exchange membrane (PEM) fuel cells and chasing its bold and brave goal of zero-emission to tackle the ever-growing concerns of carbon issue in our modern era of air travel (Mark, 2015). We find quite a few ambitious plans for manufacturing fuel cell planes for 300 to 500-mile short-haul services targeting this premium market segment by the middle of this decade (Homann, 2019). It is high time now for start-ups and larger manufacturers to come up with successful models of hydrogen fuel cell operated aircraft, and the tide should also be accompanied by quality research and investigations into the impacts of this new technology on the aircraft maintenance industry. Aviation researchers should start full swing to prepare airline businesses for the impacts on aircraft maintenance activities that are particularly relevant to hydrogen fuel cell aircraft. The authors emphasize that efforts should begin right now, at the outset of this new technology and in parallel with current fossil fuel-powered aircraft operations. Investing in and benefiting from hydrogen fuel cell technology in the aviation sector requires immediate action.

MAJOR CONCERNS AND ASSESSMENTS

Several major implications and their consequential impacts on aircraft maintenance arise simply due to aircraft design changes for the integration of hydrogen tanks and the introduction of auxiliary power units (APU), pneumatic power generation, electrical drive train technology, superconducting electrical circuits, and so on. Various maintenance safety and reliability factors are associated, particularly with the cryogenic requirements of storage, transportation and refuelling of hydrogen fuel cell aircraft. An additional maintenance burden also arises due to ongoing inspections of structural integrity and possible leakage. The awareness programs should be part of the mandatory requirements across the whole aviation industry, utilizing resources like online awareness level of training, operational hands-on level of training, the national expertise level of training, among others (Barilo, 2015). Here, the authors systematically explore the maintenance concerns to highlight relevant focus area for fruitful insights along with some estimates and assessments.

Maintenance Impacts

Ground Handling and Refueling

Aircraft ground handling and servicing is the first interface between aircraft operation and maintenance, where the concerns of hydrogen fuel's flammability hazard and certain adverse biological reactions must be addressed effectively. Aircraft refuelling and ground handling functions are the regular aspects of aircraft operation, and airport precincts must implement necessary policies, procedures, and technologies for adequate supply and distribution of hydrogen fuel. To ensure the safety of personnel directly responsible for these critical operations in airports, it is essential that they are trained in maintenance safety awareness, industry practices, and up-to-date knowledge of production, storage, distribution, and containment of liquid or compressed hydrogen required for the fuel cells (Barilo, 2015).

Careful consideration is necessary for the selection and maintenance of hydrogen containment facilities in airport areas for refuelling and ground handling operations. The presence of condensed air near a liquid hydrogen storage system can result in oxygen enrichment and explosive conditions. Inside the ground fuelling vessel, continuous evaporation of liquid hydrogen will generate gaseous hydrogen, which may result in pressure increase. Accidental air leakage into a liquid hydrogen storage vessel may introduce some moisture, condensed water, which may then lead to potential ice formation, and inadequate purging may even cause relevant instruments to malfunction. It is a challenging task to identify leaks due to the notorious feature of hydrogen seepage through the containment lines and tanks and hydrogen's ability to permeate solid materials such as polymers (Barilo, 2015).

The cryogenic temperature of hydrogen fuel storage and supply system is a serious and hazardous concern, which should mandate specialist training requirements. Potential ice formation around pumps, vents, valves, connections, and accessories may cause them to malfunction and personnel coming in touch with cryogenically cooled metals may also suffer serious injury (Schmidtche et al., 1997). Liquid hydrogen can also cause serious frostbite and hypothermia due to its extremely low boiling point. Even though hydrogen is not corrosive or poisonous, in the event of a leak there is a serious physiological concern of asphyxia because if hydrogen accumulates in enough quantity, it can potentially starve an individual of oxygen (CSIRO, 2020).

Hydrogen Tanks

Hydrogen fuel can be stored either as a compressed gas or in liquid form, although gaseous storage is normally suitable for shorter flights only. It is usually preferred in the liquid form for storage inside fuel cell-powered aircraft, as the storage tank will require a reduced volume, significantly less weight, and faster refueling times compared to the case of a gaseous hydrogen tank. Liquid hydrogen storage will be crucial for future commercial aircraft operation, particularly for short- to long-range sectors where aircraft may be carrying several tons of hydrogen per flight. The recommended design of lightweight liquid hydrogen tanks should bring about a 50 percent reduction in weight as compared to some current prototypes (Hydrogen-powered Aviation, 2020).

On the positive side, liquid hydrogen is regarded as somewhat safer than gaseous hydrogen because of its lower pressures inside the storage tanks and a lower possibility of fatigue-induced structural failure. After an aircraft crash, liquid hydrogen normally will not contaminate the water or soil of the natural environment like traditional kerosene fuel (Rondinelli et al., 2017). We also expect a relatively safer outcome in the event of a crash, as the liquid hydrogen tank is less likely to rupture because of less heat and intensity of hydrogen fire, the rigidity of liquid hydrogen tanks, buoyancy, and quick dissipation of the gaseous discharge (Brewer, 1983). The first experimental aircraft, the Tu-155, which ran on liquid hydrogen at first and liquid natural gas (LNG) later, had its hydrogen fuel tank arranged inside the air-blown rear compartment (Aviation Image Library, 2019). The Tu-155 made around 100 flights since it first flew on 15th April 1988 and is currently in storage at Ramenskoye airport near Zhukovskiy (Russian Aviation, 2021).

Generally, there are two types of hydrogen storage tank configurations in aircraft: non-integral and integral structural designs. The aircraft hydrogen tanks must cope with the aerodynamic and inertial loads as well as the fuel containment loads in the case of non-integral design, where they are usually mounted above or under the wing or on the airframe itself. For short-range aircraft, storing the hydrogen in pods below the wing could also be an option since the aerodynamic performance is not decreased much, and the maintenance and safety aspects are easier to cope with. The integral tank is a more realistic design choice for long-haul or wide-body aircraft, and the fuel tank needs to be built with lightweight material for the double-insulated tank walls and insulations. However, this may require an expanded airframe design with additional maintenance inspections to check for structural integrity at certain regular intervals (Khandelwal et al., 2013). The fuel tanks and delivery lines pose some serious risk of exposure to passengers and other personnel in the event of a leak or accident, so the tank design, installation, and proper maintenance practices must consider these concerns to make it very unlikely.

The temperature of the tanks, hydrogen embrittlement issues and permeation through the tank walls are the key factors and challenges for the cryogenic requirement of hydrogen storage inside the aircraft. The integral hydrogen storage tanks should always remain at cryogenic temperatures inside the aircraft with proper external insulations, to minimize boil-off losses and thermal cycling fatigue of the tank. When the aircraft is out of service for an extended period, or in the case of tank entry for special maintenance and inspection purposes, tank temperature should be allowed to rise only then. While carefully warming up the tank, it must be de-fuelled first and then subsequently be purged with inert gas, after which the inert gas can be replaced with air. Aircraft refuelling can then be carried out subsequently in reverse order, preceded with purging at first, then replacing of the inert gas with gaseous hydrogen and followed by a gradual chill-down before filling the tank with liquid hydrogen again (Verstraete, 2009; Brewer, 1991; Sefain, 2005).

Fuel System

Although cryogenic fuels have been designed for use in launch vehicles for space exploration and projectile propellant technologies by manufacturers like Ariane Group, liquid hydrogen has not yet been used for regular commercial air transport services (Hydrogen-powered Aviation, 2020). The first experimental aircraft, the Tu-155, carrying 18 m3 of liquid hydrogen storage tanks in the rear fuselage, took off on April 15, 1988, and successfully made several flights without any operational issues or in-flight shutdowns. The aircraft fuel section was constantly purged with air from the air conditioning system of the aircraft, along with an additional helium system to purge the pipelines and fuel system valves (Schmidtche et al., 1997; Verstraete, 2009).

For a successful deployment of the hydrogen fuel cell technology in commercial aircraft operation, it is essential to have a reliable and redundant supply of liquid hydrogen fuel with safe distribution, vaporization, and feeding mechanisms. As liquid hydrogen requires cooling down to 20 degrees Kelvin, the fuel system needs to be designed with double-insulated fuel pipes, reliable pumps, and valves to withstand such cryogenic cooling temperatures and avoid potential leakage and embrittlement of the materials of the tank and supply system (Hydrogen-powered Aviation, 2020). In the U.S., NASA is funding aviation research for a new fuel cell concept in its CHEETA (Cryogenic High-Efficiency Electric Technologies for Aircraft) program and is working on a futuristic design for a fully electric platform that uses cryogenic liquid hydrogen fuel cells (Szondy, 2019). It is a fundamental shift away from traditional jet fuel towards a more sustainable energy source with fuel cells that can develop new electrically driven propulsion systems for commercial aircraft operation.

Airframe Design

The integration of liquid hydrogen tanks, high-power fuel cell systems, and complex fuel and electricity distribution are the key challenges that need to be overcome through better aircraft designs. In contrast to familiar traditional fuel like kerosene, hydrogen fuel still requires larger volumes and storage tanks on board the aircraft, even though it has a higher gravimetric energy density. Special maintenance requirements and regular inspections must strictly adhere to the structural integrity, cracks, leakage, and operational functionality of all relevant aircraft systems. Two aircraft design concepts are currently available, namely the evolutionary and revolutionary designs, to incorporate the emerging development of hydrogen fuel cell technology in the aviation sector.

The evolutionary design is characterized by the adoption of the traditional tube and wing design, where the airframe and fuselage design are adapted for the incorporation of hydrogen storage tanks (Hydrogen-powered Aviation, 2020). The revolutionary aircraft design is characterized by a new aerodynamic concept of a blended wing-body design with better integration of liquid hydrogen storage tanks. In the evolutionary design, integral hydrogen tanks with larger volumes would result in a substantial increase in airframe length and energy demand, with costs per passenger estimated to be 40 - 50% higher. Although technically feasible for the incorporation of liquid hydrogen fuel cells, the traditional evolutionary designs are considered less suitable for the long-range sectors than the commuter and short-range sectors. The revolutionary blended-wing-body aircraft designs would be much better for long-range operation, but the forthcoming developments can be years away from entering regular commercial services (Hydrogen-powered Aviation, 2020). The aerodynamic performance of aircraft can benefit from the additional cooling capacity of liquid hydrogen, and air drags can be reduced by cooling through hydrogen for laminar flow to improve the performance of a subsonic aircraft. The cold hydrogen can be used to cool the structures of a supersonic or hypersonic aircraft exposed to aerodynamic heating (Cunnington & Parmley, 1980).

A significantly smaller amount of fuel is required for the operation of hydrogen-fueled aircraft compared to conventional fossil fuels because of the extremely high heat of combustion and lesser vibrations, unlike the traditional turbine engines. The main advantages of reduced fuel load are lower gross weight and smaller and quieter engines, even though part of the weight advantage is nevertheless lost due to heavier and bulkier hydrogen tanks. The additional requirement to absorb shocks from numerous landings, along with the required useful life left with the insulation material of the tanks, still presents some challenges (Verstraete, 2009).

Engines and Nacelle Pods

Fuel cell-driven engines can be smaller, quieter, and more efficient due to the high specific heat capacity, reduced fuel load, and cryogenic temperature of liquid hydrogen. The engine performance can be improved by using a heat sink provided with the liquid hydrogen storage system. The engine installation pods must ensure an efficient wing aerodynamics design, along with some reductions in nacelle drags and the wing maximum lift coefficient. The best positions for the engine nacelle pods are in front and below the leading edge of the wing, which must minimize the interference between the pod and the wing from a drag perspective (Verstraete, 2009). To obtain a proper centre of gravity excursion throughout the whole loading envelope, the position of the wing on the fuselage may need to be changed. Engine installations over the wing are sometimes proposed to offer noise reduction benefits, but it may influence the landing gear height by allowing a higher vertical separation between the nacelle pod and the wing. Hence, this type of installation may increase the associated aircraft maintenance costs (Verstraete, 2009).

Non-Propulsion Systems

In conventional aircraft, the APU is a small turbine engine located at the rear of the fuselage. Its main functions are to generate electrical energy and serve as an emergency backup power source in flight when the primary generator fails. Additionally, the APU provides compressed air for cabin heating and cooling, starts the primary jet engine, powers electrical equipment on board, and enables the aircraft to operate independently of other ground service equipment. The hydrogen fuel cell technology is particularly useful in generating electricity for the APU, aircraft pushback and taxiing mechanism, and various other systems and equipment on board, such as fire prevention, water storage, cabin heating systems, and the emergency Ram Air Turbine (RAT). The fuel cells can provide a more reliable, efficient, and alternative power source for aircraft operation, and Airbus is currently considering fuel cell technology for an innovative design of APU (CSIRO, 2020).

Several basic functions of aircraft operations, such as landing gear and flight control movements, normally rely on hydraulic power and pneumatic bleed air from the engine with the help of other mechanical systems. Hydrogen fuel cell operated APU can supply power to the aircraft when it is on the ground, either stationary or taxiing, and the remaining available power can also be used for aircraft heating, pressurization, galley power, and cabin entertainment systems. The water produced by fuel cells can be used as refuse water, and if it meets the required approval, it can also be used in the potable water systems for drinking purposes and in the cabin humidification system. By reducing the need for carrying water onboard, we can avoid the risks of bacterial contamination and poor water quality (Hardman, Chandan & Steinberger-Wickens, 2015; CSIRO, 2020).

For pushback and ground movements, a semi-automated system called TaxiBot was trialed previously, which raised concerns about adding congestion, CAPEX, maintenance requirements, and delays to aircraft movement. The fuel cells can be used to provide power for aircraft ground movements with the addition of an electrical drive system and motors to the wheels of the aircraft landing gears, allowing the APU to drive the aircraft during the taxiing phase. In 2013, Honeywell and Safran on a joint venture trialed an upgraded version of the taxiing system, called the "electric green taxiing system" (Hardman, Chandan & Steinberger-Wickens, 2015).

The system provides hydrogen to a proton exchange membrane (PEM) fuel cell and then converts the chemical energy from hydrogen and ambient oxygen to electrical energy for the motor providing rotational energy for the propeller. The European Marine Energy Centre is currently exploring the infrastructure options for bringing hydrogen fuel to airports and aircraft, which may have some additional maintenance impacts on the aircraft operation side. A schematic diagram of the carbon-free infrastructure layout by ZeroAvia was highlighted by Robinson (2020). Topham (2016) also illustrated a layout picture of the EasyJet hybrid energy system platform for aircraft operation.

EasyJet has unveiled plans to use hydrogen fuel cells on its aircraft to save up to 50,000 tons of fuel a year and cut its carbon emissions. The airline hopes to trial technology later this year that would allow planes to taxi to and from the runway using zero-emissions fuel. (Topham, 2016).

Assessments of the Maintenance Impacts

Ground Handling and Fueling

We do not anticipate any notable change or impact on the ground handling and servicing aspects of aircraft maintenance, as the servicing of fuel cell-powered aircraft can be handled in a similar manner to conventional fuelled aircraft. While passengers are boarding and cargo is being loaded, a liquid hydrogen aircraft may be fuelled at the gate along with other ground operations, with a similar turnaround time as kerosene-fuelled aircraft (Brewer, 1991; Sefain, 2005). Ground personnel directly involved in aircraft servicing and airport precincts need to follow rigorous safety procedures and maintenance precautions after the initial hurdles of transitioning to the new technology.

Hydrogen Tanks

The maintenance cost for an integral tank configuration will start to rise slowly due to the expanded airframe design needed to accommodate the storage tanks. This will require more maintenance checks during the initial period of introduction. The design of the integral tank configurations should last throughout the lifetime of an aircraft, with the least possible amount of necessary maintenance requirements. Hence, the overall maintenance cost for hydrogen fuel cell aircraft services is expected to decrease in the long run. Routine visual inspection of the tank structure is recommended every 8,000 hours and should be carried out from inside the tank, which is consistent with airline practice for primary structures (Verstraete, 2009). It is also recommended that the maintenance and recalibration of leak detectors be performed every 3-6 months and recorded accordingly in the relevant logs or manufacturer's recording facilities (Barilo, 2015). Inspections, troubleshooting, and repair tasks involving tank entry, defueling, and subsequent refuelling can impose additional maintenance requirements, delays, and costs, as liquid hydrogen must always be kept at cryogenic temperatures inside the tank.

Fuel System

For optimum performance and redundancy purposes, the aircraft fuel system should be built with highly durable, safe, and light components. This includes double-insulated fuel pipes equipped with cryogenic cooling features, compressors and heat exchangers equipped with leakage control and venting management, and sensors and instruments equipped for condition-monitoring capabilities. The curing process of hydrogen permeation and embrittlement in the material of storage tanks and fuel systems may also contribute to additional maintenance costs (Verstraete, 2009). As always, the maintenance cost must be kept as low as possible to make the technology commercially feasible and environmentally sustainable for commercial air travel services.

Engines and Nacelle Pods

Engine installations over the wing are proposed sometimes to take advantage of the potential noise reduction benefit, but it may then affect the landing gear height by allowing a higher vertical separation between the wing and nacelle pods. As a result, this type of installation may potentially increase the associated maintenance cost (Verstraete, 2009). For aircraft operating on short-range routes, hydrogen pods mounted below the wing could be a better design option since they do not significantly affect the aircraft's performance, and maintenance and safety aspects are easier to manage.

Maintenance Cost

The financial aspect of the aircraft maintenance business may be estimated by quantitative analysis and assessment of the total maintenance cost of the fleet, including direct labour, indirect labour, equipment, spare parts, tools, and materials required, as well as the applicable maintenance burdens. Various miscellaneous aspects of the aircraft maintenance business, such as compliance, quality, safety, and reliability, can be estimated by qualitative analysis and assessment of industry practices. This can be achieved by examining the number of compliance and safety breaches, warranty claims, aircraft incident severities, accidents, and fatalities, among others. We conduct a comparative cost analysis of the impact of fuel cell technology on aircraft maintenance by evaluating the estimates and differences between these various aspects of aircraft maintenance.

Direct Maintenance Cost (DMC)

DMC includes all relevant maintenance activities, i.e., line maintenance, base maintenance, engine overhaul, and components maintenance for a fleet of aircraft. For example, when a technician works on an aircraft to replace a part, the labour and part's cost are related to the maintenance and hence become part of DMC. A breakdown of DMC for commercial aircraft operation is shown in Figure 1 for each category, as provided by the IATA for the financial year 2013 (Fioriti, Vercella & Viola, 2018).

Figure 1. Direct maintenance cost breakdown

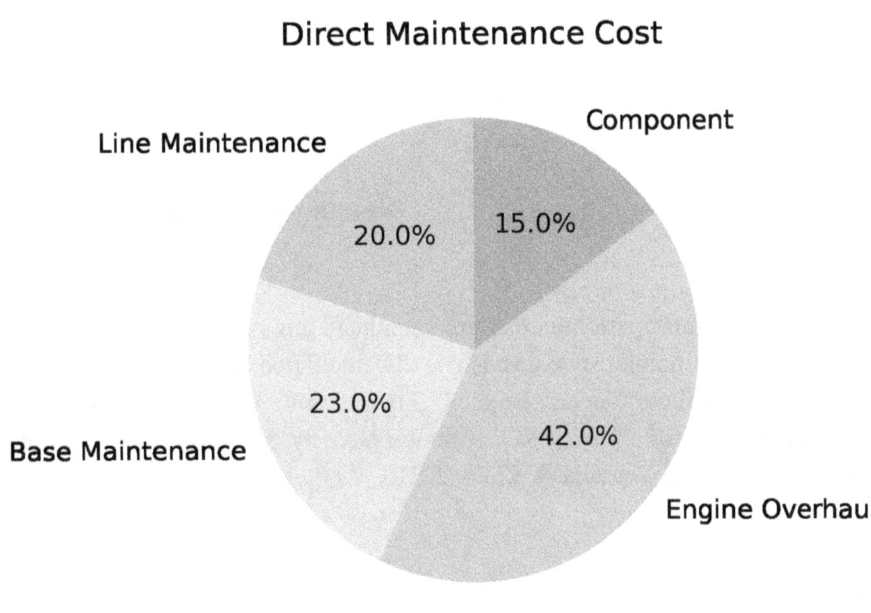

Indirect Maintenance Burden (IMB)

IMB includes the procurement expenses, allocation of the hangars and other facilities, quality control, associated engineering services etc. and is usually expressed as a percentage of DMC for a fleet of aircraft. The overhead expenses for the maintenance cost of associated equipment and tools, the establishment cost of associated building, salaries, technical facilities etc. are included in estimating the maintenance burden.

Total Maintenance Cost (TMC)

TMC is a sum of DMC and IMB, all together it contributes up to 15% of the direct operating cost of an airline. A general breakdown of the total maintenance cost for commercial aircraft operation is shown in Figure 2 for each category, as provided by IATA for the year 2013 (Fioriti, Vercella & Viola, 2018).

Figure 2. Total maintenance cost breakdown

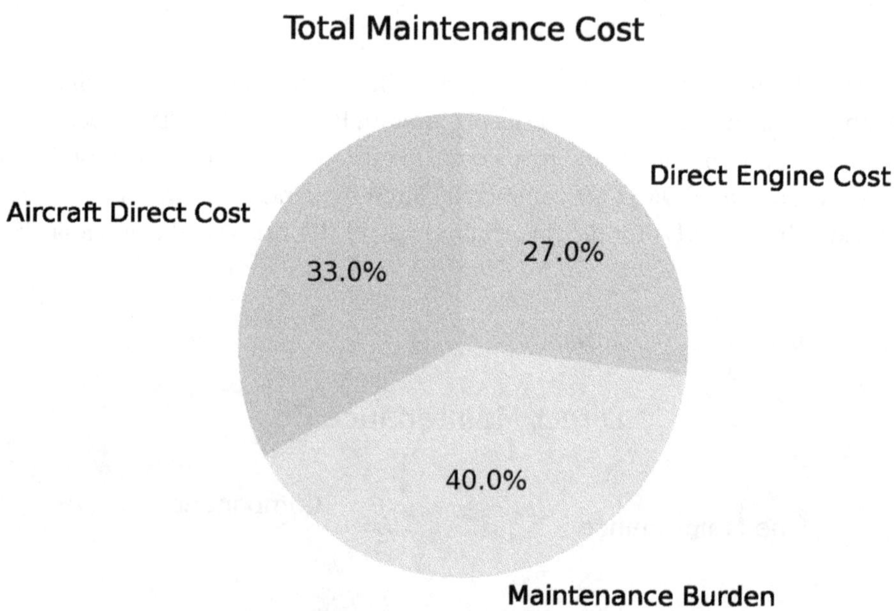

Many components of aircraft operating costs are calculated annually, such as depreciation, interest, and insurance, while the total maintenance cost is usually calculated on a per-trip basis and is typically expressed as a cost per flight hour. As can be seen, this is a quite complex method, as different cost parameters involved must be brought back to a common base by converting them to hourly costs and then summed up in total (Fioriti, Vercella & Viola, 2018).

The Overall Picture

Time Horizon

The aviation industry is targeting commercial aircraft services to be operational by 2028-2035, particularly in the commuter and short-range segment. Upon reaching the first entry-into-service milestone, the industry also has ambitious goals to replace all aircraft up to a range of 10,000 km after 2028-2038 (Hydrogen-powered Aviation, 2020). Dr. Bart Biebuyck, the Executive Director of the Fuel Cells and Hydrogen Joint Undertaking, a European public-private partnership aimed at accelerating the market introduction of these technologies, boldly stated in a jointly commissioned report on hydrogen-powered aircraft that deployment for short-range planes should be possible by 2035 (O'Callaghan, 2020). Dr. Biebuyck also said, "That means on European soil, you could connect all the big cities in Europe using hydrogen-powered planes. By 2050, the ambitious scenario is that 40% of the (European aviation) fleet would be powered by hydrogen" (O'Callaghan, 2020).

It would be highly beneficial in all segments of the aviation sector to have aircraft designs with lightweight structures and a higher level of aerodynamic efficiency, utilizing safe and reliable hydrogen components with minimum maintenance requirements, lead times, and overhead burdens (Hydrogen-powered Aviation, 2020). The new startup ZeroAvia has already flown more than a dozen flights successfully with its Piper Matrix N504EZ retrofitted with a hydrogen fuel cell (Bogaisky, 2019). The European aerospace giant Airbus has set a target of the year 2035 for zero-emission hydrogen-fueled commercial aircraft to dominate the skies with a trio of high-tech hydrogen hybrid models of new airliners (Starling, 2020). Boeing paints a picture of short-haul operation between now and 2030, medium-haul from 2030 onwards, and long-haul towards 2050, with the vision of carbon-free travel coming closer and closer in the field of commercial aviation with hydrogen technology (Gillespie, 2020). Some infographics of this bold and futuristic journey by Airbus and the aviation industry have been provided by Starling (2020) and Drăgan (2021).

Financial Estimates

Fuel cell aircraft can offer additional cost savings because of reduced maintenance, as they do not produce heat and vibration due to their limited number of moving parts, unlike the traditional turbine engine (CSIRO, 2020). ZeroAvia has initiated its Hyflyer project to supply hydrogen fuel cell drivetrains to commercial airline operators and manufacturers for 10- to 20-seater aircraft by 2022, following up successful runs with its prototype of 6-seater commuters. According to ZeroAvia, a fuel cell powertrain is going to have 75% lower fuel and maintenance costs for short-range operations up to 500 miles, to reduce the operating costs by about 50% of fossil fuel-powered aircraft (Homann, 2019).

The capital expenditure and total maintenance costs for hydrogen-powered commuters will be 10-15% less compared to conventional aircraft in the commuter segment of the aviation industry based on the total cost per available seat-kilometre (CASK). The total operating cost of fuel cell-powered aircraft operation may be higher than that for conventional aircraft due to an increase in the complexity of fuel distribution, propulsion cost, safety measures, and special maintenance checks required. The rise in maintenance costs, especially during the initial years of introduction, would be primarily due to a larger airframe size to integrate the storage tanks, but the overall maintenance cost is expected to decrease in the long run (CSIRO, 2020). The short-range segment of aviation services is forecasted to be operational by the year 2035, and the total aircraft operating cost is going to be 25% more than a kerosene aircraft adjusted for 2035 trends and technologies. A simple breakdown of this cost increase (25%) consists of higher energy costs (9%), higher capital expenditure (7%), aircraft maintenance costs (6%), and other costs (3%) (CSIRO, 2020).

Due to a range of significant differences between the traditional kerosene fuel and the forthcoming hydrogen fuel cell technology, the airframe maintenance and acquisition cost for hydrogen-fuelled aircraft may increase for the short-range and medium-range operations during the initial phase of deployment, from 10% up to 50% of the equivalent cost of kerosene-fuelled aircraft. But the direct operating costs for hydrogen-fuelled aircraft would still be competitive with a much higher fuel price, even if the maintenance and acquisition costs happen to reach the upper limit of 50% (Verstraete, 2009). For the commuter segment of the aviation industry, the aircraft operating cost would still be favourable indeed, with a reduction of 10% to 15% compared to the conventional commuters (CSIRO, 2020). It is still not viable for wide-body aircraft to be powered entirely by hydrogen fuel cells for long-range operations. A hybrid design is most likely to dominate this segment of the market, which is expected to be operational by 2050.

CONCLUSION

Since the day of the Hindenburg disaster, almost a century later, the aviation industry is now genuinely considering the use of pollution-free, clean, and energy-efficient hydrogen fuel cell technology for regular commercial aircraft operation almost a century later now. There are clear environmental benefits and higher efficiency gain of hydrogen fuel cells for all our glories, but the remaining hurdles must be overcome by appropriately preparing the airline industry for the challenges of this new technology. Certainly, there are still some uphill roads to travel for making this fantastic quasi-futuristic transition journey of hydrogen fuel cell technology ready for the aviation industry. We envisage that the aviation industry will soon overcome the challenges to make the overall maintenance profile of hydrogen-powered aircraft more favourable than the traditional turbine engine aircraft in the long run because the components of hydrogen aircraft are subject to a less extreme range of heat, pressure, and motion. The moral of the story is that hydrogen fuel cells can be used safely in the aviation industry if we diligently follow relevant safety procedures and maintenance requirements for the technology. The impacts of this forthcoming technology on the aircraft maintenance industry are reasonably within our control. The aviation world can make it happen and call it a success only if we all act together in the matter of aircraft maintenance for the overall safety, reliability, and profitability of the sustainable aviation industry.

REFERENCES

Aviation Image Library. (2018). *Tupolev Tu-155.* AirTeamImages. https://www.airteamimages.com/tpolev-tu-155_CCCP-85035_tupolev-design-bureau_354061.html

Barilo, N. (2015). *Safety Considerations for Hydrogen and Fuel Cell Applications.* Hydrogen Tools, Portal. https://h2tools.org/sites/default/files/Safety_Considerations_for_Hydrogen_and_Fuel_Cell_Applications.pdf

Bellis, M. (2019). History of Airships and Balloons. *ThoughtCo, Life Long Learning*. https://www.thoughtco.com/history-of-airships-and-balloons-1991241

Bogaisky, J. (2019). Startups bet hydrogen Fuel Cells are ready for takeoff in aviation. *Forbes Magazine*. https://www.forbes.com/sites/jeremybogaisky/2019/08/14/zeroavia-alakai-hydrogen-fuel-cell/#2cc402b918aa

Brewer, G. D. (1919). *Hydrogen aircraft technology*. CRC Press.

Brewer, G. D. (1983). An Assessment of the Safety of Hydrogen-Fueled Aircraft. *Journal of Aircraft, 20*(11), 935–939. doi:10.2514/3.48195

Caleb, A., & Kunycky, A. (2019). *Hydrogen as a Renewable Energy Carrier for Commercial Aircraft*. ResearchGate. https://www.researchgate.net/publication/336551230_Hydrogen_as_a_Renewable_Energy_Carrier_for_Commercial_Aircraft

CSIRO. (2020). *Opportunities for hydrogen in commercial aviation*. Main Report, Energy and Resources, CSIRO. https://www.csiro.au/en/Do-business/Futures/Reports/Energy-and-Resources/hydrogen-commercial-aviation

Cunnington, G. R., & Parmley, R. T. (1980). Aerodynamic Surface Cooling for Laminar Flow Control for Hydrogen-Fuelled Subsonic Aircraft. *SAE Transactions*. doi:10.4271/801155

Drăgan, O. (2021). Sustainable Aviation Needs Airports to Become Complex Hydrogen-Based Ecosystems. *Autoevolution.com*. [Photograph]. Retrieved from https://s1.cdn.autoevolution.com/images/news/sustainable-aviation-needs-airports-to-become-complex-hydrogen-based-ecosystems-163720_1.jpg

Fioriti, M., Vercella, V. & Viola, N. (2018). Cost-Estimating Model for Aircraft Maintenance. *Aerospace Research Centre* (*Vol. 55*, no. 4).

Frassinelli, M. (2012). The Hindenburg 75 years later: Memories time cannot erase. *New Jersey Real-Time News*. https:// www.nj.com/news/2012/05/the_hindenberg_75_years_later.html

Gillespie, M. (2020). *Roadmap to runway: Getting hydrogen off the ground. CSIROscope*. Blog.

Hardman, S., Chandan, A., & Steinberger-Wickens, R. (2015). Fuel cell added value for early market applications. *Journal of Power Sources*, *287*, 297–306. doi:10.1016/j.jpowsour.2015.04.056

Report, H. A. (1937). U.S. Commerce Department. *Airship.net*, Airship Blogs. Retrieved from https://www.airships.net/hindenburg/disaster/commerce-department-report/

Homann, Q. (2019). Aviation, Fuel Cell & Hydrogen Energy Association. https://www.fchea.org/in-transition/2019/11/25/aviation

Hydrogen-powered Aviation. (2020). A fact-based study of hydrogen technology, economics, and climate impacts by 2050. *Fuel Cells and Hydrogen Joint Undertakings*. Retrieved from https://www.euractiv.com/wp-content/uploads/sites/2/2020/06/20200507_Hydrogen-Powered-Aviation-report_FINAL-web-ID-8706035.pdf

Khandelwal, B., Karakurt, A., Sekaran, P. R., Sethi, V., & Singh, R. (2013). Hydrogen powered aircraft: The future of air transport. *Progress in Aerospace Sciences*, *60*, 45–59. doi:10.1016/j.paerosci.2012.12.002

Learson, D. L. (2019). NASA funds aviation research on a new fuel concept. *Physics.org*. Retrieved from https://phys.org/news/2019-05-nasa-funds-aviation-fuel-concept.html

O'Callaghan, J. (2020). *Quiet and green: Why hydrogen planes could be the future of aviation*. Horizon, The EU Research and Innovation Magazine.

Report, H. A. (1937). U.S. Commerce Department. *Airship.net, Airship Blogs*. https://www.airships.net/hindenburg/disaster/commerce-department-report/

Robinson, T. (2020). *High Time for Hydrogen*. Royal Aeronautical Society. https://www.aerosociety.com/news/high-time-for-hydrogen/

Rondinelli, S., Gardi, A., Kapoor, R., & Sabatini, R. (2017). Benefits and challenges of liquid hydrogen fuels in commercial aviation. *International Journal of Sustainable Aviation*, *3*(3), 200–216. doi:10.1504/IJSA.2017.086845

Russian Aviation. (2021). News. Retrieved from https://www.ruaviation.com/news/2021/4/15/16105/

Schmidtchen, U., Behrend, E., Pohl, H. W., & Rostek, N. (1997). Hydrogen Aircraft and Airport Safety. *Renewable & Sustainable Energy Reviews*, *1*(4), 239–269. doi:10.1016/S1364-0321(97)00007-5

Sefain, M. (2005). *Hydrogen aircraft concepts & ground support*. [Ph.D. thesis, Cranfield University, School of Engineering, Cranfield, England].

Starling, M. (2020). Future of aviation: Airbus to develop world's first zero-emission aircraft. *The Week*. https://www.theweek.co.uk/aviation/108150/future-of-aviation-airbus-zero-emission-aircraft

Szondy, D. (2019). Nasa backs development of cryogenic hydrogen system to power all-electric aircraft. *New Atlas*. Image retrieved from https://newatlas.com/nasa-cheeta-funding-aircraft-fuel-cell/59725/#gallery:3

The Hindenburg Disaster. (1997). Airships.net, Airship Blogs. https:// www.airships.net/hindenburg/disaster/

The Hindenburg Disaster. (2020). History.com. https://www.history.com/this-day-in-history/the-hindenburg-disaster

Topham, G. (2016, February). EasyJet plans to cut carbon emissions with hydrogen fuel-cell trial. *The Guardians*.

Verstraete, D. (2009, April). *The Potential of Liquid Hydrogen for long range aircraft propulsion*. [Ph.D. thesis, Cranfield University, School of Engineering, Cranfield, England].

APPENDIX

APU: Auxiliary Power Unit
AOG: Aircraft On Ground
CASK: Cost per Available Seat Kilometre
DMC: Direct Maintenance Cost
IMB: Indirect Maintenance Burden
MRO: Maintenance Repair and Overhaul
PEM: Proton Exchange Membrane
RAT: Ram Air Turbine
TMC: Total Maintenance Cost

Chapter 4
Human Resources in Aviation in the Covid-19 Era

Mustafa Kaya
https://orcid.org/0000-0002-0637-7787
İstanbul Rumeli University, Turkey

Savas Selahattin Ates
Eskisehir Technical University, Turkey

ABSTRACT

Covid-19 has adversely affected many sectors, especially aviation. Because the plane is an enclosed space and spreads the virus faster by carrying passengers between countries, aviation has been affected more than other sectors. In this process, revenues of aviation enterprises have decreased, flights and therefore the amount of required labor has decreased due to domestic and international flight bans. For this reason, the human resources units of the enterprises have taken some actions against this situation. Although almost every airline company was adversely affected by the pandemic, airlines following the right policies were able to close this process with less damage. In this book chapter, what is done in the human resources departments of aviation companies in the world will be examined.

INTRODUCTION

Oil Crisis (1973), Iran-Iraq War (1980-1988), Gulf Crisis (1990-1991), Asian Crisis (1997), September 11 Attack (2001), SARS (2002), which greatly affected the aviation industry in the last 50 years -2003), 2008 Economic Crisis and MERS(2015) have experienced many global crises. The Covid-19 pandemic, which started to be experienced as of 2019 and still continues, is one of these extraordinary situations. Aviation, which is the most important mode of transportation on a global scale, has been one of the sectors most affected by the pandemic due to its high fragility, not only because it offers travel in a closed environment on airplanes, but also because of the many restrictions imposed on both domestic and international flights. According to the Briefing Note published by McKinsey & Company at the very beginning of the pandemic, the sectors most adversely affected by the pandemic were examined.

DOI: 10.4018/978-1-6684-6835-7.ch004

The research looked at the sectors that showed the most decline in their activities. Accordingly, the commercial aviation sector economy was affected by 40%, the airline travel by 36%, the petro-gas sector by 34%, the insurance sector by 22% and the automotive sector by 21% (Craven et al, 2020).

In general, most of the international flights have been canceled after the strict bans since the start of the pandemic. Later, flights started slowly on domestic lines. Compared to 2019, there was a 60% decrease in the number of passengers carried in 2020. The number of passengers carried decreased from 350 million in January 2020 to 25 million in April 2020. Almost 90% of the passengers carried in April 2020, and the majority of the passengers carried until July 2021, were carried on domestic routes (ICAO, 2022). In this process, apart from the flight bans, due to the reluctance and fear of the passengers, the occupancy rates of the planes continued to be low for a while. In this process, airline companies started to carry out cargo flights in order to minimize the damage (Deveci et al, 2022). According to the ICAO report, in this period, the number of passengers carried on international flights decreased by nearly 100% in March 2020 compared to January, and by 85% in domestic flights (ICAO, 2022).

As can be seen in the data, there has been a great decrease in the number of passengers carried. The decline has caused great losses to the businesses, and even since the beginning of the pandemic, there have been airline companies that could not bear the costs and went bankrupt. Between March and October 2020, 16 airlines went bankrupt (Dube et al, 2021). There were a lot of layoffs during this period, which was probably the biggest crisis in aviation history and businesses lost and went bankrupt (Sun et al, 2021).

Personnel expenses, which are the biggest expenses for airline companies after fuel expenses, are very important for businesses. Due to the high decrease in the number of flights during the pandemic period, fuel expenses decreased, but personnel expenses continued in the same way whether the flight was made or not. In this period, when flights decreased, the biggest expense of airline companies was personnel expenses. Businesses that cannot provide income have also started to lay off employees. According to the news of HRWorld and calculations of Anurag Kotoky, Manish Modi and Matt Turner in July 2020, 400,000 people in the aviation industry worldwide were laid off, on leave or at risk of losing their jobs (Bloomberg, 2020). On the other hand, according to the news of BusinessToday, 10% of aviation sector employees lost their jobs between April 2020 and December 2021 (PTI, 2022).

In this book chapter, the aims and functions of human resources will be examined, and especially the selection and placement part in the aviation sector after the pandemic will be emphasized. In addition, recruitment and firing processes will be examined. In addition, during the pandemic process, the changes in the performance of the sector employees will be discussed in connection with the concerns they experience. Another issue to be examined is the global crises affecting aviation, which will be briefly examined and compared with the effects of the Covid-19 pandemic. As a result of the comparison, it will be pointed out how different the developments in the field of human resources are compared to other crises.

HUMAN RESOURCES AND AVIATION

Human Resources, which first emerged with the Industrial Revolution under the name of "Personnel Management", gained a strategic dimension in the 1990s and evolved into Strategic Human Resources Management (Bayraktaroğlu, 2011). With the beginning of the Industrial Revolution, the birth of unionism and union activities took shape as a result of important events such as world wars and the development of state protectionism (Gürüz & Yaylacı, 2004). Human Resources, which is very important for businesses, is one of the most important departments today and large investments are made by businesses. Today, businesses take steps to gain competitive advantage. It is desired to minimize the costs, to obtain cheap resources, to make improvements in the production process and marketing areas. There is a human factor in all processes (Uslu & Kedikli, 2017). Therefore, the Human Resources department is vital for businesses. Human resources basically have two purposes: estimating the human resources that the business will need in the future and determining the resources that will be required to meet this need and how the resources will be evaluated (Şimşek & Özge, 2012). In addition, this department, which has functions such as planning every process, preventing errors that may occur during production and reducing costs, serves to maximize the efficiency and income of the enterprises. In the planning process; problems and opportunities are determined, goals are determined, the starting points of the plan are determined, options are determined, the options are evaluated, the most appropriate option is decided, auxiliary action plans are created, the plans are budgeted with numerical values and the plans in hand are put into practice (Kurulgan, 2020). Today, human resources have become more important for businesses due to the change in the market due to globalization and increased competition, rapid changes in technology, the emergence of new markets, the survival needs of businesses, the change in consumer demands and the increase in employee rights (Eroğlu, 2001).

The aviation industry is a large-volume industry and has many components. The human resources department can have great effects in the aviation sector as well as in every sector. The example of Continental Airlines (Bethune, 1999), which has come to a very good place in the history of aviation with the influence of its management staff and human resources, and the example of Southwest Airlines, which still continues its success thanks to the right strategies of human resources, are important in order to understand the impact of human resources in aviation (Küçükönal & Korul, 2002). One of the most important duties of human resources is to determine the workforce needs and to recruit the appropriate people. Most of the personnel working in the aviation industry do not have the luxury of making mistakes, and if mistakes are made, the consequences can be severe. Therefore, the recruitment part in the aviation sector is more important than other sectors. Aware of this situation, Southwest Airlines is a good example. Southwest Airlines received more than 125,000 job applications during its hiring period in 1994, and recruited only 4,000 of the applicants (Halowell, 1996). Recruitment was carried out meticulously as it should be. In addition, one of the other important duties of human resources is to perform performance management and to ensure the continuity of the motivation of its employees. There are multiple ways to increase motivation, and the most important of these is wage growth. Locke et al. (1980) found in their study that individual wage incentives increase employee productivity by 30%.

After Covid-19, human resources in aviation had to adopt some changes. The change that human resources had to adopt first was the transition to distance learning and distance work at a high rate. This digital transformation has led to a leap in online technologies and has established itself permanently (ElMarady & Rahouma, 2022). It is predicted that the future of aviation will be increasingly hybrid (BCG, 2020), this will increase unemployment (Xiarewana & Civelek, 2020) and it is necessary to expand the ability to work in a fully digital environment (Agrawal et al., 2020). In summary, the increase in the use of technology in aviation and the arrival of new technologies lead to changes in employees and therefore human resources.

Mental Health and Wellbeing of Aviation Employees After Covid-19

Stress and anxiety negatively affect employees in the aviation industry, as in every business line. While stress and anxiety in the work environment cause health problems such as heart disease, increased blood pressure and depression in employees, it also causes behavioral disorders to the detriment of the enterprise, such as absenteeism, increase in work accidents, increase in compensation cases, decrease in production and increase in workforce turnover. (Robbins, Judge, amd Millet, 2015; Cooper & Dewe, 2004). In the Covid-19 period, the VUCA environment (Akdemir et al, 2021), which means variability, uncertainty, complexity and ambiguity, has emerged. The impact of the Covid-19 pandemic on employees; They analyzed them under four headings: work and employment, welfare and morale, safety behavior and security surveillance. During the pandemic period, human resources in aviation made unpaid leave, early retirement and salary reductions. According to Shin et al. (2022) in their study on cabin workers; They addressed and analyzed six psychological anxiety factors: salary reduction, perceived risk of infection, career recession, reduction in benefits, social perception, and employment insecurity. According to the findings obtained at the end of the study, it has been seen that the salary reduction harms the self-confidence of the employees, the employees have economic difficulties, they are worried about being infected, the career recession has negative effects on the work self-esteem and the job insecurity negatively affects the job self-confidence of the employees. Kim et al. (2022), on the other hand, found that work-related stressors showed a significant difference after the pandemic compared to the pre-pandemic period. According to the study, it was stated that the stressors that emerged during the pandemic period increased more than the routine stressors. At the same time, unemployment can lead to long-term general health problems (Daly & Delaney, 2013). The fear of being unemployed during Covid-19 has also been a factor that triggered this situation. Considering all these, the stress and anxiety in employees after the pandemic shows that human resources management has become difficult. It can be explained with good personnel selection, which includes stress resistance as a selection criterion in order to prevent employees' level of being affected by crisis environments (Herpertz et al., 2016).

BACKGROUND

According to Sheaffer and Negrin (2003), it is not possible for a bad human resources management to cope with crises. The human resources department always needs to make long-term plans. The success rate is very low in short-term and careless plans. During the pandemic period, businesses have been able to make wrong decisions in order to respond quickly, as they were caught unprepared. Careless planning in this process can lead to much worse results. As a result of wrong planning, the personnel working in business are negatively affected by the situation and their productivity may decrease. A good human resource management should keep the motivation of the personnel high (Persons, 1995). It is necessary to be sensitive considering the mental status of the employees. After the crisis, additional measures should be taken to eliminate the concerns of the employees. Thanks to the measures taken, there is an Increase in productivity and a decrease in costs (Braverman, 2003). However, Buhusayen et al. (2020) found in their study that business managers are overburdened by insufficient human resources. In the study conducted in Australia, it was seen that human resources were insufficient, and the underlying reasons were lack of personnel and weak leadership. In addition, it was seen that planning and internal communication were insufficient in this process. On the other hand, D'Angelo et al. (2022) stated in their study that it became difficult for human resources personnel to be physically present in the workplace in the remote working system, which was implemented after the start of the pandemic. They stated that they had to gain some skills in order to adapt to technology quickly and to continue their work professionally. In addition, it was emphasized in the study that the decisions that they wanted could not be taken because the human resources departments had to do what the senior managers said.

The pandemic has caused great damage especially to the aviation industry. On the other hand, the benefits of the pandemic were also discussed, instead of only seeing the negative sides of the pandemic. In his study, Amankwah-Amoah (2021), although he mentioned that the pandemic had negative consequences, said that thanks to the pandemic, the established routines and processes of aviation went through a mandatory change. Due to the extraordinary circumstances experienced in this process, businesses had to innovate. Innovations such as good disinfection of airplanes, contactless technologies at airports, accelerated check-in and open-middle seat technology have been gained. In the study, instead of seeing the crisis as a problem, it was argued that businesses should bring innovations to such situations and provide permanent solutions. Sun et al. (2021) supported this view in their studies and argued that the pandemic pushed aviation technology innovations. They also said that businesses should contribute to the development of the industry by sharing the data they have with each other, thanks to a common data pool. Mahfuza et al. (2020) stated that human resources processes may need to be renewed with the pandemic. Especially in the recruitment part, which is one of the most important functions of human resources, they considered it possible that the recruitment criteria will change with the change of expectations from the people to be recruited. Bilkay (2022), on the other hand, mentioned that the commitment of the personnel to the company they work for has positive results in times of crisis. He argued that this commitment should be increased together with qualified human resources management. In addition, the sustainability of commitment makes it more ready for new crises that may occur in the aviation industry. Erdağ (2021) also supported this view and stated that the human resources department is a critical department in terms of profitability and sustainability of enterprises.

Human resource management in aviation; have great impact and difficulties in three areas: airline workforce policies, costs and profitability (WowEssays, 2020). Since a large number of personnel work in the aviation industry, hiring, delegating tasks to employees and meeting employee demands is one of the main challenges for human resource management (Shanker, 2019). In addition, it is very difficult to cope with the constantly developing technologies today. It is necessary to create a strategy that will enable rapid adaptation to changing technologies (Sun et al., 2021). Many airports, especially Istanbul Airport, Singapore Changi Airport, London Heathrow Airport, Los Angeles Airport and Hong Kong Airport have started to use technologies such as smart payment, smart boarding gate, self-check-in points, virtual tower and assistants, security analysis with artificial intelligence and unmanned operation vehicles (Artar&Türkay, 2021). Apart from the exemplary technologies, the use of 3d printers in aviation is becoming widespread. Airlines such as British Airways (Gavine, 2019), KLM (Brightmore, 2020) and Qatar Airways (Carlota, 2019) and the two largest aircraft manufacturers Airbus and Boeing (Bergmann, 2022) have started to use 3d printers. For the specified technologies and more, employees should keep up with technological change and human resources in aviation have to pay attention to this area, especially in the recruitment and training.

Crises in Aviation and Covid-19

There have been many crises in the history of civil aviation, mostly fuel related. Due to the embargoes imposed by the Organization of Petroleum Exporting Countries (OPEC) in 1973, 1974 and 1979, an increase of 100 billion dollars occurred in the fuel costs of airlines (IATA, 2015). The negative effects of the crisis continued until 1987 (Strategy, 2008). However, there was no significant decrease in the demand of passengers. The Iran-Iraq War that lasted between 1980 and 1988 negatively affected the economic markets. The war of these two countries, which are very important in oil exports, reduced oil production capacity. This situation caused a new crisis in aviation and oil prices increased due to the decrease in production. Oil production returned to normal levels only 35 years after the end of the war (Saxton, 2006). In 1990, Iraq declared war on the fact that Kuwait produced oil above its production quota. The reflection of the war on aviation was again in the form of an increase in oil prices. As a result of the price instability experienced, there was a decrease in passenger mobility (Şimşek, 2021). On September 11, 2001, the terrorist organization hijacked civilian passenger planes and attacked 3 points in the USA. The event resulted in the introduction of many new measures, especially in the field of aviation security, after the attack (Türk, 2017). Although the airline employees were not affected negatively after the measures taken, a more detailed process has started for the passengers within the scope of security. After the measures, there was a decrease in the level of use of aviation by passengers. In this process, the aviation industry lost passengers and lost approximately 50 billion dollars (Şimşek, 2021). Towards the end of 2002, the SARS pandemic began in Hong Kong. Shortly after it started, there were 8422 cases and 916 deaths. In addition, the disease, which spread to 37 countries, caused more than 3 million workers to be unemployed in Asian countries (Türker, 2020). The pandemic was mainly effective in Asian countries. The crisis resulted in the decline in passenger demand for airlines and the bankruptcy of some smaller airlines. Afterwards, the Economic Crisis that started in the USA in 2007 affected the whole world negatively. The aviation industry faced the lowest passenger demand in the last 5 years due to the shock of the crisis (IATA, 2008). Due to the economic crisis, people stopped their flights because they were afraid to spend money. As a result of the crisis, more than 80 million passengers were lost in the number of passengers carried (IATA, 2019).

In the current Covid-19 pandemic, while the total number of passengers carried was 4.5 billion per year, it decreased to 1.8 billion after the pandemic.

As seen in the Figure 1, the Covid-19 pandemic has been the most negative impact in aviation history. After the pandemic, the total number of passengers transported suddenly dropped to the levels of 2004. As can be seen in the Figure 1, the crisis that affected aviation most negatively after Covid-19 occurred after the September 11 attacks. After the attacks, European airlines announced 30,000 job cuts and North American airlines 95,000 job cuts. Job cuts also extended to ground services (LSG 5,000 lay-offs, Gate Gourmet 3,000) and aircraft manufacturers (Boeing 18,000 lay-offs, Airbus 2,000)(ILO, 2002). Despite this, it is clearly seen that Covid-19 has affected aviation, which has a constantly rising trend, much more than other crises. Since the aviation industry has not experienced such a major crisis before, businesses were caught unprepared. In addition, the Covid-19 pandemic has followed a different course compared to previous crises. For example, after the increase in fuel prices due to the decrease in oil production, businesses were able to fly by adjusting their pricing accordingly. In addition, in the SARS pandemic, which is similar to Covid-19, the negative effects were mostly limited to the Asian region. SARS resulted in a 4% annual decline in Asia Pacific flights (Carlucci, 2020). In the Asian region, there were 20 percent cuts in the salaries of aviation personnel (Dombey, 2003). However, when we look at the Covid-19 pandemic, the situation seems to be different. The crisis was not limited to the increase in costs. Businesses could not fly due to flight bans. Therefore, no income was obtained. In this period, when the operations came to a standstill, the human resources departments of the enterprises made plans to survive.

Figure 1. World Passenger Traffic Evolution
Source: (ICAO, 2022)

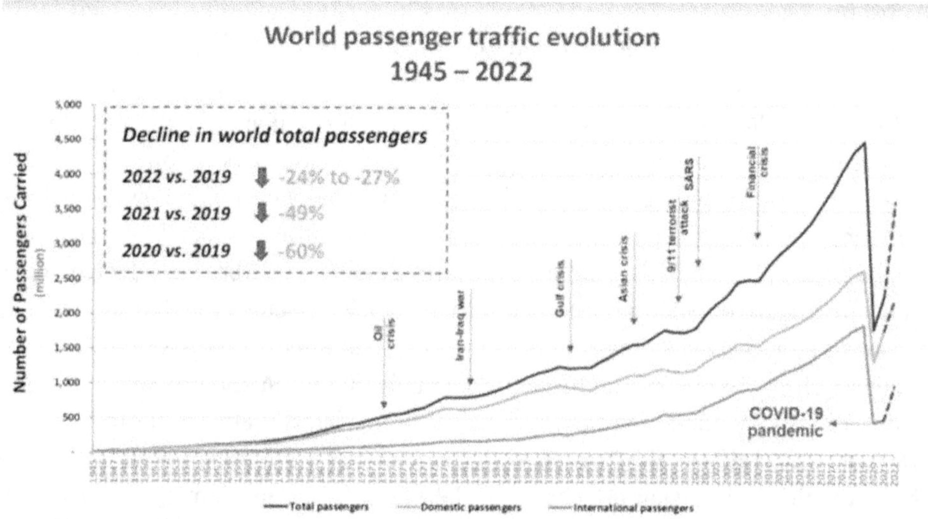

CHALLENGES IN AVIATION AND HUMAN RESOURCES IN THE COVID-19 ERA

Issues

Throughout the history of aviation, it has experienced many crises and somehow overcame them. Despite this, none of the crises experienced to date have had as negative consequences as the Covid-19 pandemic. Businesses were caught unprepared because there had never been such a major crisis before. During the pandemic process, businesses started to lose money, stress and anxiety began in the personnel, and urgent measures had to be taken against this unforeseen situation. The fact that the measures taken had to be taken in a short time affected the quality level of the decisions. Some businesses have achieved successful results as a result of the decisions they have taken and have survived the crisis with as little damage as possible. Some of the businesses that could not make the right decisions or that had insufficient financial power went bankrupt. However, one should never waste a crisis. Despite the negative tables, opportunities for businesses have been created. Opportunities have arisen for businesses in digitalization, environmentalism and cost reduction.

Problems

The global Covid-19 pandemic has brought some problems with it. The problems experienced have caused quite negative effects on businesses. The problems encountered during the pandemic period have reached levels not seen before in aviation history. Even in times of reduced confidence in aviation or economic crises, airlines were somehow able to survive with their policies.

With the onset of the Covid-19 pandemic, states have begun to take precautions. Most of the measures taken have been in the form of countries banning international flights. The fact that the flights were still continuing, which was the reason why businesses could develop new strategies in previous periods, did not happen during the pandemic period. For this reason, businesses have not followed the policy of reducing their expenses and have not been able to generate income for a long time. The problems brought by the pandemic period are as follows:

Decreased revenues: Airlines can generate revenue as long as they fly. In the pandemic, where flights have almost come to a standstill, the revenues of airline businesses have decreased considerably. During this period, when flights were not made at a high rate, both in-flight revenues and revenues from flights decreased significantly. Plans of airline companies, which plan their flights according to periods, have been completely disrupted by the pandemic period. Therefore, businesses had to make new plans. In addition, due to the economic collapse in the world after the start of flights, people took care not to spend their money except for essential needs. This has been one of the factors in the decrease in the number of passengers.

Flight bans: International flight bans, mainly imposed by the states, have put airline companies in a difficult situation. The most undesirable situation for airline companies is that the planes are on the ground. In this period, when flights were highly prohibited, planes remained on the ground. Therefore, businesses have had difficulty in generating income.

Figure 2. Seat Capacity Scenarios
Source: (ICAO, 2022)

Flights due to flight bans decreased by 95%. Even if the aviation sector started flights and gradually returned to normal, the seat capacity still has not returned to the pre-pandemic level. Even in the best-case scenario, seat capacity is negative compared to pre-pandemic. While an increase is expected in Scenario 1, Scenario 2 and Scenario 2a, a slow rise is expected after a decrease in Scenario 1a.

Seat restrictions: In the period when the flight bans that came with the pandemic began to relax gradually, some restrictions were introduced for airplanes. One of them is the seat distance rule. This rule states that passengers should sit with one seat occupied and one empty seat. In this case, the planes had to make their flights only at half capacity. Since ticket prices could not be doubled in proportion to this, businesses were able to make much less profit than the flights they made before the pandemic. Although the total cost of flights does not change much according to the number of passengers, the unit cost decreases as the number of passengers increases. This means that the less the number of passengers, the higher the unit cost and therefore the higher the ticket price reflected to the customer. The effect of this situation on human resources is; As a result of the planes flying at half capacity, there has been a decrease in the number of personnel needed. HR either took unpaid leave or laid off some of the existing employees.

Fear of passengers: Air transportation, which allows traveling in a closed environment from the beginning to the end of the flight, has reduced the rate of passengers preferring this transportation model. Passengers have given up on pleasure travel at a high rate. In addition to the decrease in flight frequency and the increase in ticket prices due to seat distance, passengers preferred not to travel at a high rate because they were afraid of being in a closed environment. This has reduced passenger rates on flights.

Tourism and online communication, online meeting: Airlines achieve high occupancy rates and make high profits on flights to touristic regions and especially in summer. In addition, business travelers are also very important for airline companies. During this period, a significant portion of tourism passengers gave up on vacation, and business travelers held their meetings via online platforms. Therefore, the aviation industry lost two of the most important passenger groups at a high rate during the pandemic period.

Personnel shortage due to layoffs: Most of the enterprises resorted to layoffs and unpaid leave due to the increase in their costs and decrease in their incomes. With the gradual start of flights, the personnel on hand was insufficient. At the beginning of the normalization process, businesses could not work with sufficient number of personnel. Since the motivation of the working personnel is low, their productivity has decreased. Afterwards, there were cancellations and delays in airline flights.

Continuation of fixed expenses: Although flights have come to a standstill and variable costs have decreased, maintenance, insurance, depreciation/rent and personnel costs have continued. Fixed expenses were quite challenging in this period when airlines could not generate revenue.

Restriction of state aids to flag carriers: The aid provided by the states in this period was limited especially to flag carrier companies. The expenses of the flag carriers were covered by the state and it was ensured that they survived the pandemic process with the least damage. The airlines, with the exception of the flag carrier companies that can survive thanks to the aid, have gone through a very difficult period. In this process, there were airline companies that could not bear the costs and went bankrupt. In this process, the USA provided a support package of 50 billion dollars, provided an additional aid of 25 billion dollars to prevent the dismissal of 750 thousand personnel working in the aviation sector for a while (Josephs, 2020), in addition, it provided 4 billion dollars to air cargo businesses and 3 billion dollars to ground handling companies (Rappeport & Chokshi, 2020). France provided support to Airbus (Horcajuelo, 2020). Germany bought a 20% stake in Lufthansa for 9 billion euros (Lotus, 2020). China, on the other hand, announced that it would support the airlines in its country (AviationVoice, 2020) and provided a maximum of 800 thousand yuan for single-aisle aircraft and 1.45 million yuan for double-aisle aircraft (Xinhua, 2020). On the other hand, the human resources departments of the airlines that were not helped had to dismiss their personnel and take more precautions during this period.

- Long-term plans lose their function: There are long-term plans made by different departments of businesses. Due to the extraordinary situation, the plans of the enterprises were interrupted to a great extent.
- Difficulty of adaptation: Employees have faced a crisis they have never seen before and processes have changed drastically. This situation required the employees to adapt to a different situation, both mentally and physically. Some of the employees had difficulties in adaptation.
- Changes in personnel skills: Changes have occurred in the recruitment process, which is one of the most important branches of human resources. Changing aviation after the pandemic asked its personnel to acquire different skills. For this reason, the criteria in the recruitment process have also changed. Some of these expectations are IT skills, critical reflection, problem solving, communication, information retrieval systems, enquiry and analysis (Lappas & Kourousis, 2016).
- Extra costs of obligations such as hygiene and seat distance: Airline operators and airports had to comply with the determined rules in order to continue their flights during the pandemic process. Especially the requirements of hygiene and social distance have caused additional costs to businesses.
- Unemployment: Businesses have suffered great losses during the Covid-19 period. Most airlines' Human Resources departments have made layoffs. Especially in the aviation sector, which was affected more than other sectors at the beginning of the pandemic, it became necessary to make layoffs or unpaid leave. This situation has led to unemployment in the aviation sector.
- Disruptions in the Supply Chain: Due to flight cancellations and disruptions, layoffs, government restrictions and economic collapse, businesses have had problems with supply. These disruptions caused the flights of the enterprises to be delayed. In this process, the failure of a supplier to supply created a dominance effect and thus disruptions occurred in the entire supply chain (Kilpatrick, 2020). Since the aviation industry is a large part of this supply chain, it was both affected and affected other industries.
- Fuel prices: Jet fuel prices hit the lowest level of the last 10 years in April 2020, and then continued to rise until June 2022, reaching the peak (IATA, 2022). Fuel prices have also increased during the pandemic period, when many negative pictures were seen. One of the other factors that put businesses in a difficult situation and increased costs was fuel.

Considering all these, human resources departments have had great responsibilities in this challenging process. The qualified human resources personnel mentioned in the previous sections were greatly needed. The human resources departments of the enterprises had to take a very important role in this period. The decisions taken are important in terms of how businesses will overcome this difficult process. Making the right decisions and making the right planning will reduce the losses.

SOLUTIONS AND RECOMMENDATIONS

In times of crisis, negative effects are experienced on employees. Employees' mental health may deteriorate and they may intend to quit their job (Vardarlıer, 2016). The pandemic is one of the clearest examples of this. In this period, layoffs and unpaid leave were experienced intensely beyond the intention to leave the job. 132.000 people working in the aviation industry in the USA were laid off between March and June 2020. United Airlines laid off 36,000 employees, almost half of its workforce, Southwest Airlines laid off 17,000, and United Airlines laid off 36,000, which is almost half of its workforce. On the other hand, more than half of JetBlue's employees and Alaska Airlines have allocated approximately 7,000 employees on unpaid leave. Delta Airlines, on the other hand, did not require layoffs or unpaid leave, thanks to the voluntary retirement of one-fifth of its employees (Guardian, 2020). In Europe, Lufthansa laid off 22.000, Tui 8.000, British Airways 12.000, Easyjet 4500, Ryanair 3.000, Airbus 15.000, Air France 7500 and Scandinavian Airlines 5.000 employees (Davies, Tidey, and Ortahamamcilar, 2020). Worldwide, airlines have lost more than 400,000 employees during the pandemic (Kelly, 2021). Apart from this, salary cuts were made except for dismissal and unpaid leave. As a result of this situation, both businesses and employees were adversely affected. Due to unemployment, unpaid leaves and reduced incomes, employees experienced a mental breakdown. This negative situation needs to be corrected in order for businesses to increase productivity in employees.

Human resources departments have functions of human resources planning, employee procurement and selection, training and development, performance management, wage management, career planning and management, workplace safety and risk management, and employee relations adjustment (Bingöl, 2013). There are moves that need to be done correctly for each function. The actions of human resources are very critical for businesses that have been adversely affected by the pandemic. If the right moves are made, it is possible for businesses to get through this process with minimum damage.

During the pandemic period, some functions have become more important. In non-crisis situations, while the supply and selection of employees is one of the most important issues, issues such as human resources planning and wage management have gained importance. Since the biggest expense for businesses after fuel is personnel expense, human resources had to focus on wage management. However, as mentioned in the previous section, wrong human resource planning can have serious consequences. As seen in the examples, after the start of the flights, there were cancellations and delays due to the lack of personnel. In addition, in order for businesses to be successful, these plans should be made for a long time and the planning should be done without rushing.

Since human resources is one of the most important departments for businesses, a qualified team is required. This department, which directly affects the success of businesses, can fail, especially in times of crisis, when it is insufficient. Although the financial strength of the business is important in crisis situations, the moves made by human resources on how to overcome the crisis are very critical. The negative effects of insufficient human resources on businesses during the pandemic period we live in are clearly seen. For this reason, it is important to create a qualified staff in the human resources department.

In the part of remuneration and rewarding, human resources decide how the remuneration will be by providing the interests of the business and all other stakeholders. At the same time, it is necessary to provide the motivation of the employee by rewarding the deserving person (Benligiray, 2020). In this way, the loyalty of the employees is increased and they work more efficiently. However, due to the economic collapse during the pandemic period, employees could not be motivated about wages. Good planning and execution in this area are effective in not reducing the motivation of employees in times of crisis. It has been easily observed during the pandemic period how high the lack of motivation in the employees is in case the planning related to the area is made incorrectly. Therefore, it is important that the moves made by human resources in the remuneration and rewarding function are correct.

The function, which was one of the most important functions of human resources before the pandemic and gained great importance after the return of aviation to normal, is the employee supply and selection function. In the employee procurement and selection function, human resources fulfill the tasks of ensuring and increasing employee-job harmony, attracting qualified and talented candidates, determining the most suitable candidate for the job, increasing the loyalty of the personnel, and retaining qualified employees (Benligiray, 2020). After the start of the pandemic period, it was especially important to increase staff loyalty and retain qualified employees. With the start of normalization, businesses had to find new employees instead of the candidates they dismissed due to the crisis. Here, the selection and placement process has taken place. As a result of the change in the expectations of the enterprises from the employees with the pandemic, the recruitment criteria have also changed. In the aviation sector, which has created a new working environment, new technologies providing contactless transactions and a different business environment have emerged compared to the pre-pandemic period. The selection of candidates by human resources according to this newly formed environment has become a process that needs to be done carefully in terms of both the smooth continuation of current affairs and the recruitment of employees who can adapt to future changes. Making the right choices by the human resources departments of the enterprises in this process will bring positive results both in the short and long term.

In Chinese, the crisis is expressed by the symbol "Weiji". Weiji means opportunity and threat (Akıncı, 2011). Here, it is seen that every crisis is a threat, but these threats can be turned into opportunities. In the pandemic process, as there are many threats, there are also opportunities. The most important of these opportunities has been that the pandemic has forced businesses to innovate. Interviews and other works during the pandemic process were done online as much as possible. The development of remote working and hybrid working models has the potential to reduce personnel needs and costs (Genç & Işıktaş, 2022). Human resources also benefited from this process. Although e-recruitment, e-training and e-competence methods (Bondarouk & Brewster, 2016) are not applications that started to be used after the pandemic, their use has become quite common during the pandemic period. In the e-training and e-recruitment process, businesses carry out the processes through electronic systems. Job postings are announced on websites for recruitment or on sites where job postings are posted. Afterwards, the candidates apply through the online system and the exam/interview processes begin. The entire process is carried out through online systems. In e-training, on the other hand, it provides access to different trainings required by the employees of the enterprise or optionally through online platforms. Training videos can be made synchronously or asynchronously (Wege et al. 2019). Electronic human resources management is a system that businesses use to store, process and store their information (Öge, 2004). All functions of human resources can be performed in this electronic environment. In the planning part, it ensures that the knowledge, skills and abilities of the employees of the enterprise are collected, stored and updated. Thanks to these systems, information about employees can be accessed quickly (Hopkins & Markham, 2003). Thanks to the database, it becomes very easy to access situations such as the performance of the employees and the financial possibilities of the organization (Doğan, 2011). In addition, training costs can be greatly reduced by conducting the trainings online. This system, which does not require a physical environment and physical presence of the educator, is quite efficient. Interviews in the aviation sector were also held online as much as possible. In this process, human resources also carried out the recruitment processes through online systems. As a result of the fact that the candidates were able to carry out the entire process online in the online application system, candidates from foreign countries also applied to the job postings opened by the enterprises. There has been an increase in job applications due to the ease of application. Although this is a factor that increases the workload of the human resources department, the chance of finding the right candidate has been increased. Due to the high number of applications, the number of candidates who meet the necessary conditions has increased and the most suitable candidate for the job has been selected. At the same time, due to the remote working environment, some jobs did not require physical relocation. This has allowed the number of candidates to increase and even to receive applications from different countries. Considering all these, the pandemic period has encouraged businesses to innovate or to use systems that are currently not used much.

At the same time, after the shock of the crisis was overcome, airlines started to search for income. Many airline companies that cannot provide passenger transportation service with airplanes have started to operate cargo flights. Airlines have made their planes suitable for carrying cargo by applying seat removal in some of their planes. Airlines that make/can do cargo flights have gained an advantage over their competitors. Thanks to the cargo flights, the negative effects could be reduced to some extent.

Figure 3. Worldwide air freight traffic from 2004 to 2022
Source: (statista, 2022)

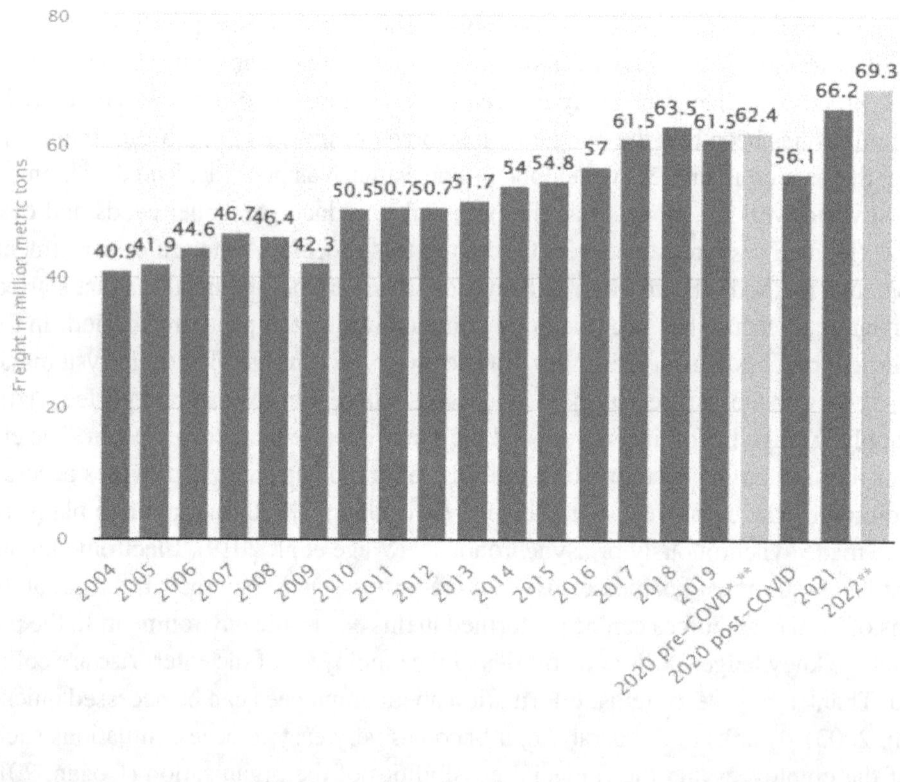

Although the amount of cargo carried after Covid-19 decreased, it exceeded the pre-Covid-19 level in 2021 and continued to increase in 2022. It is seen that the demand for cargo has increased and airlines have started to carry cargo. Businesses were able to generate income in this way. The fact that passenger carriers started to carry cargo was another good move made during the crisis.

CONCLUSION

The Covid-19 pandemic has affected aviation much more than any other crisis before, and this has been seen in a short time. In this process, businesses had to react quickly. There were intense layoffs and unpaid leave. The moves made by the enterprises were directly related to the planning and especially the competencies of the human resources departments. There have been companies that went bankrupt as a result of insufficient financial power or taking wrong decisions. At the same time, the flag carrier companies of some countries received aid from the state. Supported flag carriers were less affected by the pandemic than other businesses. However, the decisions and plans made for airlines that did not receive assistance have become even more important. The quality of decisions and planning has been influential in the level of success.

Despite all the negativities, there were also positive aspects brought by the pandemic. In particular, it has been ensured that businesses are purified from obsolete systems and systems at a high rate. It has been seen that paperwork, interviews and many other works can be done through online systems. Recruitment interviews, online realization of internal and external meetings, etc. Many jobs could be done remotely. As a result of this, both the cost of the enterprises has decreased and as a result of doing everything electronically, a more environmentally friendly formation has been realized.

In addition, the pandemic caught businesses unprepared, and as a result of the negative effects it created, it was effective in preparing businesses for major crises. Businesses that have not experienced such a great collapse before have sought ways to survive in this process. The experience of the enterprises is shedding light both for themselves and for the new airlines to be established in the future. These methods can be taken as a model in a new crisis when the right moves and successful processes are examined. In fact, the pandemic process has been a difficult test especially for aviation and has given businesses a great crisis experience. From this point of view, it is very important to see the positive aspects as well as the negative aspects of the pandemic.

Businesses need to analyze the events experienced during the pandemic period and make plans about how to react to major crises. Otherwise, it is likely that they will share the same fate with the airlines that made the wrong decisions during this period. Especially in the human resources department, it will be necessary to work very well on the steps to be taken. Businesses that make good preparations and planning will be less affected in times of crisis than their competitors. In short, businesses need to be prepared against potential crises by using this period to their advantage.

FUTURE RESEARCH DIRECTIONS

As of September 2022, the pandemic still continues. We have observed the decline and rise since the beginning of the pandemic. In this process, many steps related to aviation have been taken and continue to be taken. The results will be seen whether the steps taken in the field of human resources are made to save the day or by planning a permanent solution. It will be very useful to investigate the long-term positive and negative effects of the steps taken in future studies on the subject. In addition, it is necessary to discuss how free the human resources departments of the enterprises in this process and the positive or negative consequences of this. In addition, every crisis environment brings with it opportunities. It has been seen during the pandemic process that there have been businesses that can see the opportunities in this crisis environment and take appropriate steps. It is very critical to examine the businesses that turn the crisis into an opportunity and to put them in the literature as an exemplary model. The work to be done will ensure that the steps to be taken against global events such as Covid-19 will be more effective.

REFERENCES

Agrawal, S., De Smet, A., Lacroix, S., & Reich, A. (2020). *To emerge stronger from the COVID-19 crisis, companies should start reskilling their workforces now.* McKinsey. https://www.mckinsey.com/business-functions/organization/ our-insights/to-emerge-stronger-from-the-covid-19-crisis-companies-should-startreskilling-their-workforces-now

Akıncı, Z. (2011). *Konaklama İşletmelerinde Kriz Sürecinde İnsan Kaynakları Yönetiminin Rolü Ve Önemi* [The Role and Importance of Human Resources Management in the Crisis Process in Hospitality Enterprises.]. Süleyman Demirel Üniversitesi Vizyoner Dergisi.

Amankwah-Amoah, J. (2021). COVID-19 pandemic and innovation activities in the global airline industry: A review. *Environment International, 156*, 1–7. doi:10.1016/j.envint.2021.106719 PMID:34166874

Artar, O., & Türkay, U. İ. (2021). Havacılık sektöründe havalimanlarının dijital dönüşümü. [Digital transformation of airports in the aviation industry.] *Working Paper Series, 2*(1), 86-97.

Aviation Voice. (2020). China Offers Airlines Financial Support amid COVID-19 Outbreak. *Aviation Voice.* https://aviationvoice.com/china-offers-airlines-financial-support-amid-covid-19- outbreak-2-202003061421/

Bayraktaroğlu, S. (2011). *İnsan Kaynakları Yönetimi. [Human Resources Management.]* Sakarya: Sakarya Yayıncılık.

BCG. (2020). What 12,000 Employees Have to Say About the Future of Remote Work. *BCG.* https://www.bcg.com/publications/2020/valuable-productivity-gains-covid-19

Benligiray, S. (2020). İnsan Kaynakları Yönetimine Giriş. İnsan Kaynakları Yönetimi (s. 15). [Introduction to Human Resources Management. Human Resource Management (p. 15).] içinde Eskişehir: AÖF.

Bergmann, K. (2022). JP 9102 Bidders strengthen their industry teams. *Asia-Pacific Defence Reporter (2002), 48*(6), 30-32.

Bethune, G. (1999). *From Worst to First (1. Baskı b.).* Wiley.

Bilkay, S. (2022). *Pandemi Dönemindeki Esnek Çalışma Düzenlemelerinin İşkolikliğe Etkisinde Duygusal Bağlılığın Aracılık Rolünün Araştırılması* [*Investigating the Mediating Role of Emotional Attachment in the Impact of Flexible Working Arrangements on Workaholism in the Pandemic Period.*]. Pearson Journal Of Socıal Scıences & Humanıtıes.

Bingöl, D. (2013). *İnsan Kaynakları Yönetimi* [*Human Resources Management.*]. Beta Yayınevi.

Bloomberg. (2020). Covid 19 impact: Jobs are being wiped out at airlines worldwide, and there's worse to come. *HRWorld.* https://hr.economictimes.indiatimes.com/news/industry/covid-19-impact-jobs-are-being-wiped-out-at-airlines-worldwide-and-theres-worse-to-come/77142768?redirect=1

Bondarouk, T., & Brewster, C. (2016). Conceptualising the future of HRM and technology research. *International Journal of Human Resource Management, 27*(21), 2652–2671. doi:10.1080/09585192.2016.1232296

Braverman, M. (2003). Managing the Human Impact of Crisis. *Risk Management*, 475–488.

Brightmore, D. (2020). KLM is recycling plastic bottles to make 3D printed tools. *Manufacturing Digital.* https://manufacturingdigital.com: https://manufacturingdigital.com/technology/klm-recycling-plastic-bottles-make-3d-printed-tools

Buhusayen, B., Seetand, P.-S., & Coetzer, A. (2020). Turnaround Management of Airport Service ProvidersOperating during COVID-19 Restrictions. *Sustainability*, 1–24.

Carlota, V. (2019). Diehl Aviation worked with Airbus to create 3D printed modules for Qatar Airways. *3D Natives*. https://www.3dnatives.com/en/qatar-airways-090420194/

Carlucci, D. (2020). The Novel Coronavirus May Damage Aviation More Than SARS. *MarshMcLennan.* https://www.brinknews.com/the-novel-coronavirus-may-damage-aviation-more-than-sars/

Cooper, C., & Dewe, P. J. (2008). *Stress: A brief history*. John Wiley & Sons.

Craven, M., Liu, L., Mysore, M., & Wilson, M. (2020). *COVID-19: Briefing note*. McKinsey&Company.

D'Angelo, C. D., Gazzaroli, D., Corvino, C., & Gozzoli, C. (2022). Changes and Challenges in Human Resources Management: An Analysis of Human Resources Roles in a Bank Context (after COVID-19). *Sustainability, 14*(8), 1–14. doi:10.3390u14084847

Daly, M., & Delaney, L. (2013). The scarring effect of unemployment throughout adulthood on psychological distress at age 50: Estimates controlling for early adulthood distress and childhood psychological factors. *Social Science & Medicine, 80*, 19–23. doi:10.1016/j.socscimed.2012.12.008 PMID:23415587

Davies, P., Tidey, A., & Ortahamamcilar, B. (2020, August 18). Covid-19'un Avrupa ekonomisine etkileri: Hangi firma ne kadar işçi çıkardı? [The effects of Covid-19 on the European economy: Which company laid off how many workers?] *Euronews*. https://tr.euronews.com/: https://tr.euronews.com/2020/08/18/covid-19-un-avrupa-ekonomisine-etkileri-hangi-firma-ne-kadar-isci-c-kard

Deveci, M., Çiftçi, M. E., Akyurt, İ. Z., & Gonzale, E. D. (2022). Impact of COVID-19 pandemic on the Turkish civil aviation industry. *Sustainable Operations and Computer*, 93-102.

Doğan, A. (2011). Elektronik İnsan Kaynakları Yönetimi Ve Fonksiyonları. [Electronic Human Resources Management and Functions.] *Journal of Internet Applications and Management*, 51-80.

Dombey, O. (2004). The Effects Of Sars On The Chinese Tourism Industry. *Journal of Vacation Marketing, 10*(1), 4–10. doi:10.1177/135676670301000101

Dube, K., Nhamo, G., & Chikodzi, D. (2021). COVID-19 pandemic and prospects for recovery of the global aviation industry. *Journal of Air Transport Management, 92*, 1–12. doi:10.1016/j.jairtraman.2021.102022 PMID:36567961

ElMarady, A. A., & Rahouma, K. H. (2022). The Impact of COVID-19 on the Cybersecurity in Civil Aviation: Review and Analysis. In *2022 International Telecommunications Conference (ITC-Egypt)* (pp. 1-6). IEEE.

Erdağ, T. (2021). Havayolu Taşımacılığı Sektörü Gelişim Evrelerinin PEST Analizi ile Değerlendirilmesi:5 Dönem+Covid-19 Pandemi Süreci Dönemi Kapsamında Bir İnceleme. [Evaluation of the Development Stages of the Air Transport Sector by PEST Analysis: 5 Semesters + A Review within the Scope of the Covid-19 Pandemic Process Period.]. *The International Journal of Social Sciences (Islamabad)*, 422–461.

Eroğlu, U. (2001). *İnsan Kaynakları Yönetiminde Yeni Yönelimleri Hazırlayan İki Güç: Sanal Organizasyonlar Ve Stratejik Tabanlı Düşünce* [*Two Forces Preparing New Directions in Human Resource Management: Virtual Organizations and Strategically Based Thinking.*]. ISGUC The Journal of Industrial Relations and Human Resources.

Gavine, A. (2019). British Airways is exploring 3D printing for cabin efficiencies. *Aircraft Interiors International.* https://www.aircraftinteriorsinternational.com/news/airline-news/british-airways-is-exploring-3d-printing-for-cabin-efficiencies.html

Genç, M., & Işıktaş, S. (2022). Covıd-19 Pandemi Sürecinin, Türk Havacılık Sektörü Çalışanlarına Ve Müşterilerine Olan Etkileri: Swot Analizi. [The Effects of the Covid-19 Pandemic Process on Turkish Aviation Sector Employees and Customers: Swot Analysis.] *International Journal of Arts & Social Studies,* 44-65.

Gürüz, D., & Yaylacı, Ö. G. (2004). *İletişimci gözüyle insan kaynakları yönetimi* [*Human resources management through the eyes of a communicator.*]. MediaCat Kitapları.

Halowell, R. (1996). Southwest Airlines: A Case Study Linking Employee Needs Satisfaction And Organizational Capabilities to Competitive Management. *Human Resource Management, 35*(4), 513–534. doi:10.1002/(SICI)1099-050X(199624)35:4<513::AID-HRM5>3.0.CO;2-Z

Herpertz, S., Nizielski, S., Hock, M., & Schütz, A. (2016). The relevance of emotional intelligence in personnel selection for high emotional labor jobs. *PLoS One, 11*(4), e0154432. doi:10.1371/journal.pone.0154432 PMID:27124201

Hopkins, B., & Markham, J. (2003). *E-Hr: Using Intranets to Improve the Effectiveness of Your People.* Gower Publishing Limited.

Horcajuelo, G. (2020). Airbus, France Discuss State Support Amid COVID-19-CEO. *Teletrader.* https://www.teletrader.com/airbus-france-discuss-state-support-amid-covid-19 ceo/news/details/52014346?ts=1595118512588

IATA. (2008). *Airlines to Lose US $ 5.2 Billion in 2008 - Slowing Demand and High Oil to.* IATA.

IATA. (2015). *ACMG Report.* Airline Airline Cost Management Group. IATA.

IATA. (2019). *Economic Performance of the Airline Industry. 2019 End Year Report.* IATA.

IATA. (2022). iata.org. https://www.iata.org/en/publications/economics/fuel-monitor/

ICAO. (2022). *Effects of Novel Coronavirus (COVID-19) on Civil Aviation: Economic Impact Analysis. Montreal.* ICAO. -

İnal, İ. H., Akdemir, A., & Cihan, S. (2021). Pandemi sonrası oluşan VUCA ortamının çalışan insan kaynakları kaygı düzeyi ve verimliliği üzerine etkisi [The effect of the post-pandemic VUCA environment on employee human resources anxiety level and productivity]. *Yönetim Bilimleri Dergisi*, *19*(40), 347–374.

International Labour Office (ILO). (2003). *The impact of the restructuring of civil aviation on employment and social practices.* Submission to the Worldwide Air Transport Conference: Challenges and Opportunities of Liberalization, Geneva, ILO

Josephs, L. (2020, March 16). *US airlines seek more than $50 billion in government assistance as coronavirus roils business.* CNBC. https://www.cnbc.com/2020/03/16/us-airlines-seek-more-than-50-billion-in-aid-as-coronavirus-roils-business.html

Kelly, J. (2021). Airlines Lost Over 400,00 Workers—United Airlines Announced Another 14,000 Jobs May Be Lost. *Forbes.* https://www.forbes.com/sites/jackkelly/2021/02/01/airlines-lost-over-40000-workers-united-airlines-announced-another-14000-jobs-may-be-lost/?sh=74443cd024b3

Kilpatrick, J. (2020). COVID-19 Managing supply chain risk and disruption. *Deloitte.* https://www.deloitte.com/global/en/services/risk-advisory/analysis/covid-19-managing-supply-chain-risk-and-disruption.html

Kim, S., Wong, A. K. F., Han, H., & Yeung, M. W. (2022). Airline employees' stress amidst the COVID-19 pandemic and its job-related consequences. *Asia Pacific Journal of Tourism Research*, *27*(1), 30–47. doi:10.1080/10941665.2021.1998158

Küçükönal, H., & Korul, V. (2002). *Havayolu İşletmelerinde İnsan Kaynakları Yönetimi* [*Human Resources Management in Airline Operations.*]. Sosyal Bilimler Dergisi.

Kurulgan, M. (2020). *İnsan Kaynakları Planlaması. İnsan Kaynakları Yönetimi (s. 75). [Human Resources Planning. Human Resource Management (p. 75).]* İçinde Eskişehir: Anadolu Üniversitesi.

Lappas, I., & Kourousis, K. I. (2016). Anticipating the need for new skills for the future aerospace and aviation professionals. *Journal of Aerospace Technology and Management*, *8*(2), 232–241. doi:10.5028/jatm.v8i2.616

Locke, A. E., Feren, B. D., McCaleb, M. V., Shaw, N. K., & Denny, T. A. (1980). The relative effectiveness of four methods of motivating em-ployee performance. *Changes in working life*, 363-388.

Mahfuza, J., Nawrasb, N., Jassimc, A. G., Hafazd, N. A., Hayatule, S., & Imanf, B. (2020). The effect of after COVID-19 human resource management approaches on organizational entrepreneurship. *Systematic Reviews in Pharmacy, 11*(10).

Öge, S. (2004). Elektronik İnsan Kaynakları Yönetimi (E-HRM)'nde İnsan Kaynakları Enformasyon Sistemi (HRIS)'nin Önemi ve Temel Kullanım Alanları. [The Importance of Human Resources Information System (HRIS) in Electronic Human Resources Management (E-HRM) and Main Usage Areas.] Ulusal Bilgi, Ekonomi ve Yönetim Kongresi, (s. 109-117). Eskişehir.

Persons, W. (1995). Crisis Management. *Career Development International*, 27.

PTI. (2022). COVID-19 impact: 10% jobs in aviation sector lost between Apr 2020 and Dec 2021, says govt. *BusinessToday*. https://www.businesstoday.in/industry/aviation/story/covid-19-impact-10-jobs-in-aviation-sector-lost-between-apr-2020-and-dec-2021-says-govt-327667-2022-03-28

Rappeport, A., & Chokshi, N. (2020). Crippled Airline Industry to Get $25 Billion Bailout, Part of It as Loans. *The New York Times.* https://www.nytimes.com/2020/04/14/business/coronavirusairlines-bailout-treasury-department

Robbins, S. P., Judge, T. A., & Millett, B. (2015). *OB: the essentials*. Pearson Higher Education AU.

Saxton, J. (2006). Iran's Oil and Gas Wealth. *Joint Economic Committee Research Report.*

Shanker, M. (2019). Recruitment process and its impact on retention of commercial pilots in Indian aviation industry. *Business Process Management Journal*, *26*(3), 736–751. doi:10.1108/BPMJ-12-2018-0376

Sheaffer, Z., & Rita, M.-N. (2003). Executives' Orientation As Indicators of Crisis Management Policies and Practices. *Journal of Management Studies*, *40*(2), 577. doi:10.1111/1467-6486.00351

Shin, Y., Choi, J. O., & Hyun, S. S. (2022). The Effect of Psychological Anxiety Caused by COVID-19 on Job Self-Esteem and Job Satisfaction of Airline Flight Attendants. *International Journal of Environmental Research and Public Health*, *19*(7), 4043. doi:10.3390/ijerph19074043 PMID:35409725

Şimşek, H. (2021). Havayolu Taşımacılığında Kriz Yönetimi. [Crisis Management in Air Transport.]. *Journal of Aviation Research*, *3*(1), 21–40. doi:10.51785/jar.777273

Şimşek, M. Ş., & Özge, H. S. (2012). *İnsan Kaynakları Yönetimi* [Human Resources Management.]. Eğitim Yayınevi.

Statista. (2022). Worldwide air freight traffic from 2004 to 2022. *Statista.com*. https://www.statista.com/statistics/564668/worldwide-air-cargo-traffic/

Strategy, A. (2008). *The full-blown fuel crisis*. Aviation Economics.

Sun, X., Wandelt, S., & Zhang, A. (2021). Technological and educational challenges towards pandemic-resilient aviation. *Transport Policy*, *114*, 104–115. doi:10.1016/j.tranpol.2021.09.010 PMID:34566279

The Guardian. (2020, Ekim). Major US airlines to lay off thousands of workers as Covid-19 support expires. *The Guardian.* https://www.theguardian.com/: https://www.theguardian.com/business/2020/aug/14/us-airlines-layoffs-covid-19-support-delta-united-american

Türk, A. (2017). 21. *Yüzyılda Terör Konsepti Havacılık Güvenliği. [Century Terror Concept Aviation Security.]* [Thesis of SHGM Havacılık Güvenliği ve Daire Başkanlığı Uzmanlık. Ankara].

Türker, G. Ö. (2020). Covid-19 Salgını Turizm Sektörünü Nasıl Etkiler? Turizm Akademisyenleri Perspektifinden Bir Değerlendirme. [How Does the Covid-19 Outbreak Affect the Tourism Sector? An Evaluation from the Perspective of Tourism Academics.] *International Journal of Social Sciences and Education Research*, 207-224.

Uslu, D. Yeter, & Kedikli, E. (2017). Sürdürülebilirlik Kapsamında Yeşil İnsan Kaynakları Yönetimine Genel Bir Bakış. [An Overview of Green Human Resources Management in the Context of Sustainability.] *Üçüncü Sektör Sosyal Ekonomi, 52*(3), 66-81.

Vardarlıer, P. (2016). *Strategic approach to human resources management during crisis*. 12th International Strategic Management Conference, (pp. 464-472), Antalya, Turkey. 10.1016/j.sbspro.2016.11.057

Wege, L. G., Ngige, C. D., & Dieli, O. J. (2019). Electronic Human Resource Management (E-Hrm) and Organizational Performance In Nigeria. *International Journal of Management and Entrepreneurship*, *1*(1), 53–69.

WowEssays. (2020). Research Paper On HRM Challenges In Aviation. *WowEssays*. https://www.wowessays.com/free-samples/research-paper-on-hrm-challenges-in-aviation/

Xiarewana, B., & Civelek, M. E. (2020). Effects of covid-19 on China and the world economy: Birth pains of the post-digital ecosystem. Journal of International Trade. *Logistics and Law*, *6*(1), 147–157.

Xinhua.net. (2020). China to Support Air Transport Enterprises Amid Fight Against COVID19. *Xinhua*. http://www.xinhuanet.com/english/2020-05/27/c_139091729

Chapter 5
Aviation and Climate Change: Becoming a Climate-Neutral Industry

Anil Padhra
University of West London, UK

Salim Kurnaz
https://orcid.org/0000-0002-8060-5151
Süleyman Demirel University, Turkey & Kazimiero Simonaviciaus University, Lithuania

ABSTRACT

The aviation industry is currently on a path to becoming climate-neutral and various stakeholders such as airlines, airports, and regulatory bodies are setting targets to achieve this goal. The purpose of this chapter is to explain aviation's contribution to climate change and the existing efforts of the industry to reduce and mitigate climate impact. The chapter covers recent evidence suggesting that our changing climate is beginning to impact the industry, largely in a negative manner. Therefore, the industry will need to plan and adapt to its changing environment while developing new technologies and approaches. The challenges and opportunities of alternative aircraft fuel and energy source are presented, along with modifications to infrastructure that will accelerate the transition to a carbon and climate-neutral industry.

INTRODUCTION

In 2019, the aviation industry contributed 2.1% of all human-induced carbon dioxide emissions and approximately 12% of all carbon emissions from the transport sector (ATAG, 2020). As the aviation industry begins its recovery in the post-pandemic world, it's contribution to carbon emissions and climate change will increase relative to other transport modes. This is in part, due to the rapid expansion of aviation expected in the post-pandemic period which will see more passenger carriage and aircraft operations. It is also due to higher power and energy requirements of aircraft operations which makes it difficult to transition to alternative energy sources that are able to deliver the payload and range performance achievable with existing fossil fuel sources. At the same time, the rapid electrification of surface transport modes such as cars and heavy goods vehicles reduces their contribution to climate change,

DOI: 10.4018/978-1-6684-6835-7.ch005

resulting in a proportional increase in the climate impact of aviation. However, the aviation industry has had a successful history of developing new technologies and operational practices to drive efficiency and reduce environmental impact. During periods of high fossil-fuel prices, airlines have placed greater emphasis on introducing technology and developing practices to reduce fuel burn, helping to reduce operational cost and emissions. According to Schafer and Waitz (2014), aircraft fuel burn reduced by 70% between 1960 and 2000.

Although improvements to aircraft, engines and operational procedures have enabled aviation to become relatively more efficient and environmentally friendly, the continued rate of development is unlikely to fast enough to meet decarbonisation targets. Instead, alternative fuel and energy sources provides the best hope for radical improvements. The most promising of these technologies is electric-battery aircraft and hydrogen-powered aircraft. However, these sources have their own challenges particularly in terms of airport infrastructure developments and logistics to accommodate new aircraft.

From a commercial perspective, there are also pressures for the industry to become greener. In the last decade, investors and shareholders have been demanding more sustainable practices by airlines and airports, and passengers are becoming more environmentally conscious when travelling. In some countries, there has been evidence of declining air travel demand. In the first three quarters of 2019 average monthly demand for domestic air travel in Sweden fell by 8.7% compared to 2018 (Gossling et al. 2020).

Aviation's contribution to climate change is not just limited to carbon dioxide emissions. Aircraft engines also emit oxides of nitrogen (NO_x) which have a net warming impact, in the short-term, due to the formation of atmospheric ozone (Lee, 2018). In addition, water vapour and soot emissions from engines may form condensation trails (contrails) depending on atmospheric conditions. These line-shaped clouds reflect incoming solar radiation (cooling effect) but also trap outgoing radiation (warming effect). The net effect is that of warming and contributes to climate change; an effect which is further enhanced if contrails evolve into long-lasting cirrus clouds.

Furthermore, the period of aviation's growth and expansion over the last half century has coincided with a period of rapid climate change with average surface temperatures increasing by 0.2°C per decade (Allen et al., 2019). The associated extreme weather events, sea-level rises and changes to other weather phenomenon has and will continue to impact aviation operations. For example, coastal airports may need to build sea defences and airlines may need to plan for more frequent and severe in-flight turbulence. Therefore, not only does aviation impact to climate change, but a changing climate also impacts the aviation industry. Thus, climate change is a significant component for the operational and commercial sustainability of the industry.

The objectives of this chapter are to:

- Explain how the aviation industry contributes to climate change through carbon and non-carbon emissions.
- Explore the efforts to reduce and mitigate climate impacts by various aviation stakeholders including aerospace manufacturers, airlines and airports.
- Assess the impact of climate change on aviation in terms of aircraft operations and passenger demand and commercial impact.
- To explore the development of a range of alternative fuel and energy sources which have the potential to accelerate decarbonisation and enable to the industry to become climate-neutral.

CLIMATE CHANGE

The issue of climate change is receiving more and more attention in society. It's impact is widespread affecting almost every part of the globe. In almost every human activity there is a climate footprint and therefore the work of all industries and their organisations is becoming more scrutinised. The aviation industry has developed a reputation of being amongst the most contributing to climate change, despite the evidence suggesting that the contribution is small in relation to total human-induced greenhouse-gas emissions.

Climate change is not a newly discovered environmental issue. At the end of the 19th Century, the Swedish Chemist and Nobel laureate, Svante Arrhenius suggested that higher concentrations of carbon dioxide (CO_2) in the atmosphere would result in warmer atmospheric temperatures (Arrhenius, 1896). It is worth noting that changes to the Earth's climate due to natural causes such as volcanic eruptions and changes in ocean circulation patterns have been occurring for millions of years. However, significant changes to global climate due to human activity has only occurred since the start of the industrial revolution. The United Nations, Framework Convention on Climate Change considers the role of human activity in their definition of Climate Change as follows:

...a change of climate which is attributed directly or indirectly to human activity that alters the composition of the global atmosphere and which is in addition to natural climate variability observed over comparable time periods (UN,1992).

Human activity causes the release of greenhouse gas emissions, which trap heat from solar energy and have a warming impact on the atmosphere. The primary source of emissions has been the burning of fossil-fuels such as oil and gas. The main greenhouse gases include carbon dioxide, methane, nitrous oxides, fluorinated gases and water vapour. Not all gases are the same and their warming potential, concentration and lifetime in the atmosphere vary. For example, compared to carbon dioxide, methane has a greater ability to trap heat but it's lifetime in the atmosphere is approximately 10-12 years, whereas the lifetime of carbon dioxide is of the order of tens of thousands of years. Over a 100-year period, methane has a global warming potential approximately 25 times greater than carbon dioxide. However, the rate of release and the long lifetime of carbon dioxide has meant that it has become the main focus of attention in the effort to reduce greenhouse gas emissions.

Probably the most accurate record of recent carbon dioxide concentrations in the atmosphere are those from the Mauna Loa Observatory in Hawaii where daily continuous measurements have been taken since 1958. Figure 1 shows the measured data with a steady increase of CO_2 concentrations from 313 parts per million in 1958, passing 400 parts per million in 2013 to the current level of 420 parts per million more recently.

Figure 2 shows the correlation between CO_2 concentrations (measured from ice core samples) and atmospheric temperature over the last 400,000 years. The fluctuations align almost perfectly and indicate the sensitivity of the Earth's climate system to atmospheric composition. If the current trend in CO_2 concentrations continue the average global surface temperature will be 1.5°C warmer than pre-industrial times by 2035 and 2-3°C warmer by 2100 (IPCC, 2021).

Figure 1. Carbon dioxide measurements since 1958 taken from the Mauna Loa Observatory in Hawaii, US. [Source: Global Monitoring Laboratory – Earth System Research Laboratories, National Oceanic and Atmospheric Administration, United States Department of Commerce, https://gml.noaa.gov/ccgg/trends/]

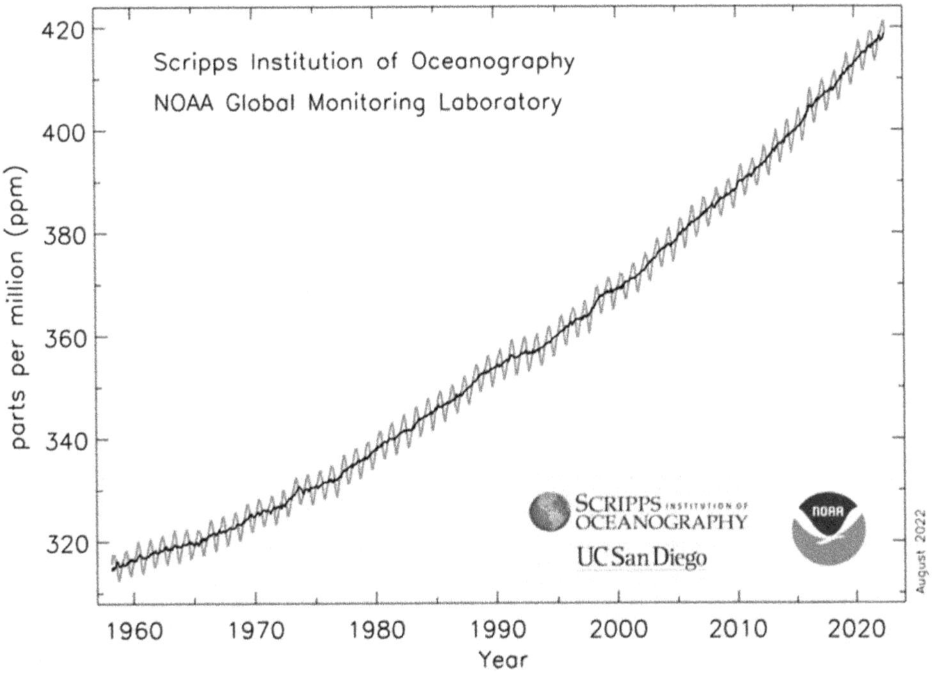

Figure 2. Correlation between atmospheric carbon dioxide concentration and temperature
Source: Petit et al. (1999)

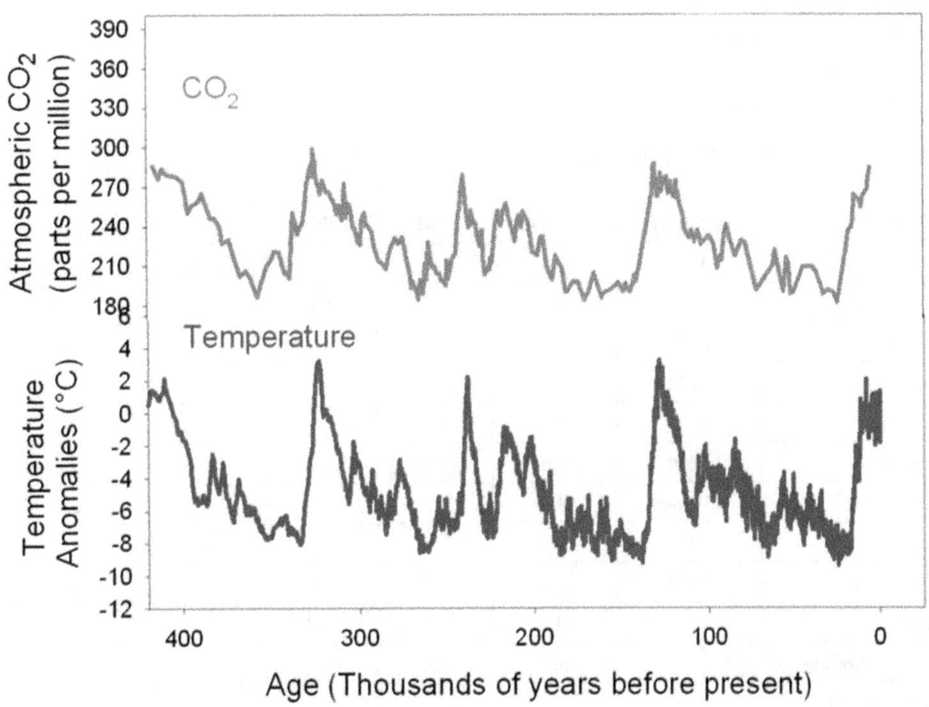

AVIATION CARBON EMISSIONS

Aviation's contribution to climate change is primarily from the release of carbon dioxide emissions from a variety of sources. These can be broadly described as being direct and indirect sources. Direct sources include the combustion of jet fuel in aircraft engines, emissions associated with the production and assembly of aircraft and the emissions associated with airport operations including terminal buildings and cargo facilities. Indirect emissions include for example, the transport of passengers travelling to and from the airport, the construction of infrastructure and the broader supply chain that serves the industry.

Direct aircraft emissions are primarily from the main engines and the auxiliary power unit (APU) located towards the rear of the aircraft which supplies electrical power during ground operations. There is a linear correlation between engine thrust and fuel consumption and between fuel consumption and carbon dioxide emissions. The most energy intensive phase of operation is during take-off and initial climb when engine thrust is approximately 75-100% of maximum thrust. During the take-off run, fuel burn can vary depending on aircraft and engine type, weight, meteorological conditions, airport elevation and engine thrust settings. Fuel burn values have been observed to vary between 39kg for an Airbus A321 aircraft to 102kg for a Boeing 747-400 aircraft at full thrust (Koudis et al. 2017). This phase of flight is relatively short compared to the cruise phase where engine thrust is typically 35-50% of maximum thrust. A typical narrow-body aircraft engine operating in the cruise will burn 2,700kg of jet fuel per hour. Every kilogram of jet fuel burned releases 3.16 kilograms of carbon dioxide, resulting in 8,500 kilograms of carbon dioxide per hour. There are also other greenhouse gas emissions including 3,300kg of water vapour, 30kg of nitrogen oxides, 2.5kg of sulphur dioxide as well as carbon monoxide, hydrocarbons, and particulate matter (EASA, 2019). The descent, approach and landing phases of flight are relatively less energy intense and fuel burn for most flights is between 1-7% of the total. Table 1 shows the percentage of fuel burn by flight phase for a range of routes from London. The longer the flight distance, the larger the proportion of fuel burn in the cruise phase. For shorter flights, fuel burn at the airport during taxi-in and taxi-out could be as much as 16% and possibly more at congested airports where taxi-times are longer.

Table 1. Fuel burn percentage by phase of flight for various routes and aircraft type

Route	Aircraft Type	Taxi-Out	Take-Off	Climb	Cruise	Approach	Taxi-In
London to Dubai	**A380**	1.8	0.6	1.5	94.5	1.0	0.6
London to Hong Kong	**A350**	1.1	0.4	1.1	96.3	0.7	0.4
London to New York	**B777**	1.7	0.6	1.4	94.6	0.8	0.9
London to Athens	**A320**	3.6	1.2	3.1	89.0	2.0	1.2
London to Helsinki	**A321**	3.7	1.6	3.9	87.3	2.4	1.1
London to Madrid	**A320**	6.0	1.9	4.8	82.5	3.1	1.8
London to Lisbon	**A320neo**	3.6	1.7	4.4	86.4	2.7	1.2
London to Geneva	**A320**	8.5	2.8	7.3	75.4	4.7	1.5
London to Edinburgh	**A319**	11.4	3.5	8.9	68.1	5.8	2.3
London to Paris	**A319**	11.2	4.0	10.3	62.3	6.7	5.5
Average		**5.3**	**1.8**	**4.7**	**83.6**	**3.0**	**1.6**

Source: Robertson (2022)

In absolute terms, the energy requirements of aircraft operations are such that a single-aisle aircraft operating a one-way short-haul trip between Paris, in France and Madrid, in Spain would generate the same total carbon dioxide emissions generated by 3 people in a whole year. Similarly, a single 13-hour flight operating from Paris to Singapore would generate the same levels of carbon dioxide emissions as one person would in their entire life.

In-flight fuel burn is heavily dependent on aircraft weight. In the cruise phase, heavier aircraft require more lift generation. The two main mechanisms for generating lift are to increase the aircraft speed or increase the flight angle of the aircraft. Both mechanisms increase the drag force which would slow the aircraft down. To maintain constant speed, the engines need to generate more thrust by increasing the fuel flow rate, thereby causing more emissions. To fly the same distance, the higher fuel flow rate for heavier aircraft means that more fuel needs to be carried which in turn increases the weight of the aircraft. Later in the chapter, the efforts of aircraft designers and airlines to reduce weight is discussed.

The main aircraft emissions at the airport occur during the taxiing phase and the turnaround process. During taxi, the aircraft engines are close to idle during which thrust is approximately 7% of maximum. An Airbus A320 aircraft taxiing with both engines at idle for 10 minutes typically burns 140kg of fuel and emits 450kg of carbon dioxide. In comparison, the four-engine Airbus A380 burns 630kg of fuel and releases nearly 2000kg of carbon dioxide over the same time period (ICAO, 2022). Turnaround emissions are more difficult to quantify and depend on turnaround time which can vary from less than 30 minutes for short-haul low-cost operator to more than 3 hours for a long-haul operator. As soon as an aircraft arrives on stand, the APU is switched on and the main engines are switched off. Unless an external power and cabin air source are available, the APU will remain switched on during the duration of the turnaround. Padhra (2018) observed the average turnaround time for a short-haul European airline was 42 minutes, during which the APU fuel burn was on average between 40kg (if external power was available) and 67kg (if external power was not available). The corresponding carbon dioxide emissions were between 126kg and 212kg per turnaround.

In addition to climate impact, aircraft engine emissions also contribute to poor air quality in and around an airport. To account for aircraft emissions that take place outside the airport boundary and contribute to poor air quality around the airport, the International Civil Aviation Organisation (ICAO) has developed a standard Landing and Take-Off (LTO) cycle. The LTO cycle includes emissions up to 3000ft above airport level and assumes that the aircraft engines operate at 100% thrust (take-off) for 0.7 mins, 85% thrust (initial climb) for 2.2 mins, 30% thrust (approach) for 4 mins and 7% thrust (idle) for 26 mins. This standard approach enables fuel burn and emission comparisons between different aircraft and engines. The LTO cycle carbon dioxide emissions for a Boeing 787 Dreamliner aircraft (Trent-1000 engines) is 5473kg while the similar size Boeing 767 aircraft (RB211 engines) has an LTO emission of 6055kg, a difference of 10%.

The other main airport emissions include those from passengers and airport staff travelling to and from the airport and ground service vehicles such as fuel and catering trucks. Figure 3 shows the carbon footprint breakdown for London Heathrow Airport in 2017. The carbon footprint is approximately equal for aircraft and non-aircraft sources.

Figure 3. Percentage carbon footprint share for London Heathrow Airport in 2017
Source: Heathrow (2018) and Walker (2018).

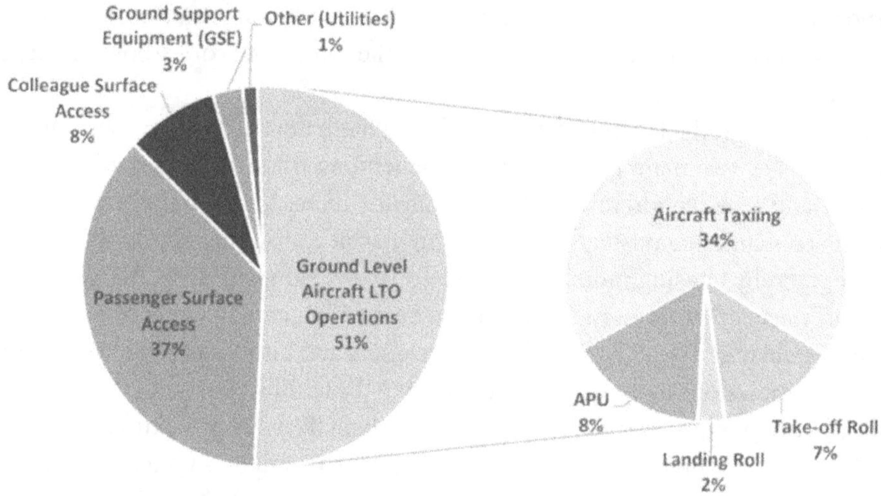

MITIGATION OF AVIATION CARBON EMISSIONS

Efforts to mitigate aviation carbon emissions have focused on three main areas; improvements in aircraft and engine design, improvements in operational practices and developing greener airports and infrastructure.

The most effective method to reduce aviation's climate impact is to improve the efficiency of aircraft and engines such that fuel burn and emissions are reduced. Quantifying improvements is challenging due to the large variations in aircraft passenger capacity and range. The metric most often quoted for comparison purposes is the fuel consumption per passenger per km. Using this metric, aircraft manufactured in 2014 were 45% more fuel efficient than those produced in 1968 (Kharina & Rutherford, 2015). It is expected that from now until 2045, aircraft fuel efficiency will improve by 0.98% per annum (ICAO, 2019). The vast majority of design improvements have come from the engines. When Boeing produced the first commercial jet aircraft, the Boeing 707 in 1958, the Pratt and Whitney JT3D engine consumed 53kg per hour per kilonewton of thrust at maximum thrust setting. By 1994, when the Boeing 777 was introduced, it's General Electric GE90 engines consumed 33kg per hour per kilonewton. The main design feature that has contributed to the improvement is the increase in the bypass ratio of the engines. The bypass ratio of an engine is the ratio of the mass of air that bypasses the core of the engine to that which passes through the core of the engine and is mixed with fuel for combustion. The higher the by-pass ratio, the greater the propulsive efficiency. Earlier engines such as the Pratt and Whitney JT3D engine had a bypass ratio of 1.4:1. More recent engines such as the Pratt and Whitney PW1100G used to power the Airbus A320neo aircraft have a bypass ratio of 12.2:1. Recent improvements in engine efficiency have come from the use of gears which enable the fan blades to rotate at a different speed to the low-pressure compressor which enables optimisation and improved propulsive efficiency.

Additional fuel burn and emissions reductions come from aerodynamic improvements which help to reduce drag. Lift generation from the wings causes wingtip vortices which reduce their ability to generate lift. Additional lift is achieved by flying at a higher angle of attack which also induces more drag. Reducing the induced drag can be achieved by installing wingtip devices such as winglets and sharklets.

According to Guerrero et al. (2020), winglets can achieve fuel burn reduction of 4-6%. The inclusion of winglets does require strengthening of the wings which add weight to the aircraft. For short flights, the additional weight cancels out the fuel saving benefits of the winglet and therefore they are more effective on longer flight sectors.

Advances in aerospace materials have enabled stronger and lighter aircraft structures resulting in lower fuel burn and emissions as explained earlier. In particular, the higher proportion of composite materials has enabled aircraft structural weight to reduce by 20%. The Boeing 787 Dreamliner, shown in Figure 4, consists of 80% composite materials by volume and 50% by weight (Boeing, 2022). This weight reduction leads to a 20% reduction in carbon emissions compared to an aluminium aircraft. Estimates suggest that by 2050, if the global fleet of aircraft transitioned to composite aircraft, carbon emissions could be reduced by 14-15% relative to a fleet that maintains its existing aluminium-based configuration (Timmis et al. 2015).

Figure 4. Diagram of the Boeing 787 Dreamliner aircraft showing the use of composite materials by weight.
Source: Aly (2017)

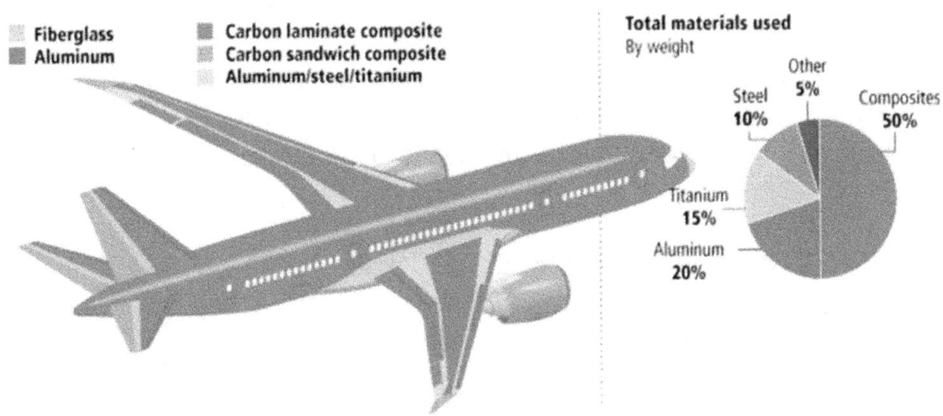

While aircraft and engine improvements have resulted in the greatest reductions in fuel burn and emissions, further reductions can be achieved once the aircraft is delivered to the operator, which is most cases is an airline customer. Two of the same aircraft operated by different airlines can have different climate impacts depending on their business model, routes, maintenance operations and internal operating policies. A common practice among all operators is to reduce aircraft operating weight. This can be achieved by, for example, reducing catering weight, installing lighter seats and carpet, converting in-flight books, magazines, flight charts, manuals and documents to digital format, reducing galley weight and reducing passenger baggage allowances. In any case, such initiatives must maintain flight safety and protect commercial revenue. A general rule-of-thumb for most aircraft is that for every 100kg of weight reduction, the aircraft consumes 3kg less fuel per hour (de Moor, 2020). In addition to emission reductions, this can result in significant reductions in direct operating costs due to lower fuel carriage and maintenance costs.

Flight operations procedures can also lead to significant emission reductions. For example, conducting reduced engine taxi between the runway and stand at the airport can reduce fuel burn and carbon emissions by up to 33% (Airbus, 2004). Modern aircraft systems are now able to compute the optimum speed and altitude inflight, to minimise fuel burn and emissions. This information is displayed on the flight management system in the cockpit and can be requested by the flight crew to air traffic controllers. Indeed, air traffic management provides some of the best opportunities to improve flight efficiency. Although aircraft fly from point to point (waypoints) in the airspace, air traffic controllers are sometimes able to offer shorter routes reducing flight distance, flight time and emissions. Parts of the airspace may also be permanently closed for commercial air traffic due to military airspace allocation. In Europe in recent years, the implementation of Flexible Use of Airspace (FUA) has removed the permanent segregation of civil and military airspace and instead airspace segregation is temporary, based on real-time usage within a specific time period. This enables airlines to operate more efficient flight plans with reduced in-flight emissions.

The vertical trajectory of aircraft is also managed by air traffic controllers. Providing uninterrupted climbs and descents can reduce the need to apply additional engine thrust, helping to reduce emissions. These climb and descent profiles are known as Continuous Climb Operation (CCO) and Continuous Descent Operation (CDO). Robinson and Kamgarpour (2010) conducted a study of CDOs in the US and found that the estimated savings were less than 25kg for over 45% of the flights, and less than 100 kg for over 87% of the flights. In Europe, a Eurocontrol study estimated that the theoretical maximum benefit of CCO and CDO operations were up to 350,000 tonnes per year (over 1m tonnes of carbon dioxide emissions). However, it was noted that the achievement of 100% CCO and CDO across the European network was not possible for a number of reasons including safety, weather, capacity and controller workload.

CLIMATE IMPACT OF NON-CARBON EMISSIONS

While the focus of governments, industry bodies and aviation stakeholders has been on reducing carbon dioxide emissions, the aviation industry is also beginning to better understand the climate impact of non-carbon effects. The main effects are due to emissions of sulphate and soot aerosols, the emissions of nitrogen oxides and the formation of condensation trails (contrails). The climate impact of these effects is often quantified in terms of the 'Radiative Forcing' (RF) metric. Radiative forcing is the net energy flux in the atmosphere caused by each of these effects and is measured in terms of Watts per metre squared (W/m^2). A positive net RF represents warming effect in the atmosphere, while a negative net RF represents a cooling effect.

The radiative forcing associated with the emissions of nitrogen oxides is complicated to estimate due to multiple secondary effects. Nitrogen oxides present in the atmosphere are associated with the formation of ozone which is a greenhouse gas and therefore causes a warming effect. On the other hand, nitrogen oxide contributes to the destruction of methane which is a potent greenhouse gas and therefore has a cooling effect on the atmosphere. The overall impact is estimated to result in a positive RF (warming effect). The emission of sulphate aerosol particles helps to reflect solar radiation back to space and therefore cause a cooling effect. Soot particles emitted from aircraft engines are often small, non-volatile black particles that trap infrared radiation leading to warming. Both sulphur and soot particles are likely to have an indirect effect of cloud formation which can have both a warming and cooling effect depending on the altitude and characteristics of the clouds.

Aircraft condensation trails (contrails) form another significant non-carbon impact of aviation. On clear days, aircraft flying in cold and humid conditions can be seen forming line-shaped clouds referred to as contrails. Contrails were first observed in 1915 (Ettenreich, 1919) and the theoretical framework for their formation was first presented in 1941 (Schmidt, 1941). Contrails form when water vapour particles emitted from the engines latch onto soot particles, emitted by engines or naturally-occurring in the atmosphere. If the ambient temperature is less than approximately -40°C the water freezes to form ice crystals. The physical characteristics of the contrail depend on water vapour and soot concentrations which can vary depending on the fuel type. Many of the contrails observed have a short lifetime and disappear within a matter of minutes. However, if aircraft are flying in atmosphere where the relative humidity with respect to ice is greater than 100% (ice-supersaturated), contrails can persist for hours and may evolve into cirrus clouds. Like clouds, contrails reflect incoming solar energy back out to space creating a cooling effect, but they also absorb outgoing surface radiated energy creating a warming effect. The net effect is usually a warming of the atmosphere. In fact, some studies suggest that the global warming impact of contrails could be as much as a third to a half of the total impact from all the carbon emissions ever emitted by aircraft. For persistent contrails, recent studies suggest a radiative forcing value between 4 and 16 mW/m^2. When including the impact of contrails which have evolved into cirrus cloud, estimates of radiative forcing are between 12 and 86 mW/m^2. The large variations in radiative forcing estimates occur because the science of contrails is not very well understood. Net radiative forcing is strongly influenced by the physical characteristics of contrails. These characteristics depend on the following variables:

- The amount of water vapour and particles present in the atmosphere occurring naturally or added by the jet engine exhaust
- The temperature, humidity, and pressure of the ambient atmosphere
- The wind and turbulence characteristics of the atmosphere
- The wingspan and size of the aircraft
- The propulsive efficiency of the jet engines
- The type and blend of fuel used to power jet engines

All these factors determine the size and shape of ice-crystals present in the contrails, the width and depth of contrails, the evolution of the contrail dimensions over time and the lifespan of the contrails.

Figure 5 summarises the radiative forcing due to aviation from multiple components from pre-industrial times to 2005. It should be noted that the level of scientific understanding for many of the non-carbon effects is low resulting in high levels of uncertainty.

Since the net effect of contrails is of warming, a promising approach to reduce their impact on the climate is to fly aircraft in regions of the atmosphere where contrails are less likely to form. This can be achieved by flight planning the vertical and horizontal route of aircraft to avoid contrail forming regions. Several studies have shown that even small changes in cruise altitude could prevent contrail formation (Avila et al. 2019; Teoh et al. 2020). Due to the large volume of traffic in the airspace, it is more challenging to deviate from the planned trajectory in-flight. Therefore, climate-optimal flight trajectories need to be pre-planned as part of the flight-planning process. The first step is to forecast parts of the atmosphere that are cold and humid enough for contrails to form so that they can be avoided. The dynamic nature of the atmosphere means contrail conditions vary on an hourly-basis and over a few hundred metres. Seasonal changes to the atmosphere mean that at any specific location, the frequency of contrail conditions will vary throughout the year. Figure 6 shows the average percentage coverage of ice supersaturated regions in US airspace between November 2014 and September 2015 at different flight level altitudes. Contrails are most likely to form between flight levels 320 and 350 and in the months of June, July, August, and September.

Figure 5. Radiative forcing components in Watts per square metre from global aviation as evaluated from preindustrial times until 2005. Error bars represent the 90% likelihood range for each estimate. The geographic spatial scale of the radiative forcing from each component and the level of scientific understanding (LOSU) are also shown on the right.
Source: Lee (2018)

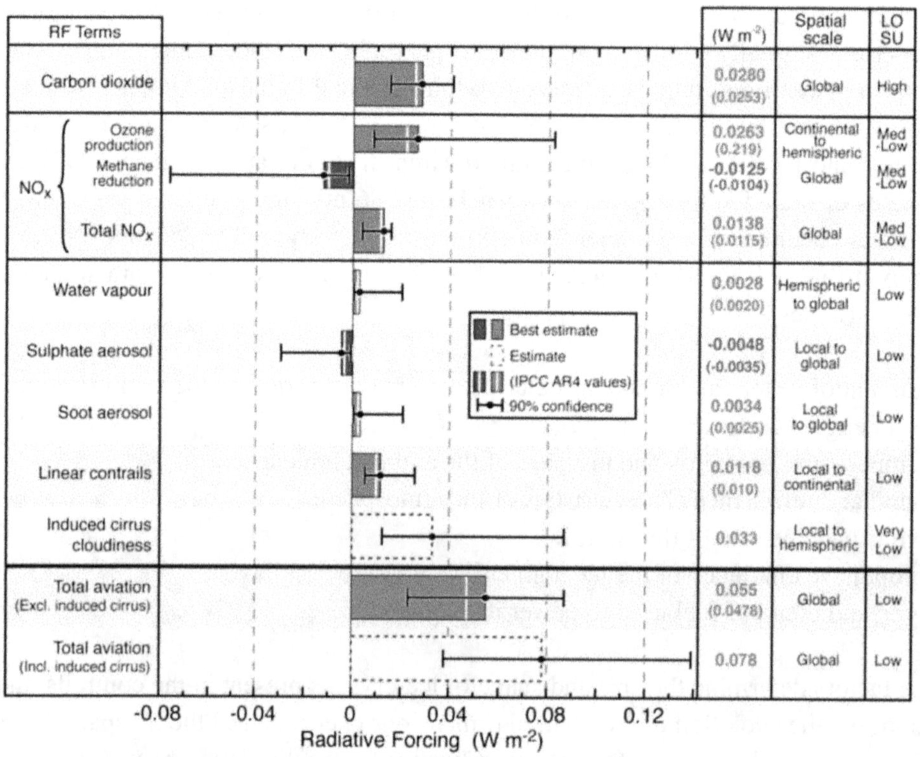

CLIMATE IMPACT AND ADAPTATION

The impact of aviation-related emissions on climate change is now widely acknowledged and relatively well understood. However, our knowledge of how the changing climate impacts the aviation industry is less well understood. Since the medium in which aircraft operate is our atmosphere, any changes, for example due to climate change, will have an impact on aviation. The main climate impacts for aviation are at airports, where risk to operations and infrastructure are the main concerns, and in-flight where a more energetic atmosphere may pose a risk to flight safety. A comprehensive summary of the impacts is provided by Ryley et al. (2020) and Gratton et al. (2022).

Figure 6. Average percentage coverage of Ice-Supersaturated regions (ISSR) in US airspace from flight level 220 to flight level 380 between November 2014 and September 2015.
Source: Avila and Sherry (2016)

Flight Level	Nov-14	Dec-14	Feb-15	Mar-15	Apr-15	May-15	Jun-15	Jul-15	Aug-15	Sep-15
380	0	0	0	0	0	0	2	4	4	4
370	2	1	0	0	2	3	9	13	14	13
360	1	1	0	0	2	2	9	12	11	12
350	5	6	3	2	6	9	18	23	23	19
340	10	14	8	6	15	17	26	28	30	26
330	2	3	2	2	4	7	11	11	11	11
320	12	22	15	12	21	24	24	26	24	23
310	8	11	9	5	10	10	9	9	10	7
300	4	6	3	2	5	7	5	5	6	4
290	8	13	10	5	14	15	9	5	7	6
280	3	7	4	2	7	7	4	1	2	2
270	4	8	5	3	9	5	3	1	1	2
260	2	3	2	1	3	2	1	0	0	1
250	3	5	4	2	4	1	0	0	0	1
240	2	2	2	1	2	1	0	0	0	0
230	1	1	1	0	1	0	0	0	0	0
220	1	1	1	0	1	0	0	0	0	0

In a warmer atmosphere, air particles move further apart and therefore the density of the air reduces. Density is an important physical parameter for aircraft operations. The forces of lift and drag generated by an aircraft are dependent on the density of air such that if the density of air is reduced, the ability of the aircraft to generate lift reduces and vice versa. Engine performance is also negatively impacted by lower air density since thrust force is reduced. During take-off, the reduced ability of the aircraft to generate lift and thrust results in a longer take-off distance on the runway. Some of the largest airports around the world have very long runways and therefore an increased take-off distance is less of a concern. However, at smaller regional airports and those located on some islands, runway length could be limited and aircraft may not be able to depart safely within the regulated limits. A study by Gratton et al. (2020) considered the impact of climate change on take-off performance for an Airbus A320 aircraft at some Greek airports. Figures 7a and 7b show the average minimum daily temperature and average wind speed trend at Chios Airport, one of the airports studied. Between 1974 and 2017, on average, the temperature increased by 0.75°C per decade and wind speed decreased by 0.23 knots per decade. The increase in temperature and reduction in wind speed resulted in the take-off distance increasing by an average of 8m per year for an Airbus A320 operating at maximum take-off weight. When an aircraft is unable to take-off in the runway distance available, airlines have to reduce the weight of the aircraft by offloading payload such as cargo, baggage and passengers. A lower weight reduces the lift force requirement of the aircraft, enabling aircraft to take-off in a shorter distance. For the island of Chios, the required reduction in payload was equivalent to 38 passengers and their baggage for the period between the A320's entry into service in 1988 and 2017. Offloading payload reduces airlines' ability to generate revenue and profit leading to a direct negative impact on airlines ability to be financially sustainable on some routes. A degradation in aircraft take-off and climb performance is also expected to lead to further negative impacts such as higher noise levels in the vicinity of airports due to shallower climb paths and slower climb to cruise altitude resulting in more fuel burn, though there are currently no studies that have quantified these impacts.

At some airports climate change is resulting in very warm temperatures, heavy precipitation and convective activity. These extreme weather events may cause airfield flooding, ground subsidence, inundation of underground infrastructure, loss of utilities provision such as electrical power, heat damage to the airport surface and disruption to surface access (Burbidge, 2018). Infrastructure damage can pause operations, delay and/or cancel flights and cause disruption to passenger and freight journeys. Airports may end up with large financial costs to repair and maintain critical infrastructure and potential loss of revenue. A warmer climate is also resulting in higher sea-levels due to the melting of sea ice and thermal expansion of the oceans. This increases the risk of airport flooding for coastal airports. A recent study assessed more than 14,000 airports worldwide and concluded that a 2°C rise in global temperature would place 100 airports below mean sea level (Yesudian and Dawson, 2021). This could happen by year 2100 and the highest risk airports were identified as Suvarnabhumi Airport in Bangkok, Thailand, and Shanghai Pudong Airport in China. Coastal airports have four main adaptation choices. They can:

1. Protect the airport by constructing sea defences
2. Raise the airfield to move operations above mean sea-level
3. Relocate the airport away from coastal areas and
4. Reclaim or float the airport to remain above mean sea-level

All of these adaptation measures involve very expensive engineering work and could cost hundreds of millions of dollars. For example, the estimated cost of building a sea defence wall, 1km long and 1 metre high could be as high as $70million. Some cities and countries have the space and finance available to invest in adaptation measures, move airports inland or absorb flights into neighbouring airports. However, many locations and especially those most reliant on airports for tourism, food and medical supplies will struggle to find alternative airport locations or investment to keep pace with the sea level rise.

In-flight operational impacts due to climate change are likely to lead to flight delays and inefficiencies. For example, a higher frequency of convective weather leading to thunderstorms could halt aircraft refueling and cause flight routes to be extended to avoid such weather. The North Atlantic air traffic region is one of the most studied in terms on climate impact. A key meteorological feature of this region is the jet stream, a conveyor belt of strong wind that moves air flow from west to east and is located at an altitude similar to the cruise altitude of aircraft. A series of jet streams are found in the mid-latitudes and occur because of the rotation of the earth and the difference in temperatures between the tropical equatorial regions and the cold polar regions. Flights operating from west to east such as from New York to London benefit from the presence of a strong jet stream because the winds carry the flight across the Atlantic at a faster ground speed reducing flight times by a few hours in some cases. However, flights travelling from east to west often encounter strong Jetstream headwinds which increase flight times. Due to climate change, there is evidence that the strength and shape of the Jetstream is changing which will inevitably impact aircraft operations and flight routings. Williams (2016) showed that if carbon dioxide levels were to double, the strengthening of the jet stream winds across the Atlantic are likely to cause shortening of eastbound flights and lengthening of westbound flights. For a round-trip across the Atlantic the reduction in flight time in one direction and the increase in flight time in the opposite direction would not cancel out and the net impact would be an increased flight time of more than 90 seconds. This may not sound like much but prior to the Covid-19 pandemic there were more than 300 round trips per day along the North Atlantic Corridor. If each flight took 90 seconds longer the collective additional flight time per year would be more than 2000 hours and would add $22 million to the fuel cost for airlines.

An even more concerning impact of changes to the Jetstream in on flight safety. At the boundary of fast-moving winds within the Jetstream and the slower moving airflow around it, vertical wind shear can be generated. This wind shear causes instability in the atmosphere known as clear-air turbulence. When flying through turbulent airspace, passengers experience a bumpy flight. In some severe cases of turbulence, injuries to passengers and crew and damage to aircraft have occurred. Studies have shown that if future carbon dioxide concentrations are double that of the pre-industrial period, changes to the jet stream would not only increase the frequency of turbulence but also the strength (Williams and Joshi, 2013). Figure 7 shows the difference in turbulence intensity in the North Atlantic flight corridor for a double CO2 scenario compared to pre-industrial CO2 concentrations for the winter season. The rectangular box between Ireland and Canada represents the area of airspace most likely to be taken by flights between Europe and North America.

Recent scientific research is shedding light on many other impacts of climate change on aviation. In addition to the operational impacts, changing passenger demand and travel patterns is also likely to have a commercial impact on airports and airline operators. Therefore, the need to understand how aviation will be impacted in the future is crucial to the sustainability of the aviation industry.

ALTERNATIVE ENERGY SOURCES AND DECARBONISATION

Since the dawn of commercial aviation, aircraft have been powered by fossil-fuels, most notably kerosene-based fuels. In recent decades, the climate impact of using such fuels has led scientists and technologists to develop alternative energy sources. Here we consider three alternative fuel and energy sources that are likely to replace fossil-fuel sources in the medium to long term.

In aircraft engines, kerosene can be replaced, in part, by synthetically-produced liquid fuels, collectively known as sustainable aviation fuels (SAF). Where the raw source of the fuel is biomass derived from used cooking oil, household and business waste and from plants and trees, SAFs are commonly referred to as biofuels. Biofuels are drop-in fuels since very little, if any, modification to existing aircraft and engines is required.

Figure 7. Spatial patterns of North Atlantic flight-level winter clear-air turbulence in a changing climate. The quantity shown is the median of variant 1 of Ellrod's turbulence index, computed from 20 years of daily-mean data in December, January, and February at 200 hPa.
Source: Williams and Joshi (2013)

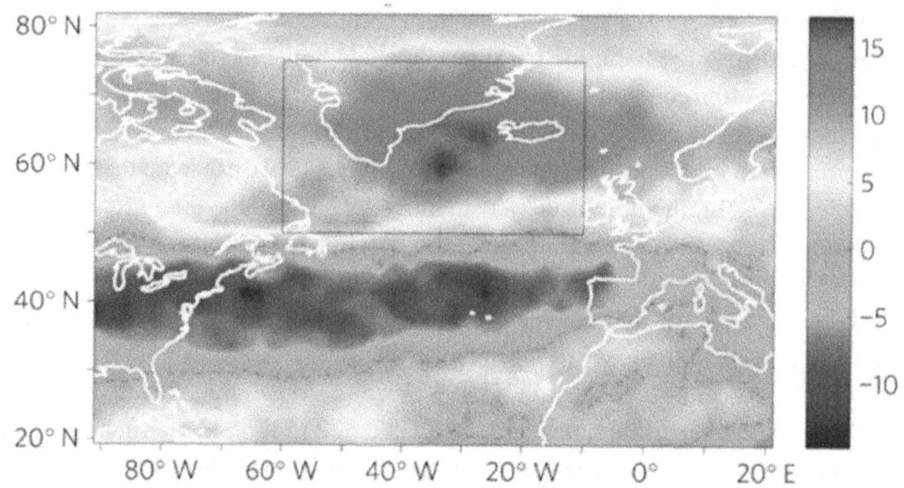

Figure 8 shows the process of converting jatropha fruit into jet fuel. Biofuels are considered as being sustainable because plants grown as feedstock have already absorbed carbon dioxide from the atmosphere, before being released back into the atmosphere during combustion and therefore the net carbon emissions are zero. However, the production process, transportation of feedstock and distribution of the processed fuel to airports all involve the emission of greenhouse gases and therefore biofuels are not completely carbon neutral. Carbon emission reductions are usually 20-90% compared to kerosene-based jet fuel. Current regulations mean that biofuels needs to be blended with traditional kerosene-based jet fuels up to a maximum percentage of 50%. The operational feasibility of using biofuels has resulted in multiple airlines around the world trialing their use in commercial flights. In 2008, Virgin Atlantic Airlines in the UK became the first airline in the world to operate a commercial flight powered by a biofuel blend. By 2019, 150,000 flights had made use of biofuels. Greater usage of biofuels is partly due to airports making the fuel available through an efficient supply chain and storage facilities. By 2021, 5 airports including Oslo, Stockholm, Bergen, Los Angeles and Brisbane airport had a regular supply of biofuels ready to be used by airlines serving those destinations.

Additional benefits of biofuels include lower emissions of other gases and particles which can help to improve air quality and may even reduce the formation of contrails. Biofuels generally contain less impurities and therefore produce less emissions such as sulphur dioxide. Less impurities also means lower emissions of soot particles which contribute to the formation of contrails. A study by Brauer et al. (2021) involved the measurement of engine emissions in the cruise phase of flight for an aircraft powered by a 49% biofuel blend. The measurements showed a 40% reduction in ice particle numbers, halving the optical depths of contrails and therefore reducing climate impact. While the use of aviation biofuels are considered to be friendlier for the environment there are some challenges and limitations. Firstly, the infrastructure for the production and distribution of biofuels is less well established than that for crude oil and therefore the current cost of producing biofuels is too high to stimulate large scale demand. Sec-

Figure 8. Flow diagram of the integrated process for converting jatropha fruit into jet fuel.
Source: Alherbawi et al. (2021)

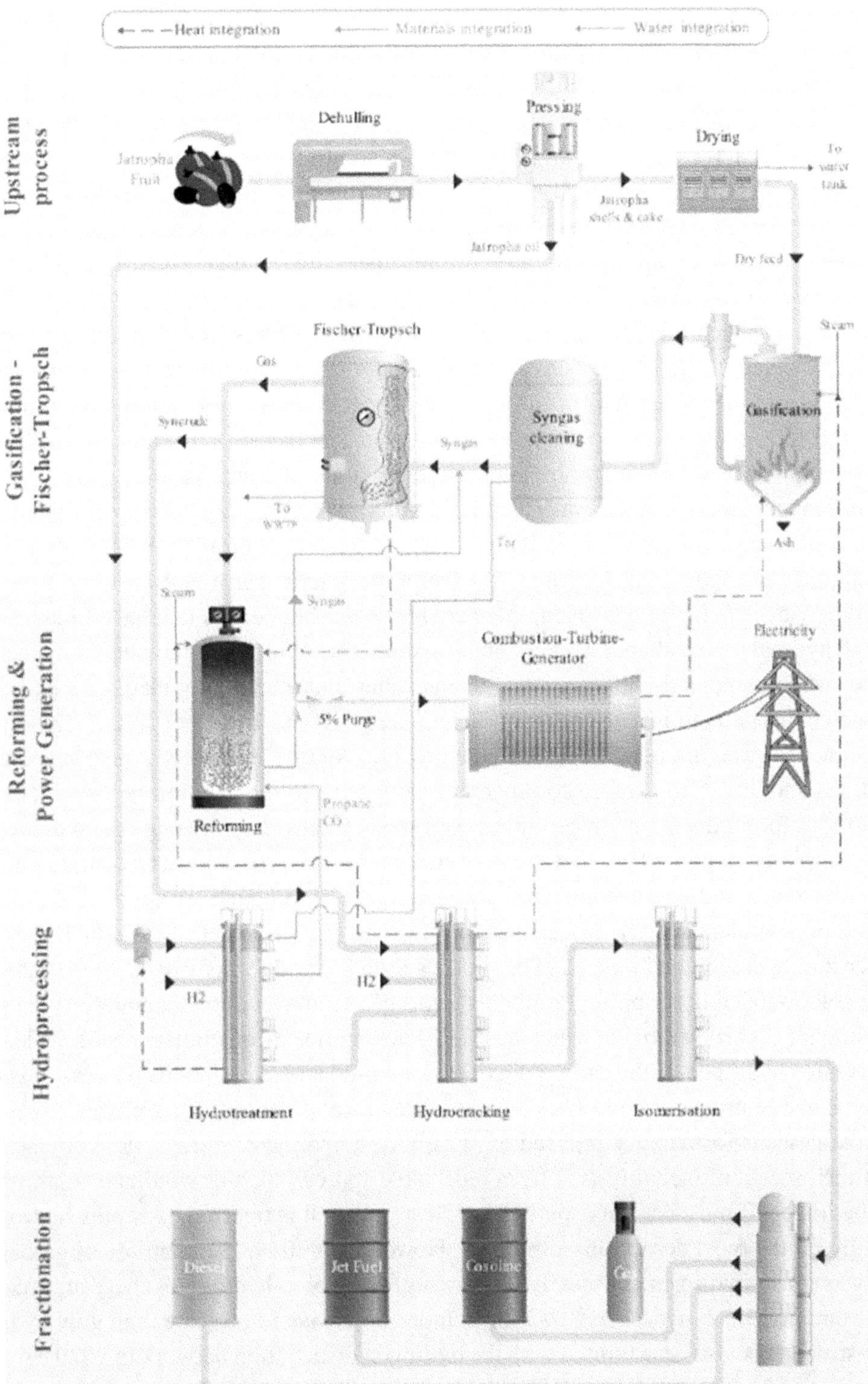

ondly, a large amount of biomass is required to produce a small quantity of biofuel. Currently, a tonne of biomass can at best produce 100 litres of biofuel. To meet the high demand for jet fuel, biomass would have to be produced especially for the purpose of creating biofuels. This would add significant pressure on land use and global food supply creating a moral and ethical issue in relation to global food poverty. So the current emphasis is on harvesting biomass from existing waste products to convert into jet fuel. Until radically new aircraft technology is developed and implemented, biofuels will provide one of the pathways to net zero carbon emissions in the short- to-medium term.

A more promising, though less well-established technology is the use of hydrogen as a fuel and energy source for aircraft. Hydrogen is the most abundant chemical element, and its low mass is ideal for aviation. Hydrogen in its liquid form can be combusted directly in an aircraft jet engine. However, current practice for prototype aircraft is for hydrogen to be used as a compressed gas in a fuel cell which generates electricity which then powers propellers to generate a thrust force. In either case, hydrogen combines with oxygen in the atmosphere and emits water vapour and there are no carbon emissions at all. An even greater benefit of hydrogen-powered aircraft could be in relation to contrail formation. Existing aircraft form contrails due to the water vapour and soot particles that are emitted from engines. While hydrogen aircraft will emit significantly large amounts of water vapour, there will be little or no soot emissions and therefore the water vapour emitted would mainly need to combine with lower concentrations of background particles in the atmosphere to form contrails. The resulting contrails are likely to be thinner and short-lived leading to a lower atmospheric warming impact.

Despite the many benefits of hydrogen, there are also some challenges that have limited widespread use. Although hydrogen is an abundant element, it's natural form is often combined with other elements. For example, when combined with oxygen it forms water and when combined with carbon it forms methane. The challenge is to isolate hydrogen in an energy-efficient manner so that carbon emissions or other greenhouse gases are not generated in the process. An efficient process also keeps the production costs down and therefore is more commercially attractive to the aviation industry. The four most common methods for producing hydrogen involve (i) steam methane reforming where hydrogen is produced from the mixing of natural gas and steam from heated water; (ii) coal gasification in which coal is combined with oxygen and steam leading to hydrogen as the by-product; (iii) pyrolysis which involves the separation of natural gas into hydrogen and solid carbon and (iv) electrolysis which uses electricity to split water into hydrogen and oxygen. The process used to produce hydrogen and its energy source determine a colour-coded description. Producing hydrogen from steam methane reforming produces a significant amount of carbon dioxide which has well-known impacts on climate change. This method is referred to as Grey hydrogen. If the carbon dioxide is captured and stored in the process, it is referred to as blue hydrogen. The process of pyrolysis also produces carbon but in the more convenient solid form rather than the gaseous form and is referred to as turquoise hydrogen. If the source of electricity used to carry out the process of electrolysis is from renewable sources such as wind and solar, it is referred to as green hydrogen. If the electricity source is nuclear power, it is referred to as pink hydrogen. Green hydrogen is by far the most decarbonised process. However, in 2019, less than 5% of global hydrogen production was green hydrogen and nearly 95% was grey or blue hydrogen. There are many barriers to this. For example, Green hydrogen is 2-3 times more expensive to produce than grey hydrogen. The current infrastructure is also very limited with the hydrogen distribution network in 2020 being only one fifth of a percent of that for natural gas. In addition to the production challenges, there are also further challenges when storing hydrogen. Hydrogen gas is highly flammable and liquid hydrogen has a very low temperature below minus 200 degrees Celsius. Therefore, high-pressure or heavily insulated tanks are required which prevent the fuel from being stored in aircraft wings like conventional kerosene-based fuel. Instead, the storage tanks would need to be located in the fuselage of the aircraft posing an additional challenge. Figure 9 shows a comparison of energy density by volume against the energy density by mass of a group of materials and technologies.

Figure 9. Comparison of the volumetric versus gravimetric energy density of a group of materials and technologies.
Source: Ehteshami and Chan (2014)

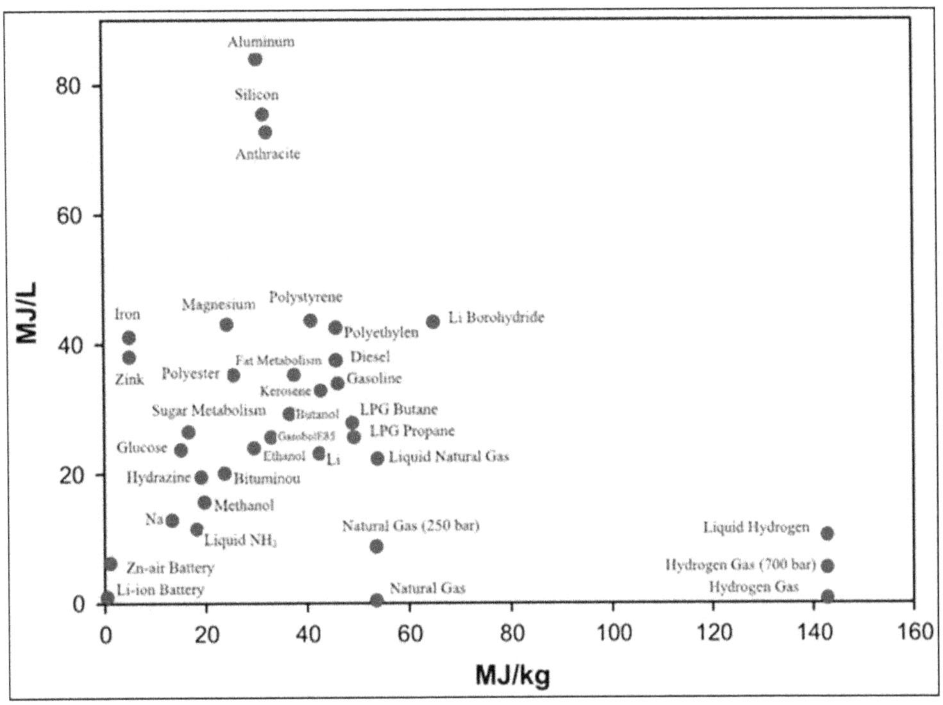

While the energy content per kilogram of hydrogen is three times greater than conventional kerosene-based jet fuel, it's energy content per cubic metre is three times less. Therefore, aircraft fuel tanks would need to be significantly larger than current tanks to achieve the same flight range. The larger tanks would reduce space for payload making hydrogen less attractive for airline revenue generation.

Assuming that economically sustainable hydrogen aircraft will be available in the next 30 years, the next challenge will be the planning and development of airport transportation and storage infrastructure to enable aircraft refuelling. Many of the world's airports have direct pipelines for kerosene-based fuel but there is little or no infrastructure for direct liquid hydrogen supply. Therefore, in the first cases, trucks are likely to be required to supply airports. A single truck could carry 5 tonnes of hydrogen which is a quarter of the energy content of one truck carrying kerosene-based jet fuel. This could lead to congestion around airports and additional carbon emissions if the trucks are not carbon-neutral. The storage of liquid hydrogen would also take up 4 times more space per unit of energy than current kerosene-based fuels requiring airports to make more space available on or close to the airfield.

The combustion of gaseous or liquid fuels to provide an energy source will inevitably lead to some form of emissions. Energy that comes from an electrical source has the benefits of zero in-situ emissions. For aviation, this has significant positive impacts on local air quality around airports. Zero emissions also means that contrails cannot form and therefore the non-carbon impacts of aviation are mitigated. If the electrical power has been generated from renewable sources such as wind and solar, then even the carbon impacts are eliminated. These benefits have encouraged designers of modern commercial aircraft to slowly transition towards becoming more electric. The Boeing 787 Dreamliner, for example, houses 6 electrical generators on board compared to traditional aircraft which have 3 generators. These generators are used to provide power for aircraft systems and lighting and also recharge the 2 sets of lithium-ion batteries on board the aircraft. Recent developments in electrical technology have enabled all-electric transportation modes, most notably in the automotive and rail industry. While several small prototypes of electric aircraft currently exist, it will be sometime before the introduction of all-electric commercial aircraft similar in size and range to existing regional aircraft. Battery technology is still not advanced enough to provide the high energy requirements of relatively high-payload, long-range flying.

Figure 10. Maximum energy and power onboard a range of air, sea and land vehicles.
Source: Epstein and O'Flarity (2019)

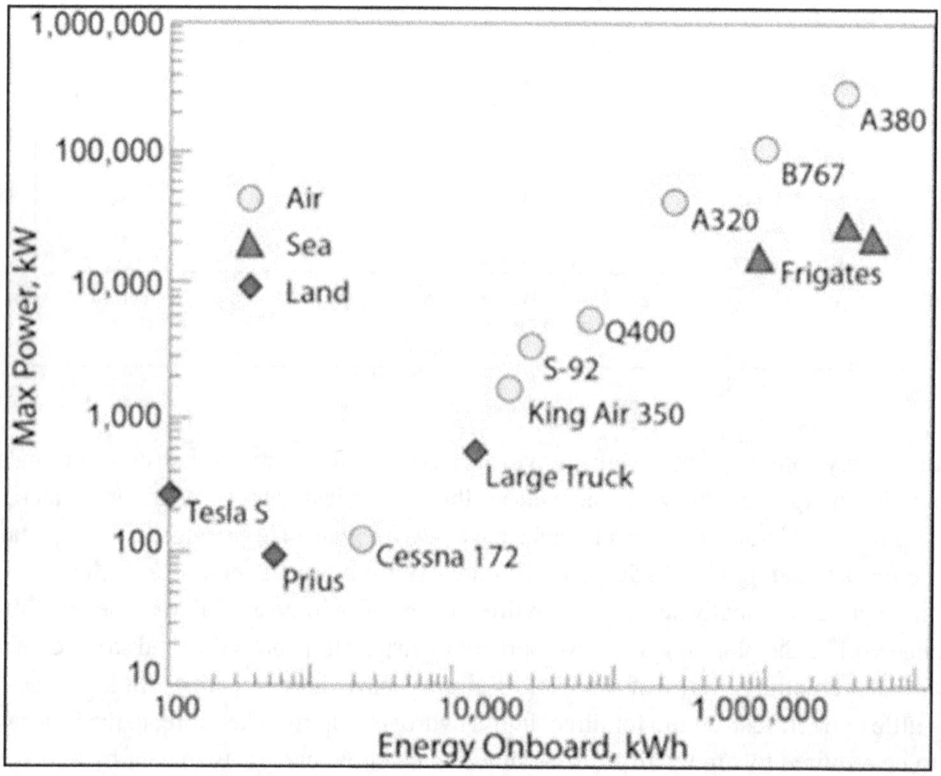

Figure 10 shows the maximum power requirements for several land, sea and air transport vehicles against the energy carried on board these vehicles. Existing electric cars have relatively small power requirements of less than 200kW and carry the equivalent of 1000kWh of electrical energy. Small aircraft such as the Cessna 172 are of similar size to cars and many of the all-electric aircraft prototypes that currently exist are of this size. However, for larger commercial aircraft such as Q400 which conducts regional flights, the maximum power requirements at take-off are 30 times greater; a value that current batteries cannot provide in the space available on aircraft. A useful metric to consider is the energy per kg that an electric battery could deliver. In 2020, the best lithium-ion battery was able to provide 250Wh per kg. In comparison kerosene-based aviation fuel delivers 12,500 Wh per kg. The minimum requirement for a short-range electric aircraft would be 750-2000kWh per kg. The current rate of battery development is a doubling of energy per kg every 17 years. Therefore, it could be 30 years by the time the first all-electric commercial aircraft are introduced into service. The next major challenge for all-electric aircraft is the large amount of electricity that would need to be produced from renewable sources to power a global fleet of regional all-electric aircraft in a sustainable manner. Based on 2015 figures, this would equate to approximately 1% of the total worldwide electricity consumption. A more likely application of electrical power in commercial aircraft is through a hybrid-electric approach. Hybrid technology involves a combination of conventional fuel and electrical power to propel an aircraft forward. For example, energy intensive phases of flight such as take-off and climb could be conducted using jet engines powered by conventional fuel, while the taxi, cruise and descent phase could take place with electric power alone. Up until April 2020, Airbus, Rolls Royce and Siemens were working on a hybrid-electric aircraft concept based on the existing BAE 146 aircraft. The aircraft named E-Fan X would have one of the four jet engines replaced by an electrical motor. Hybrid-electric aircraft could also be flown strategically to reduce climate impact. For example, aircraft flying in ice-supersaturated regions of the atmosphere where contrail formation is likely could be powered by electrical means to prevent climate warming.

MARKET-BASED MEASURES FOR CLIMATE CHANGE

Many of the operational solutions for achieving a carbon-neutral aviation industry have already been implemented. Radical technological solutions are reliant on significant progress quickly or come with additional challenges such as infrastructure development. Another driver for minimising the climate impact of aviation is to introduce market-based measures (MBM). MBM's are economic instruments that penalise airlines and airports for their carbon emissions and/or incentivise activities to reduce carbon emissions. There are three main types of MBMs, namely, taxes and levies, carbon offsetting and emissions trading. Taxes and levies are revenues collected by the government in response to certain activity. For example, the government of a country may introduce or increase fuel taxes, navigation charges or airport charges to deter airlines from expanding their operations resulting in lower fuel consumption and carbon emissions. They also encourage airlines and airports to invest in methods to reduce carbon emissions such as newer aircraft. Since aviation is an international industry, taxies and levies are not popular as it creates an unlevel playing field in the market and may discourage foreign airlines from operating flights to a particular airport or country. Carbon offsetting involves the reduction of emissions in another sector or an increase in carbon storage to offset the emissions in your own sector. For example, the carbon emissions of a specific flight may be estimated and then the airline may invest in a voluntary carbon offset project such as reforestation that plants trees to absorb the equivalent carbon emissions. Offsetting is usually a more cost-effective method than an organisation reducing its own emissions. An emissions trading scheme (ETS) is a system whereby the government or a central authority issues a limited number of permits (caps) that enables an organisation or sector to emit the equivalent amount of emissions. Organisations can then trade permits with other organisations by selling excess permits (if emissions have been reduced) and buying additional permits if more emissions will take place. The scheme allows emissions targets to be met and adjusted as new technology enters the market. The European EU Emissions Trading Scheme has included the aviation sector since 2012. Under the scheme, all airlines operating in Europe regardless of their registration country are required to monitor, report and verify their emissions, and to surrender allowances against those emissions. They receive tradeable allowances covering a certain level of emissions from their flights per year.

The system has so far contributed to reducing the carbon footprint of the aviation sector by more than 17 million tonnes per year. Since the scheme covers airlines operating in Europe, but registered in non-European countries, it has been seen as unilateral and unfair. In particular, countries such as China, the USA, Canada, Russia and Japan have strongly opposed the scheme claiming that it disadvantages developing countries and adds to airline operating costs. In response, in 2016, member states of the International Civil Aviation organisation (ICAO) adopted a global market-based method for aviation emissions termed, Carbon Offsetting and Reduction Scheme for International Aviation (CORSIA). The implementation of CORSIA is over three phases – two initial, voluntary phases (2021-2023 and 2024 – 2026) and a mandatory phase that would take place from 2027. During the initial phases, CORSIA only applies to international flights between states that have volunteered to take part. International flights to and from states that have not volunteered will be exempt. During the mandatory stage, which begins in 2027, CORSIA will cover all international flights, including those travelling to or from states that had not volunteered for the early phases. For fairness, the least developed nations, small island nations and nations with a very small share of international aviation traffic will be exempt. It is estimated that CORSIA will mitigate around 2.5 billion tonnes of CO2 between 2021 and 2035, which is an annual average of 164 million tonnes of CO2. This is equivalent to the total annual CO2 emissions from the Netherlands across all sectors.

SUMMARY

Aviation's impact on the environment, and in particular climate change is relatively small compared to other transport sectors such as road vehicles. However, as the industry begins its recovery in the wake of the Covid-19 pandemic, aviation traffic will increase rapidly and its contribution to climate change will increase. The primary contribution is through carbon emissions associated with aircraft operations, passenger and freight movement by surface transport to and from airports and emissions associated with airport operations, terminals and buildings. There are also non-carbon impacts due to the emission of nitrogen oxides at high altitude that cause an imbalance in radiative forcing and the emissions of water vapour and soot that can form contrails. The impact of the latter is less well-understood though likely to have a significant climate impact in addition to carbon emissions. The industry has recognised that operating more efficiently not only reduces climate impact but also reduces operating costs which contribute to the financial stability of the sector. Thus, airlines, airports, airspace managers and aircraft designers and manufacturers have placed a considerable amount of effort in developing and implementing environmental initiatives. Despite, year-on year improvements, the industry will still not achieve the target of net-zero carbon emissions by 2050 unless radical alternatives are found to kerosene-based fuels. Sustainable aviation fuels such as biofuels are a good short-term approach as net carbon emissions are lower. However, biofuels still result in tailpipe emissions and there are doubts as to whether sufficient feedstock is available to produce the volumes of biofuels required by the industry. The transition to electric aircraft is promising but the development of new battery-technology needs to be rapidly progressed to achieve financially sustainable commercial flights that can be operated with large payload and range. This is unlikely in the short-to-medium term given the historical rate of progress. The use of hydrogen as a fuel and energy source is gaining traction in the aviation industry with several trials and prototypes expected within the next decade. The large-scale green production of hydrogen fuel and the associated infrastructure required for storage and distribution in a safe and economical manner remain challenges. Alongside operational and technological developments, market-based measures such as the ICAO CORSIA initiative will be mandated from 2027 for which the industry must prepare now. Accelerating decarbonisation initiatives and achieving climate-neutrality are becoming more and more important as several recent studies have begun to highlight the negative impact of a changing climate on the industry. Rising sea-levels, heatwaves and more frequent and severe thunderstorms and in-flight turbulence are already impacting the industry and adaptation measures are being considered to maintain the sustainability of the industry.

REFERENCES

Air Transport Action Group (ATAG). (2020). *Aviation Benefits Beyond Borders*. ATAG. https://aviationbenefits.org/environmental-efficiency/ climate-action/waypoint-2050/

Airbus. (2004). Getting to Grips with Fuel Economy. *Airbus, 4,* pp. 82.

Alherbawi, M., McKay, G., Mackey, H. R., & Al-Ansari, T. (2021). A novel integrated pathway for Jet Biofuel production from whole energy crops: A Jatropha curcas case study. *Energy Conversion and Management, 229*, 113662. doi:10.1016/j.enconman.2020.113662

Allen, M. R., Dube, O. P., Solecki, W., Aragón-Durand, W., Cramer, S., Humphreys, M., & Mulugetta, R. (2019). Framing and Context. In: Global Warming of 1.5 C. An IPCC Special Report on the impacts of global warming of 1.5 C above pre-industrial levels and related global greenhouse gas emission pathways, in the context of strengthening the global response t.

Aly, N. M. (2017). A review on utilization of textile composites in transportation towards sustainability. [). IOP Publishing.]. *IOP Conference Series. Materials Science and Engineering, 254*(4), 042002. doi:10.1088/1757-899X/254/4/042002

Arrhenius, S. (1896). On the influence of carbonic acid in the air upon the temperature on the ground. *Philosophical Magazine, 4*, 237–276.

Avila, D., & Sherry, L. (2016, April). Method for analysis of Ice Super Saturated Regions (ISSR) in the US airspace. In 2016 Integrated Communications Navigation and Surveillance (ICNS) (pp. 10B2-1). IEEE.

Avila, D., Sherry, L., & Thompson, T. (2019). Reducing global warming by airline contrail avoidance: A case study of annual benefits for the contiguous United States. *Transportation Research Interdisciplinary Perspectives, 2*, 100033. doi:10.1016/j.trip.2019.100033

Boeing. (2022). *787 Dreamliner by Design: Advanced Composite Use.* Boeing. https://www.boeing.com/commercial/787/by-design/#/advanced-composite-use

Bräuer, T., Voigt, C., Sauer, D., Kaufmann, S., Hahn, V., Scheibe, M., Schlager, H., Huber, F., Le Clercq, P., Moore, R. H., & Anderson, B. E. (2021). Reduced ice number concentrations in contrails from low-aromatic biofuel blends. *Atmospheric Chemistry and Physics, 21*(22), 16817–16826. doi:10.5194/acp-21-16817-2021

Burbidge, R. (2018). Adapting aviation to a changing climate: Key priorities for action. *Journal of Air Transport Management, 71*, 167–174. doi:10.1016/j.jairtraman.2018.04.004

de Moor, S. (2020). A Fuel Efficiency Masterclass: Part Two of Three. *Aircraft Analytics*. https://aircraft-analytics.com/insights/a-fuel-efficiency-masterclass-part-two-of-three

EASA (European Union Aviation Safety Agency). (2019). European Aviation. *Environment Reporter, 2019*, 112.

Ehteshami, S. M. M., & Chan, S. H. (2014). The role of hydrogen and fuel cells to store renewable energy in the future energy network–potentials and challenges. *Energy Policy, 73*, 103–109. doi:10.1016/j.enpol.2014.04.046

Epstein, A. H., & O'Flarity, S. M. (2019). Considerations for reducing aviation's CO2 with aircraft electric propulsion. *Journal of Propulsion and Power, 35*(3), 572–582. doi:10.2514/1.B37015

Ettenreich, R. (1919). Wolkenbildung über einer Feuersbrunst und an Flugzeugabgasen. [Cloud formation over a conflagration and aircraft exhaust fumes.]. *Meteorologische Zeitschrift (Braunschweig), 36*, 355–356.

Gössling, S., Humpe, A., & Bausch, T. (2020). Does 'flight shame'affect social norms? Changing perspectives on the desirability of air travel in Germany. *Journal of Cleaner Production, 266*, 122015. doi:10.1016/j.jclepro.2020.122015

Gratton, G., Padhra, A., Rapsomanikis, S., & Williams, P. D. (2020). The impacts of climate change on Greek airports. *Climatic Change, 160*(2), 219–231. doi:10.100710584-019-02634-z

Gratton, G. B., Williams, P. D., Padhra, A., & Rapsomanikis, S. (2022). Reviewing the impacts of climate change on air transport operations. *Aeronautical Journal, 126*(1295), 209–221. doi:10.1017/aer.2021.109

Guerrero, J. E., Sanguineti, M., & Wittkowski, K. (2020). Variable cant angle winglets for improvement of aircraft flight performance. *Meccanica, 55*(10), 1917–1947. doi:10.100711012-020-01230-1

Heathrow. (2018). *Heathrow Carbon Footprint 2017*, p. 5. Heathrow Airport Ltd.

ICAO. (2019). *ICAO Environmental Report 2019: Destination Green – The Next Chapter.* ICAO.

ICAO. (2022). *ICAO Aircraft Engine Emissions Databank.* ICAO, Doc 9646. https://www.easa.europa.eu/domains/environment/icao-aircraft-engine-emissions-databank

IPCC. (2021). Summary for Policymakers. In *Climate Change 2021: The Physical Science Basis. Contribution of Working Group I to the Sixth Assessment Report of the Intergovernmental Panel on Climate Change* [Masson-Delmotte, V., P. Zhai, A. Pirani, S.L. Connors, C. Péan, S. Berger, N. Caud, Y. Chen, L. Goldfarb, M.I. Gomis, M. Huang, K. Leitzell, E. Lonnoy, J.B.R. Matthews, T.K. Maycock, T. Waterfield, O. Yelekçi, R. Yu, and B. Zhou (eds.)]. (pp. 3–32). Cambridge University Press. doi:10.1017/9781009157896.001

Kharina, A., & Rutherford, D. (2015). Fuel Efficiency Trends for New Commercial Jet Aircraft: 1960 to 2014. *The International Council on Clean Transportation, White Paper*, p. 27.

Koudis, G. S., Hu, S. J., Majumdar, A., Jones, R., & Stettler, M. E. (2017). Airport emissions reductions from reduced thrust takeoff operations. *Transportation Research Part D, Transport and Environment*, *52*, 15–28. doi:10.1016/j.trd.2017.02.004

Lee, D. S. (2018). *The current state of scientific understanding of the non-CO2 effects of aviation on climate*. UK Department for Transport, Manchester Metropolitan University.

Padhra, A. (2018). Emissions from auxiliary power units and ground power units during intraday aircraft turnarounds at European airports. *Transportation Research Part D, Transport and Environment*, *63*, 433–444. doi:10.1016/j.trd.2018.06.015

Petit, J. R., Jouzel, J., Raynaud, D., Barkov, N. I., Barnola, J. M., Basile, I., Bender, M., Chappellaz, J., Davis, M., Delaygue, G., Delmotte, M., Kotlyakov, V. M., Legrand, M., Lipenkov, V. Y., Lorius, C., PÉpin, L., Ritz, C., Saltzman, E., & Stievenard, M. (1999). Climate and atmospheric history of the past 420,000 years from the Vostok ice core, Antarctica. *Nature*, *399*(6735), 429–436. doi:10.1038/20859

Robertson, B. (2022). *Which part of a flight uses the most fuel?* OAG. https://www.oag.com/blog/which-part-flight-uses-most-fuel

Robinson, J., III, & Kamgarpour, M. (2010). Benefits of continuous descent operations in high-density terminal airspace considering scheduling constraints. In *10th AIAA aviation technology, integration, and operations (ATIO) conference* (p. 9115). 10.2514/6.2010-9115

Ryley, T., Baumeister, S., & Coulter, L. (2020). Climate change influences on aviation: A literature review. *Transport Policy*, *92*, 55–64. doi:10.1016/j.tranpol.2020.04.010

Schäfer, A. W., & Waitz, I. A. (2014). Air transportation and the environment. *Transport Policy*, *34*, 1–4. doi:10.1016/j.tranpol.2014.02.012

Schmidt, E. (1941). Die entstehung von eisnebel aus den auspuffgasen von flugmotoren. [The formation of ice mist from the exhaust gases of aircraft engines.] Schriften der Deutschen Akademie der Luftfahrtforschung, Verlag R. Oldenbourg, München, Heft 44, 5(44), 1-15.

Teoh, R., Schumann, U., & Stettler, M. E. (2020). Beyond contrail avoidance: Efficacy of flight altitude changes to minimise contrail climate forcing. *Aerospace (Basel, Switzerland)*, *7*(9), 121. doi:10.3390/aerospace7090121

Timmis, A. J., Hodzic, A., Koh, L., Bonner, M., Soutis, C., Schäfer, A. W., & Dray, L. (2015). Environmental impact assessment of aviation emission reduction through the implementation of composite materials. *The International Journal of Life Cycle Assessment*, *20*(2), 233–243. doi:10.100711367-014-0824-0

UN (United Nations). (1992). United Nations Framework Convention on Climate Change. UN.

Walker, C. (2018). Heathrow Airport 2017 Emission Inventory. *Ricardo Energy and Environment, Report number ED 11486* (1) 38. Cheshire, UK.

Williams, P. D. (2016). Transatlantic flight times and climate change. *Environmental Research Letters*, *11*(2), 024008. doi:10.1088/1748-9326/11/2/024008

Williams, P. D., & Joshi, M. M. (2013). Intensification of winter transatlantic aviation turbulence in response to climate change. *Nature Climate Change*, *3*(7), 644–648. doi:10.1038/nclimate1866

Yesudian, A. N., & Dawson, R. J. (2021). Global analysis of sea level rise risk to airports. *Climate Risk Management*, *31*, 100266. doi:10.1016/j.crm.2020.100266

Chapter 6
Rethinking Managerial Practice on Low-Cost Airline Organizational Structures:
The Lesson-Learned Amid the Pandemic - Case in Thailand

Kanittha Charernnit
https://orcid.org/0000-0002-3016-9451
College of Hospitality Industry Management, Suan Sunandha Rajabhat University, Thailand

Kannapat Kankaew
https://orcid.org/0000-0003-1127-5627
College of Hospitality Industry Management, Suan Sunandha Rajabhat University, Thailand

Rojanard Waramontri
College of Hospitality Industry Management, Suan Sunandha Rajabhat University, Thailand

Korawin Kungwol
College of Hospitality Industry Management, Suan Sunandha Rajabhat University, Thailand

ABSTRACT

This chapter investigates the managerial practices of low-cost carriers during the pandemic enhancing its operations to remain in track. The managerial practices in the study were high competence workforce, organizational culture, risk management, and strategic human resources management. The multiple regression was applied. The results found that LCCs highlighted on risk management, followed by high competence workforce, and strategic human resource management affecting to organizational structure. Overall, there seems to be some evidence to indicate that the discrepancies and risk management plan for any circumstances of uncertainty must be implemented. While the training and development of employees should be conducted continuously especially during the period of non-operating. A further issue suggested to practice the strategic human resource management (SHRM), that aligned organization objectives fit with the current situation both internally and externally, since the SHRM would eventually affect to organizational structure as a whole.

DOI: 10.4018/978-1-6684-6835-7.ch006

INTRODUCTION

The widespread of COVID-19 is severely derailed airlines' operations alike other businesses. Additionally, the airline industry is highly vulnerable to external environment namely; political, economic, social, technology, and environmental issues. Furthermore, the destructive competition, the threat of new entrants, the customer's buying power, the threat of substitution, and the bargaining power of suppliers which is known as five forces indicated the airlines are defenselessness (Kankaew, 2016, 2022). In addition, in the era of fourth industrial revolution (IR 4.0) all kind of organizations must learn to transform themselves respectively to the changes (Kankaew et al., 2021) that placing more tough managerial issue in organizational structure. In the antecedent of the COVID 19, airlines were competing strongly by proposing its service quality and employing the price mechanism attracting their customers. Specifically, the price mechanism was heavily administered as a pull factor attracting passenger to purchase the ticket. Yet, at that time there were customers flying enhanced the airlines' s financial stability. Logically, customers are the major veins that survive the airline interim of funding and operations. Therefore, some airlines are still kept in track on its businesses and services in such fluctuated crises (Kim & Sohn, 2022). Howbeit, with the penetration of the COVID 19, its dreadful impact has ruined the tourism and transport industry because of the stringent restriction and prohibition travelling across frontier in myriad countries (Scarlett, 2021) caused people ceasing to fly. As in Thailand, the Ministry of Interior (2021) has set up Coronavirus Disease 2019 Epidemic Administrative Center issuing various measurements to control the virus dissemination, such as set up curfew, quarantine travelers flying from both domestic and international, fine people who ignore of wearing mask, and distribute vaccinations throughout the country. Over and above, the COVID-19 delta variant exposed in the country the government has implemented stricter regulations. This was resulted to seize the air operation in certain cities, decrease its flights and change the type of aircraft for save the operation fortune. The Civil Aviation Authority of Thailand-CAAT (2021), then, has announced the notification forbidden to fly from July 21st 2021 onward until further notice when the pandemic was relieved. The CAAT, at the same time, regulated the airlines to maintain social distancing onboard the aircraft. This regulation means that airlines could carry passengers only 50 percent of its aircraft capacity. Hence, it is barely for air carriers to run the business especially low-cost airlines that the ticket prices are rather cheap.

The pandemic costs a lot of fortune to various businesses around the world (Amankwak-Amoah et al., 2021). Particularly, airline companies where the cost of operation as well as non-operation cost are high. And, to the airline business when the aircrafts are on ground during the pandemic, it costs a lot of administrative expenses to the organization to maintain the aircrafts in good condition, parking fee, and employees' salary. Therefore, most of the airlines and related business have offered early retire packages for their employees, reduce salary, and layoff workforces (Seanyen et al., 2020). Moreover, the infectious disease is not only affected to the airlines, likewise it derailed the tourism industry and the country's economic as a whole. Chiefly, the tourism industry is one of primary keys industry which generate revenue to Thailand. According to the Ministry of Tourism and Sport (2021) disclosed that the income from visitors in 2020 minus 81.16 percent from 2019. As evidence, it can be assumed that the disease impacting on social appalled, economic growth rate, and health care awareness has led the airlines in Thailand turmoil and freeze. This situation has resulted the airlines lacking fund flow causing the airlines' operational stagnation. The airlines, although, implemented various measurements such as leave without pay package offering, take turns on flying schedule, reduce salary up to 75 percent. Peculiarly, low-cost carriers that are notably competing purely on price which the profit margin is less and depending on the numbers of passenger. Addedly, as prior mentioned the strong competition in air transport industry has gradually periled the air carriers. Notably, the low-cost airlines were highly affected. Therefore, 7 Thai CEOs from airlines' companies including low-cost plea for soft loan from the government to cope with unpleasant situation (Thairath, 2021). For these reasons, this study focused on how low-cost carriers practice to survive amid the pandemic. What are required for well-organized processes in handling the tough situations. Thence, this study aims to assess the managerial factors that low-cost carriers highlight during COVID-19 to carry on the business.

BACKGROUND

Today, it sounds like the pandemic was relieved even though the new COVID-19 variant BA.4 and BA.5 were discovered. Anyhow, the symptom of the disease seems to be not severe due to the herd community and vaccination protection plan have been done. Therefore, to set back the economic activities were put in place especially the tourism and travel industry. For instance, the abolishment of travelers' health quarantine procedures to Thailand was implemented (Royal Thai Government, 2021). Nevertheless, the Amazing Thailand Safety and Health Administration or SHA has also been enforced to assure travelling health safety in Thailand. Addedly, the government has promoted the tourism industry. As refers to The Bureau of Information Office of The Permanent Secretary (2022) reported the Deputy Prime Minister and Minister of Public Health held the event called "Thailand Moving Together: hug, eat, fly, travel and get closer again". This is to arouse the travelling to Thailand and cherishing tourism industry. As a consequence, the government expected the tourism and air transport industry would generate a huge revenue to the country. The airlines industry, certainly, is benefited from this policy. It is expected that there will be a grand amount of folks of tourists visiting the country after two years suffering from the pandemic. Attributable to this, it revitalizes the airline industry not only in Thailand but in many countries where the restriction was relieved. As appears in the news and social media, there were many airlines lack of manpower handling flight operation and passengers. That, unfortunately, resulting to the flight cancellation and undelivered bags left over in the airport (Bangkokbiznews, 2022). Somehow, another shockwave is attacking global economic worldwide due to the consequence of the pandemic and the geopolitical unrest war. These situations make economic even worst and fluctuated, where the inflation rises significantly around the world (Desilver, 2022). On top of that, the fuel prices roar up leading to the airfares soaring as the aviation experts predicted that the fuel price and ticket fee will be higher (Chokshi and Krauss, 2022). Alike the Department of Transportation's Bureau of Transport Statistics (BTS) (2022) reported the fuel price in air transport expenditure rose up 57.8 percent in April 2022 comparing with April 2019. It results to the unit operating cost of the air carrier getting higher. Literally, scholars (Santos-Reyes and Badillo-Portillo, 2022) studied the choice of transportation mode in Mexico, they found people prefer using bus rather than air transport even though the air transport is faster and more comfortable. In contrast, the air transport is considered too expensive and was perceived unreliable to the force majeures that may occur. On the other hand, Onianwa (2022) suggested the global collaboration and idea-sharing for reshaping the air transport in new era by using the lesson-learned from current pandemic situation should be considered, since the air transport relates with various stakeholders in the operations. Principally, as Lakshmanan (2022) cited the pandemic has stagnated the air transport, but in re-operating has put many challenges to the aviation for example; human resources and assets, safety and security issues. Hence, this study we emphasize the organizational issues. And, we have identified four managerial issues affecting the airlines' organizational structure resulting to high-performance organization. The four managerial factors including risk management (RM), organizational culture (OC), strategic human resource management (SHRM), and high competence workforce (HCW).

Risk Management (RM)

The corona virus is a disaster that collapse airline financially and organizationally. The disease is reckoned as a crisis to organization. Where, the crisis was defined as an abrupted incident that wrecking the daily operation system. Organization needs quick and clear instructions to mitigate its effects (Coombs, 2015). Organization responses to crises differently depending on the problems they confronted. The risk management model namely COSO enterprise risk management (ERM) is widely applied as a guideline (Park et al., 2021) for investigating, analyzing and prognosticating potential risks. The model composes of five internal control issues including environment, risk assessment, activities, information and communication, and monitoring (Ma & Ma, 2021). Whereas, Hayne & Free (2014) stated the COSO model is representing an entire comprehension of organization in managing risk. The COSO enterprise risk management was combined in 3D box align with internal environment precisely, objective setting, event identification, risk assessment, risk response, control activities, information and communication, and monitoring. According to the study of Park et al. (2021) found organization ignored to practice COSO framework has confronted with various problem issues such as resources, financial, and internal controlling. On the other hand, Demek et al., (2018) revealed the insufficient implementation of formal ERM on social media use in organization. For this reason, it can be assumed that risk management is an institutionalized process creating to control, response the occurrences of risk in timely manners. However, the effectiveness of risk management implementation requires forceful power (Tagod et al., 2021). In contrast, Durst, Hinteregger, & Zieba (2019) proposed the knowledge of risk management for the organization performance is as a soft measurement. They identified that knowledge of risk management as an application of techniques and tools to specify, analyze, respond to the risk by formulizing, applying and retention organization's knowledge. Since, the knowledge is critical in today's knowledge-based economy era. They, also, stated the soft measurement of organizational performance including: innovative, responsive, sustainable, and agile. In sum, it can be seen that risk management was embedded and practiced in organization's structure, processes, and employees' knowledge. It was brought into action by leadership and knowledge management. In concordance to the study of Crovini et al. (2021) found the absence of processes and strategies were related to lack of the leaders' knowledge, and attitude toward risk. Whilst Adeleke et al. (2018) found rules and regulations were the factors influencing the risk management in construction business. Also, there were risk management studies in the airline industry that excel on safety risk, flight operation (Li et al., 2020; Elkhweldi & Elmabrouk, 2015; Qiu et al., 2020) employees' fatigue (Bourgeois-Bougrine, 2020) and air traffic control management risk (Dedek et al., 2020). It is noticeably that in airline business the scholars have shed the light mainly on risk management in safety and operations. Due to the primary concern of airlines' service is safety. Yet, there is lack of the study on organizational structure risk management in airline companies. Thence in this study, we highlight risk management practice in airline's organization rather than the safety during the pandemic affecting organizational structure.

Organizational Structure (OC)

The organizational culture is the shared values, beliefs, and expectation that captivate employees and system within organization. The organizational culture enhances the identity, sense of oneness, commitment, and communication. It motivates organization's social energy to inspire employees achieving the firm's goals and objectives (Kankaew & Treruttanaset, 2020). Thus, the appropriate culture has to be cultivated and socialized throughout organization for the belief and behavior change of people. The organizational culture was counted as a core competency and capital of organization encouraging organization's capabilities and performance through employees' values (Azeem et al., 2021). The organizational culture is described into four categories including; adhocracy, clan, hierarchy, and market culture. It can be explained that the dynamic and innovative environment resulted from adaptability, creativity and flexibility was named adhocracy. While, a stress-free, supportive, and warm environment enhances teamwork, collaboration, and unity called clan culture. On the other hand, the organizational culture stresses on working standard, formalities, controlling, and regulations named hierarchy. The hierarchy culture is the efficiency seeking, stability and long-term goals achievement. The last type is market culture that emphasis on external environment intact with suppliers and customers to competitiveness (Ližbetinová et al., 2016). Further stated by Farine et al. (2021) the organizational culture is a primary factor of organization identity and performance. It is the human resource professional responsibility to incorporate employees' interests and organizations' interests for high performance. Found from their study, the organizational culture has had a positive toward employees' performance. Another study from Summerill et al (2010) discovered a positive organizational culture in water safety plans as such togetherness, proactive, service-minded, accountability, continuous improvement. In Thailand, the organizational culture system is hierarchical, high-power distance, patronage, face-saving, personal connection, and bureaucratic in public organization (Kankaew & Treruttanaset, 2020). However, LCCs in Thailand are private organization that is rather differed from public organization. Still, some characteristics of the Thais' culture exist in people belief and thoughts. Therefore, in this research we are investigating whether LCCs put the effort on organizational culture amid the pandemic from employees' perspectives. It is based on the concept that organizational culture is one of the key factors for competitiveness in service sector (Oriade et al., 2021). Resulting from Spicer (2020) suggested organization should reform its culture parallelly to the environmental changes during the outbreak. Organization, by doing so, should examine its existing cultural values for instance team engagement and employee's psychology state toward the environmental changes. Henceforth, the organizational culture of LCCs toward organizational structure is focused in this study.

Strategic Human Resource Management (SHRM) and High Competence Workforce (HCW)

The concept of SHRM is organization adopt its practice to environmental fit both internally and externally. That is the human resource activities must align with organization strategies as well as environmental changes. On the other hand, SHRM is enhancing the role of line manager as human resource professional. As a consequence, the nature of daily operations in human resource duties will rely online manager. Since, the line manager works closely and intimately with the employees than human resource professional. Hence, line managers could understand, motivate, control and resolve the problem issues immediately (Nisada, 2011; Way & Johnson, 2005). Positively, Adiguzel et al. (2020) cited that the strategic human resource management is salient to encourage employees toward job satisfaction and performance. However, the SHRM in fluctuated environment has to be vigorous and flexible to amend. The quicker change of SHRM would flourish organization opportunity to refurnish accommodating the threats from uncertainty and risk. Therefore, it is important for human resource (HR) professional to understand and swiftly change policies and practices by integrating to organization strategies. In addition, HR professional should be developed and train skills set to concrete organizational capabilities (Ahammad et al., 2020). Confirming from, Zehir et al. (2016) stated SHRM is the integration of organization's strategies both vertically and horizontally. Where, the vertical integration refers to HR implement firm's strategies into practice throughout. While, the horizontal integration is the HR incorporate firm's strategies into practice in its department (Werbel & DeMarie, 2005). These integrations encompassed strategy to recruit, select, manpower planning, and training and development etc., The type of fit has been proposed by Werbel & DeMarie (2005) separated into three types namely; person-job fit, person-group fit, and person-organization fit. The person-job fit defines as placing propre employees' knowledge, skills, and abilities with the task. Secondly, a person-group fit is to match employees' knowledge, skills, and abilities complementing and supporting the working group. And lasty, a person-organization fit refers to harmonize employees' values, interests and needs with organizational culture. These three fits would create organizational competencies, such as task-based competencies, team-based performance leading to innovation, and shared-values that lead to engagement and firm's identity. The person-organization fit, actually, is creating specific organizational culture in multi dimension like customer service.

As a consequence, SHRM is the holistic management of organization's human resource in aligning to strategies both internal and external environment. HR professional should develop human resources' skills, knowledge, and abilities in line with strategies that led by the market environment. It is to assist organization creating high-competence workforce (HCW). At the same time, the cultivating organizational culture fit with the market demand is essential (Harris & Ogbonna, 2001). Since the intellectual of HCW that embedded in a person known as know-how allow people manifest on the job performance. So that, organization could reap its investment on employees striving to its objectives. Alike other businesses, human resources in airline companies are recognized as a primary key competence resource (Appelbaum & Fewster (2003). Logically, the employees are driving organization's strategies and operation activities. There were studies on HR practices collation between foreign airlines and inland airlines (Ogochukwu et al., 2008), reengineering of HRM system (Roy, 2013), HRM in airlines as global business (Bertillo & Salando, n.d.), turn over intention of airline's workforces (Kampitak & Yoopetch, 2018). Yet, there is found no study SHRM practices amid the COVID-19 crisis. Thereby, this study aims to measure the degree of LCCs toward SHRM practices during the infectious disease crisis.

Organizational Structure (OS)

The COVID-19, catastrophically, augmented the complication of both micro and micro environment to management team and human resource professional (Hamouche, 2021). Certainly, it affects to the whole organizational structure in term of daily operations and its architectural system. This is resulting to firm's require re-structuring, reshaping working procedures, chain of command, and controlling. Bai et al. (2017) explained the organizational structure is a formal organizing of working system, and a holistic aligning of work and activities within organization. On the account of incongruous of organizational structure could lead to poor firm's performance. Aside from Schulman (2020) the formal organizational structure relates with roles, rules, communication channel, supervision and distribution of assignments. While, the informal structure is hidden in each activity. Thusly, Will et al. (2019) separated the organizational structure into three types wit; polyarchic, hierarchical, and hybrid. The polyarchic is the organizational structure where employees are communicating with each other, the decision-making on the innovation executed autonomously. The hierarchical structure refers to the organizational structure giving authority to manager in each department with specific hierarchical level, and allow specific decision-making power. Thirdly, hybrid structure is the combination of hierarchical and polyarchic. The combination structure allows employees to connect with similar team from different departments, empowering communication and decision-making. Conversely to (Ahmady et al., 2016; Ekutu et al., 2020) the organizational structure has two types including; mechanic and organic structure. The mechanic structure is defined clearly the division, roles, controlling, and its system. On the other hand, the organic structure is resilient, enriching employees' rotation, making decision based on their capabilities. Apart from today, the working environment is changed and the organizational structure is not flat like in the past. The main cause is from the pandemic that has pushed new normal life span, as such social distancing and work from home policy. In fact, this impact of life span change could affect mentally to the employees (Kropp, 2021). Though, the contingent approach should be applied in organizational structure. Besides, the other managerial factors should be considered for organization performing for more effective in the rough environments too. Especially, in the vulnerably industry like airlines business, so in this study we explore the organizational structure from the airlines' employee's aspect.

In sum, the managerial factors in this research expressly; the high competence workforce refers to eager employees who willing to learn and continuously developing oneself. Additionally, these workforces are congregated on teamwork, recognizing well of their airlines' products and process, and services. They are able to perform the task without close supervision, and committed to the result. Awhile, organizational culture is the airlines where people are engaged with teams, departments, and organizations. Subordinates are motivated and supported by leaders. There are clear rules and regulations in the innovative atmosphere. After all, risk management refers to the organization has measurement procedures and follow up the risk that is possibly occur in organization. Airlines, also, should have contingency plan to cope with internal and external crisis. In addition, the risk management plan must keep all workforces informed with lean communication. And the organization should be able to control the risk occurred to admitted level. Last managerial factor, the SHRM refers to the airlines that excel on line manager's role, pay per performance system, and 360-degree employees' assessment system are implemented. The workforces' development plan is in line with airlines' strategies.

METHODOLOGY

The quantitative approach was employed in this study. The data was collected by administering questionnaires along with the purposive sampling applied. There are four low-cost airlines registered under the state of Thailand. This research focuses on employees of the four low-cost airlines operating to and from Suvarnabhumi International Airport (BKK) which is primary airport of the capital city of the country, and Don Mueang International Airport (DMK), the secondary airport in the capital city of Thailand. Hence, researchers distributed the questionnaires to the employees of the four anonymous low-cost carriers. The questionnaires' reliability has been tested with the value of Cronbach's Alpha 0.94-0.95 of each item. The abbreviations were employed instead of using the name of low-cost carriers namely; VA, TA, AA and NA. There were 150 respondents equally from each airline, thence the total number of respondents were 600 persons. The multiple regression analysis was used to test the casual relationship of dependent and indigenous variables. The significance level based on 5 percent, which means the significance of statistic result has to be lower than 0.05. The anticipated results of this study would be useful for the airlines considering the findings to apply for managerial practices and administration for the effective and efficient operations. Ultimately, this lesson-learned amid the pandemic crisis would crystalize how should the organization adapt and strive to achieve sustainability in the swiftly change of external environment. Also, it is important to organization implementing the continuous learning and development of its workforces, reshaping the organizational culture, and risk management planning.

RESULTS

The result of respondent's number both male and female were 302 and 293 persons. It is accounted of 50 percent in average of male and female. The year of working of respondents in the airline company represented in ascending order are; 25 percent for 1-5 years, 26 percent for 6-10 years, 33 percent for 11-15 years, and 16 percent for over 15 years. The majority of respondents hold the position of senior officers' 41 percent, followed by 32 percent as operation officers, 15 percent work as airline's supervisors, and managers 12 percent. The data was concluded in figure 1.

Figure 1. Respondent Demographics

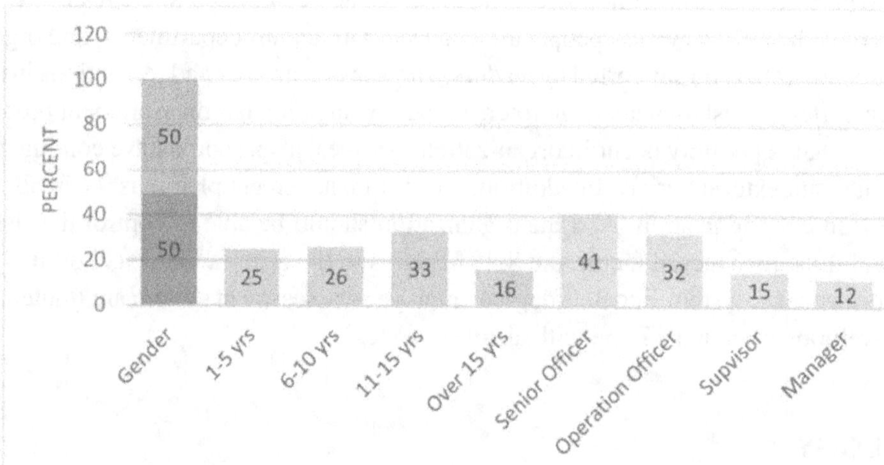

Simultaneously, the descriptive statistic disclosed that low-cost carriers (LCCs) in Thailand were focus attention on high competence workforce, abide by organizational culture, risk management, and strategic human resource management. The mean values of each factor are 3.70 for HCW, 3.66 for OC and RM, and 3.65 for SHRM. The standard deviation of all variables was not much different as shown in figure 2.

Figure 2. Mean Values and Standard Deviation of Factors Affecting LCCs Organizational Structure

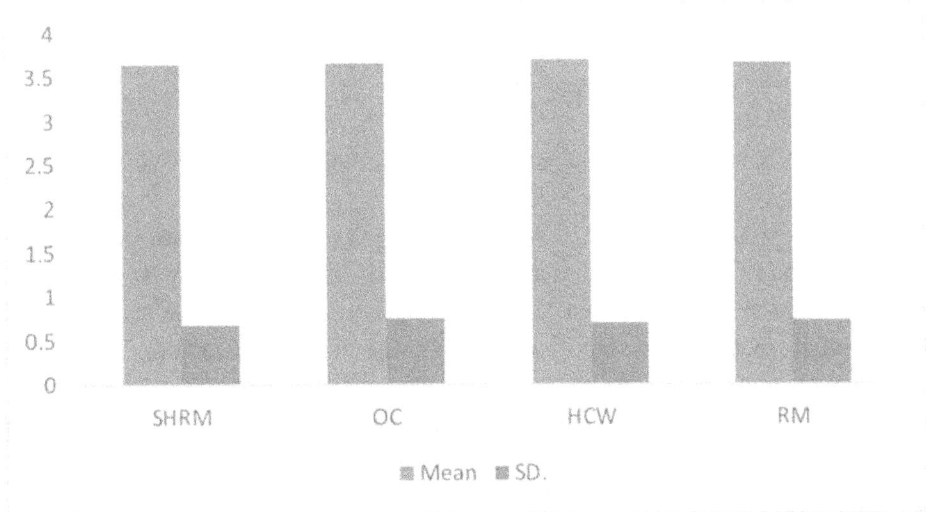

From the mean values result, it is noteworthy that LCCs in Thailand are weighted the importance to all factors. As a consequence of the mean values and standard deviation are not distinct. However, we can interpret that the high competence workforce has highest attention amid the outbreak. Yet, the airlines are not ignoring the importance of organizational culture, risk management, and strategic human resource management. For that reason, the multiple regression has been examined the correlation of four managerial factors affecting organizational structure. That, finally, would create high-performance organization. The result of multiple regression analysis of all variables is demonstrated in table 1.

Table 1. Multiple regression analysis of managerial factors affecting to organizational structure

Model	Unstandardized Coefficients		Standardized Coefficients	t	Sig.
	B	Std. Error	Beta		
(Constant)	0.546	0.118		4.613	0.000
SHRM	0.178	0.046	0.164	3.880	0.000
OC	0.085	0.049	0.087	1.745	0.081
HCW	0.283	0.060	0.268	4.703	0.000
RM	0.311	0.047	0.310	6.663	0.000

* Dependent Variable: Regression R = 0.75, R-Square = 0.56, Adjusted R-Square = 0.55, Std. Error of the Estimate = 0.49

According to table 1, it can be inferred on the basis of a foretasted analysis that: the significant linkage between SHRM and organizational structure is (0.000), the organizational culture has no relation to organizational structure because of its significance level is (0.081) that above 0.05, high competence workforce is linked to organizational structure at significant level (0.000), and risk management is related to organizational structure at significant level (0.000). From the regression model equation set:

$$Y = a + \beta 1 + \beta 2 + \beta 3$$

Organizational Structure = 0.55 + 0.164 Strategic Human Resource Management + 0.268 High Competence Workforce + 0.310 Risk Management

This can be explained that:

If every 1% increase in strategic human resource management affects to organizational structure by 16.4%,

If every 1% increase in high competence workforce affects to organizational structure by 27%, and

If every 1% increase in risk management affects to organizational structure 31%.

In consequence, amid the COVID-19 LCCs workforces' perspectives belief that the implication and practices of risk management is most important among other managerial factors. The high competence and strategic human resource management are in series afterwards. Oppositely, the employees view that an existing organizational culture of the airlines has no significance to the organizational structure during the pandemic. That is the leaders should excel on handling the risk that ruining organization, and taking care of high competence employees by not letting them demotivated. Surely in the rough times, airlines have to cut off their flights so that a well-planned of organizational in terms of manpower, strategies and risk communication throughout the company are crucial. Nevertheless, it is deniably that the strategic human resource management play a key role. The SHRM has to work closely with management team to flourish and enhance the organization performance. Whereas, the organizational culture has to be reframed in accordance with the external factors that rapidly and suddenly changed.

To be more specific, we have examined the managerial factors affecting organization for competitiveness during the COVID-19 from the workforces' viewpoints. The regression model was set as follow;

$$Y = a + \beta 1 + \beta 2 + \beta 3 + \beta 4$$

Organization's competitiveness = 0.60 + (-0.066) Strategic human resource management + 0.305 Organizational culture + 0.313 High competence workforce + 0.240 Risk management

The multiple regression result was summarized in Table 2 as follow…

Table 2. Multiple regression analysis of managerial factors affecting to organization's competitiveness during COVID-19

Model	Unstandardized Coefficients		Standardized Coefficients	t	Sig.
	B	Std. Error	Beta		
(Constant)	0.597	0.127		4.697	0.000
SHRM	-0.076	0.049	-0.066	-1.535	0.125
OC	0.315	0.052	0.305	6.038	0.000
HCW	0.350	0.065	0.313	5.412	0.000
RM	0.255	0.050	0.240	5.080	0.000

* Dependent Variable: Regression R = 0.74, R-Square = 0.54, Adjusted R-Square = 0.54, Std. Error of the Estimate = 0.52

In term of organization's competitiveness during the pandemic, the results shown that airlines' employees underline on high competence workforce, followed by organizational culture, and risk management. On the flip side, the airlines' employees are not prioritized on strategic human resource management. In harmony with table 2 revealed that if the organization increase 1% in SHRM practices, the competitiveness will minimize 6 percent. As opposed to high competence workforce, organizational culture, and risk management would upgrade firm's competitiveness to 31.3 percent, 30.5 percent, and 24 percent respectively in the condition of increasing 1 percent of management to each factor.

CONCLUSION

To this end, keeping business stable and progress in the intense shake of environment is not facile. Specifically, amid the immense impact from COVID-19 that shut down people to travel and cracked all airlines. To be survive, it is required a strong financial of the airlines themselves to brace its business entity. Besides financial issue, the good organizational structure is vital for organization to cope with the crisis. To highlight the result of descriptive statistic in this study, the managerial factors of LCCs in Thailand in terms of high competence workforce, organizational culture, risk management, strategic human resource management are not different significantly influencing the organizational structure. Then, we examined these issues from the employees' perspectives on current practices (during the pandemic) with inferential statistic. Thereafter, the multiple regression results showed that risk management has greater impact on organizational structure during the pandemic. It is followed by high competence workforce and strategic human resource management. While the organizational culture is not significant impact on organizational structure. Logically, the risk management is a primary issue for practicing during the crisis. That is all organizations should exist discrepancies plan and risk management plan to cope with uncertainty might occur. Furthermore, when the airline companies are forced to cease their flights, as a consequent it is inevitably to cut the cost incurred. In such way, air carriers are reducing salary and number of employees. As though Bangkokbiznews (2021) reported the International Air Transport Association-IATA survey 55 percent of airlines plan to lay off its employees, whereas 45 percent of airlines were already laid off the employees. Regarding the layoff issue, it is resulting to the airlines select and keep only high competence workforces. These employees are those who able to perform multi tasks. They are having both tacit and explicit know-how that are embedded in them as intellectual capital of airlines. The high competence employees could respond and deliver value-added service standard to customers, which would influence the customer choice (Worasuwannarak & Kankaew, 2022). Nevertheless, the continuous training and development of existing employees should be conducted during the non-flight operations. Lastly, the strategic human resource management is important part on both communication, practices of risk management, retain high competence workforces, and develop talents. Moreover, the HR professional has to work closely with leaders and department head as business partners, planning and integrating business objectives to fit with environment. Howsoever, it can be assumed that the airlines' employees view the existing organizational culture misfit with the external environment. As the result indicated, the organizational culture has no influence on organizational structure. So that, the resilient adapting to the changes of organizational culture is suggested. This is consistent to Imam (2020) cited the roles of human resource management in airlines in the course of crisis is to deal with organization changes through the leadership and organizational culture.

Finally, the managerial factors in term of competitiveness during the COVID-19 are including: high competence workforces, organizational culture, and risk management. While the strategic human resource management is not significant to the competition from employees' perspectives. However, the airlines should recognize that their products and services especially LCCs are rather the same, even the ticket price is slightly different. Despite that, the high competence workforces are those who could deliver quality services that distinct from its rivals. As well as the unique service culture has to be implemented in organizational culture. And the measurement of risk management should continuously be conducted. We recommended, from this study, the airlines should apply all managerial factors and adapt according to the contingent of current environment.

REFERENCES

Adeleke, A. Q., Bahaudin, A. Y., Kamaruddeen, A. M., Bamgbade, J. A., Salimon, M. G., Khan, M. W. A., & Sorooshian, S. (2018). The Influence of Organizational External Factors on Construction Risk Management among Nigerian Construction Companies. *Safety and Health at Work*, *9*(1), 115–124. doi:10.1016/j.shaw.2017.05.004 PMID:30363069

Adiguzel, Z., Ozcinar, F. M., & Karadal, H. (2020). Does Servant Leadership Moderate the Link Between Strategic Human Resource Management on Rule Breaking and Job Satisfaction? *European Research on Management and Business Economics*, *26*(2), 103–110. doi:10.1016/j.iedeen.2020.04.002

Ahammad, F. M., Glaister, W. K., & Gomes, E. (2020). Strategic Agility and Human Resource Management. *Human Resource Management Review*, *30*(100700), 1–3.

Ahmady, A. G., Mehrpour, M., & Nikooravesh, A. (2016). Organizational Structure. *Social And Behavioral Sciences.*, *230*, 455–462.

Amankwak-Amoah, J., Khan, Z., & Osabutey, E. L. C. (2021). COVID-19 and Business Renewal: Lessons and Insights from the Global Airline Industry. *International Business Review*, *30*, 01802. PMID:36568574

Appelbaum, H. S., & Fewster, M. B. (2003). Global Aviation Human Resource Management: Contemporary Compensation and Benefits Practices. *Management Research News*, *26*(7), 59–71. doi:10.1108/01409170310783592

Azeem, M., Ahmed, M., Haider, S., & Sajjad, M. (2021). Expanding Competitive Advantage Through Organizational Culture, Knowledge sharing and Organizational Innovation. *Technology in Society*, *66*, 101635. doi:10.1016/j.techsoc.2021.101635

Bai, W., Feng, Y., Yue, Y., & Feng, L. (2017). Organizational Structure. *Cross-Functional Integration and Performance of New Product Development Team.*, *174*, 621–629.

Bangkokbiznews. (2021). dœ̄nthāng pī mai thūa lōk pūan! yoklœ̄k kwā sī phan thīeo bin - khāt nakbin - lūkrūa [In Thai: A New Year Travelling Chaos: Over four Thousand Flights were cancelled due to the lack of pilots and crew]. *Bangkok Biz News.* https://www.bangkokbiznews.com/news/892592

Bangkokbiznews. (2021). IATA chī lūkčhāng ‘ sāi kān bin ‘ sūan yai sọ̄ tokngān nai sipsọ̄ng dūan [In Thai: IATA says most airline workers are likely to lose their jobs in 12 months]. *Bangkok Biz News.* https://www.bangkokbiznews.com/news/892592

BertilloJ. B.SalandoJ. (n.d.) Human Resource Management Practices in Airline Industry: The British Airways Global Perspective. Available at: SSRN: doi:10.2139/ssrn.2292797

Bourgeois-Bougrine, S. (2020). The illusion of Aircrews' Fatigue Risk Control. *Transportation Research Interdisciplinary Perspectives.*, *4*, 100104. doi:10.1016/j.trip.2020.100104

Bureau of Information Office of the Permanent Secretary. (2022, June 16). *Meet & Greet Thailand Moving Together*. Bureau of Information.www.pr.moph.go.th

Bureau of Transportation Statistics. (2022, June 06). *U.S. Airlines' April 2022 Fuel Consumption Down 9.5% from Pre-Pandemic 2019; Aviation Fuel Cost per Gallon Hits Record High.* BTS. www.bts.gov.us

Chokshi, N., & Krauss, C. (2022, April 15). Fuel Prices Send Airfares Higher, but Travelers Seem Ready to Pay. *The New York Times.* www.nytimes.com

Civil Aviation Authority of Thailand- CAAT. (2021). *CAAT NEWS 28/2564*. CAAT. www.caat.or.th

Coombs, T. W. (2015). *Ongoing Crisis Communication: Planning, Managing, and Responding* (4th ed.). SAGE.

Crovini, C., Ossola, G., & Britzelmaier, B. (2021). How to Reconsider Risk Management in SMEs? And Advanced, Reasoned, and Organized Literature Review. *European Management Journal*, *39*(1), 118–134. doi:10.1016/j.emj.2020.11.002

Demek, C. K., Raschke, I. R., Janvrin, J. D., & Dilla, N. W. (2018). Do Organizations Use a Formalized Risk Management Process to Address Social Media Risk? *International Journal of Accounting Information Systems*, *28*, 31–44. doi:10.1016/j.accinf.2017.12.004

Desilver, D. (2022, June 15). *In the U.S. and around the World, Inflation is High and getting Higher.* Pew Research Center. www.pweresearch.org

Dudek, E., Krzykowska-Piotrowska, K., & Siergiejczyk, M. (2020). Risk Management in (Air Transport with Exemplary Risk Analysis Based on the Tolerability Matrix. *Transport Problems*, *15*(2), 143–156. doi:10.21307/tp-2020-027

Durst, S., Hinteregger, C., & Zieba, M. (2019). The Linkage Between Knowledge Risk Management and Organizational Performance. *Journal of Business Research*, *105*, 1–10. doi:10.1016/j.jbusres.2019.08.002

Ekutu, A. C., Edeh, O. F., Alamina, P. U., Fern, S. Y., Kumari, P., & Eder, E. J. (2020). Effects of Organizational Structure on Employee Resilience. *Journal of Business School.*, *3*(2), 75–85.

Elkhweldi, A. M., & Elmabrouk, Kh. S. (2015). *Aviation Risk Management Strategies*. In the 2015 International Conference on Industrial Engineering and Operations Management, Dubai, United Arab Emirates.

Farine, F. K., Aditia, R., & Rahmadana, F. M. (2021). An Empirical Study of Leardership, Organizational Culture, Conflict, and Work Ethic in Determining Work Performance in Indonesia's Education Authority. *Heliyon, 7*(7), e07698. doi:10.1016/j.heliyon.2021.e07698 PMID:34386638

Hamouche, S. (2021). Human Resource Management and the COVID-19 Crisis: Implications, Challenges, Opportunities, and Future Organizational Directions. *Journal of Management & Organization*, 1–16. Advance online publication. doi:10.1017/jmo.2021.15

Harris, C. L., & Ogbonna, E. (2001). Strategic Human Resource Management, Market Orientation, and Organizational Performance. *Journal of Business Research, 51*(2), 157–166. doi:10.1016/S0148-2963(99)00057-0

Hayne, C., & Free, C. (2014). Hybridized Professional Groups and Institutional Work: COSO and the Rise of Enterprise Risk Management. *Accounting, Organizations and Society, 39*(5), 309–330. doi:10.1016/j.aos.2014.05.002

Imam, H. (2020). New Normal: An HRM Perspective of Post COVID-19 in the Airline Industry. Preprints. 2020120723.

Kampitak, T., & Yoopetch, C. (2018). The Study of Employee Retention in Thailand's Airlines Industry. *Dusit Thani College Journal., 12*(2), 193–208.

Kankaew, K. (2016). Thai Airways International: The economy Crisis Resolutions. *Actual Problems of Economic., 2*(176), 261–265.

Kankaew, K. (Ed.). (2022). *Global Air Transport Management and Reshaping Business Models for the New Era*. IGI Global Publishing. doi:10.4018/978-1-6684-4615-7

Kankaew, K., Ekachat, T., Rojanard, W., Nisara, P., Kumgwol, K., Sitikarn, B., & Charernit, K. (2021). Coping with the Changes that Challenge Business and Education Sectors in Thailand 4.0. *International Journal of Asian Business and Information Management, 13*(2), 1–11. doi:10.4018/IJABIM.287588

Kankaew, K., & Treruttanaset, P. (2020). The Organizational Culture Affecting Job Performance of Newly Hired Employees: A Case Study of the Customs Bureau at Bangkok Suvarnabhumi International Airport, Thailand. In A. Bejaoui (Ed.), *Corporate Leadership and Its Role in Shaping Organizational Culture and Performance*. IGI Global Publishing.

Kim, M., & Sohn, J. (2022). Passenger, Airline, and Policy Responses to the COVID-19 Crisis: The Case of South Korea. *Journal of Air Transport Management, 98*, 102144. doi:10.1016/j.jairtraman.2021.102144 PMID:34539103

Kropp, B. (2021). Trends that will shape work in 2021 and beyond. *Harvard Business Review*. www.hbr.org

Lakshmanan, R. (2022). Air Transport Safety and Security. In K. Kankaew (Ed.), *Global Air Transport Management and Reshaping Business Models for the New Era*. IGI Global Publishing. doi:10.4018/978-1-6684-4615-7.ch004

Li, J., Wu, J., Wang, G., Wang, W., Wu, M., & Xu, J. (2020). Architecture and Key Technologies for Airline Operational Control System Based on Risk Preprocessing. *IEEE 4th Information Technology, Networking, Electronic and Automation Control Conference*. IEEE. 10.1109/ITNEC48623.2020.9085064

Ližbetinová, L., Lorincová, S., & Caha, Z. (2016). The Application of the Organizational Culture Assessment Instrument (OCAI) to Logistics Enterprises. *International Journal of Maritime Science & Technology, 63*(3), 170–176.

Ma, J., & Ma, C. (2011). Factors Analysis Based on the COSO Framework and the Government Audit Performance of Control Theory. *Procedia Engineering, 15*, 5584–5589. doi:10.1016/j.proeng.2011.08.1036

Ministry of Interior. (2021). *Coronavirus Disease 2019 Epidemic Administrative Center*. MOI. https://www.moicovid.com/

Ministry of Tourism and Sport. (2021). *Tourist Statistic 2020*. MOTS. https://www.mots.go.th/more_news_new.php?cid=592

Nisada, W. (2011). *Strategic Human Resource Management*. Golden Time Printing.

Ogochukwu, U., Chinonye, U., & Callistus, I. (2008). Benchmarking Human Capital Strategies of Domestic and Foreign Airlines: Evidence from Nigeria. *International Journal of Development and Management Review., 13*(1), 36–55.

Onianwa, I. O. (2022). COVID-19 Pandemic and Air Transport Business in Nigeria. In K. Kankaew (Ed.), *Global Air Transport Management and Reshaping Business Models for the New Era*. IGI Global Publishing. doi:10.4018/978-1-6684-4615-7.ch002

Oriade, A., Osinaike, A., Aduthene, K., & Wang, Y. (2021). Sustainability Awareness, Management Practices and Organizational Culture in Hotels: Evidence from Developing Countries. *International Journal of Hospitality Management, 92*, 102699. doi:10.1016/j.ijhm.2020.102699

Park, K., Qin, J., Seidel, T., & Zhou, J. (2021). Determinants and Consequences of noncompliance with the 2013 COSO Framework. *Journal of Accounting and Public Policy, 40*(6), 106899. doi:10.1016/j.jaccpubpol.2021.106899

Qiu, S., Yao, D., Wang, Z., & Zhao, G. (2020). Research on Airline Operational Risk Management and Control Method. *IOP Conf. Series. Materials Science and Engineering*. IOP. 10.1088/1757-899X/715/1/012053

Roy, S. (2013). Flying Low: HR Challenges in the Airline Industry. *IPBJ, 5*(1), 49–75.

Royal Thai Government. (2021, October 2021). Prime Minister's Statement "Open the Country for Tourists Entering without Detention". Royal Thai Government. www.thaigov.go.th

Santos-Reyes, J., & Badillo-Portillo, A. (2022). Planning for a Trip: Selecting the Mode of Air Transport. In K. Kankaew (Ed.), *Global Air Transport Management and Reshaping Business Models for the New Era*. IGI Global Publishing. doi:10.4018/978-1-6684-4615-7.ch001

Scarlett, G. H. (2021). Tourism recovery and the economic impact: A panel assessment. *Research in Globalization., 3*, 100044. doi:10.1016/j.resglo.2021.100044

Schulman, R. P. (2020). Organizational Structure and Safety Culture: Conceptual and Practical Challenges. *Safety Science, 126*, 104669. doi:10.1016/j.ssci.2020.104669

Seanyen, T., Wentbap, W., Chuayprasit, W., Srayudh, K., & Hasoh, A. (2020). Analysis of Dealing with COVID-19: The Impact on Aviation Industry. *Journal of Multidisciplinary in Humanities and Social Sciences*, *3*(2), 209–220.

Spicer, A. (2020). Organizational Culture and COVID-19. *Journal of Management Studies*, *57*(8), 1737–1740. doi:10.1111/joms.12625

Summerill, C., Pollard, J. T. S., & Smith, A. J. (2010). The Role of Organizational Culture and Leadership Implementation for Improved Risk Management. *The Science of the Total Environment*, *408*(20), 4319–4327. doi:10.1016/j.scitotenv.2010.06.043 PMID:20624635

Tagod, M., Adeleke, A. Q., & Moshood, D. T. (2021). Coercive Pressure as a Moderator of Organizational Structure and Risk Management: Empirical Evidence from Malaysian Construction Industry. *Journal of Safety Research*, *77*, 139–150. doi:10.1016/j.jsr.2021.02.011 PMID:34092304

Thairath. (2021). *Ang sīang sa'ư̄n čhāk phanakngān sāi kān bin wọ̄n rat hai ngœ̄nkū Soft Loan tọ̄ lomhāičhai* [In Thai: Listen to the sobbing from the airline staff. Ask the state to give soft loans per breath]. Thairath. www.thairath.co.th

Way, A. S., & Johnson, E. D. (2005). Theorizing about the impact of strategic human resource management. *Human Resource Management Review*, *15*(1), 1–19. doi:10.1016/j.hrmr.2005.01.004

Werbel, J. D., & DeMarie, S. M. (2005). Aligning Strategic Human Resource Management and Person-Environment Fit. *Human Resource Management Review*, *15*(4), 247–262. doi:10.1016/j.hrmr.2005.10.001

Will, G. M., Al-Kfairy, M., & Mellor, B. R. (2019). How Organizational Structure Transforms Risky Innovations into Performance- A Computer Simulation. *Simulation Modelling Practice and Theory*, *94*, 264–285. doi:10.1016/j.simpat.2019.03.007

Worasuwannarak, B., & Kankaew, K. (2022). The Image Value of Southeast Asia Airlines: A Study of Attribute that Led to Image Value of Choosing Southeast Asia Airlines by Mean-End Theory Approach. In K. Kankaew (Ed.), *Global Air Transport Management and Reshaping Business Models for the New Era*. IGI Global Publishing. doi:10.4018/978-1-6684-4615-7.ch009

Zehir, C., Gurol, Y., Karaboga, T., & Kole, M. (2016). Human Resource Management and Firm Performance: The Mediating Role of Entrepreneurial Orientation. *Procedia: Social and Behavioral Sciences*, *235*, 372–381. doi:10.1016/j.sbspro.2016.11.045

ADDITIONAL READING

Bejaoui, A. (Ed.). (2021). *Corporate Leadership and Its Role in Shaping Organizational Culture and Performance*. IGI Global. doi:10.4018/978-1-5225-8266-3

Davenport, H. T., & Harris, G. J. (2017). *Competing on Analytics: The New Science of Wining*. Harvard Business Review Press.

Frei, F., & Morriss, A. (2012). *Uncommon Service*. Harvard Business Review Press.

Hamel, G., & Zanini, M. (2020). *Humanocracy: Creating Organizations as Amazing as the People Inside Them*. Harvard Business Review Press.

Kankaew, K. (Ed.). 0 (2022). Global Air Transport Management and Reshaping Business Models for the New Era. Hershey: IGI Global.

Taneja, N. K. (2011). *The Passenger has gone Digital and Mobile*. ASHGATE.

Chapter 7
What Are Airlines' Possible Strategic Responses to Overcome Unexpected Catastrophic Events Post COVID-19?

Mehmet Şahin Durak
https://orcid.org/0000-0001-8761-7131
Kırklareli University, Turkey

Gökhan Tanrıverdi
https://orcid.org/0000-0001-9863-3620
Erzincan Binali Yıldırım University, Turkey

ABSTRACT

Events that deeply affect the world and the airline industry, such as terrorist attacks or disease outbreaks, are expressed as unexpected events since they are not predicted or desired events. The responses employed by airlines are of great importance for the future of airline companies for their survival. In this context, this study aims to reveal a strategic response repertoire that airlines can employ to sustain their existence in case of facing an unexpected events for the post-Covid-19 era. Studies on strategic responses during unexpected events were collected from WoS and Sciencedirect databases. These studies were analyzed and discussed under the themes of preparedness/resilience efforts, resume, exit, adaptation, and action, determined by the authors. The study is expected to guide airline companies on how to respond to various unexpected events post-Covid-19 period.

DOI: 10.4018/978-1-6684-6835-7.ch007

INTRODUCTION

Most crises experienced in the world negatively affect many things around them. However, it may also present new opportunities. Crises such as the September 11 terrorist attack and the 2008 financial crisis have severely reshaped consumer buying behavior (Mathurin et al., 2021) and the industry environment (Sheng et al., 2020). In this direction, as crises occur, questions arise about how businesses will act/what actions they will take in crisis environments, how they will respond to crises, and what strategic responses they will provide to survive (Amankwah-Amoah et al., 2021).

The Covid-19 outbreak, which emerged in Wuhan, China in December 2019, turned into a pandemic in a short time by affecting the whole world adversely (WHO, 2020). The air transportation industry has been recorded as one of the industries most affected by this catastrophic event, along with the tourism industry since the industry has been shown as the main factor behind the rapid spread of the pandemic. While almost every industry has felt some impact, few have been hit as hard as the airline and travel industry. The sudden and sharp decline in travel demand was recorded as much worse than seen after September 11, 2001, and the 2008 financial crisis (Curley et al., 2020). The report published by Eurocontrol states that 99% of the available capacity offered by airlines landed on the ground in 2020 compared to the data of March 2019 (Eurocontrol, 2020). The airline industry faced a cost of around 492 billion dollars by the end of 2020 since hygiene and social distance requirements in flights increased costs, and demand remained very low to fly owing to mysophobia in the summer times (Bouwer et al., 2021).

"Unexpected catastrophic events" are defined as disruptions that are difficult to predict and cause unexpected effects that seriously hinder the activities of companies (Meyer, 1982). When faced with unexpected events, businesses have to successfully manage these crisis periods to survive. Crisis periods are times when businesses feel responsible for the external and internal environment and are under pressure. Although the responses developed by businesses are shaped according to the severity of the crisis, the effectiveness of these responses can be questioned. Indeed, almost all developed strategic responses were ineffective during the pandemic period (Suk & Kim, 2021). As one of the most suffering stakeholders in the aviation industry, airlines have also responded to the danger of extinction in various ways in difficult times such as the 11/09 terrorist attacks, the 2008 economic crisis, and the current Covid-19 pandemic crisis. While most airlines stopped their operations partially by ceasing their high-cost operations and reducing personnel costs, some of them get converted their passenger planes to phreighters and attempted to resume their operations with cargo transportation. However, airlines can have limited maneuverability during crises due to their obligations arising from contracts before crises (Budd et al., 2020a). In this context, this study aims to reveal a strategic response repertoire that airlines can employ to sustain their existence in case of facing unexpected events in the post-Covid-19 era by drawing them from published studies on past and current crises and airlines' responses. In the extended literature, no study presents all strategic responses as a repertoire. Existing studies only reveal the strategic responses given by the airlines for the relevant crisis. Therefore, the study is original, and it is expected that the study will guide airline managers in case of possible crises.

The study continues as follows: In the second section, the relationship between air transportation and crises is discussed. In the third section, the methodology and data collection process are introduced, while in the fourth section, the strategic response repertoire is presented, and the studies published on the strategic responses of airline companies in the extant literature are examined. Last and the fifth section, the conclusions are drawn.

IMPACT OF COVID-19 PANDEMIC AND PREVIOUS CRISES ON AIR TRANSPORT

Air transport facilitates the economic growth and development of countries by accelerating the connections between continents and countries and making serious contributions to the global economy (The World Bank, 2021). However, the airline industry is characterized by high fixed costs, and it is immediately affected by crises. As the last and most important example of this, the new coronavirus, COVID-19, has hit most sectors in the world and has led to many industries coming to a standstill. Until September 2022, approximately 613 million cases of COVID-19 have been seen worldwide, with 6.5 of the cases. More than one million deaths resulted (Hopkins University, 2022). The pandemic has led to restrictions on movement and travel bans. As a result of these restrictions, the transport sector especially air transportation has been impacted severely (Serrano & Kazda, 2020). Moreover, air transport has never been affected by an outbreak of this size, and undoubtedly the transport mode most affected by the pandemic is air transport (European Commission, 2020).

The first effects of the COVID-19 pandemic on the airline industry were the closure of borders, travel restrictions imposed by governments, grounding of aircraft due to insufficient demand, and quarantine measures for international travelers (Amankwah-Amoah, 2020). After April 2020, when the effects of the pandemic became evident, 90% of airline passenger traffic decreased. From February to April, the impact of the pandemic on air transport was felt globally, as the COVID-19 pandemic spread, passenger traffic decreased by 94% year-on-year, and industry figures bottomed out (IATA, 2020; IATA, 2021). Looking at the picture only in the USA for that period, it was estimated that travel expenses will decrease by approximately 400 billion dollars for 2020 and the cumulatively increasing loss will reach 900 billion dollars. These figures mean seven times more than the damage experienced in the 9/11 crisis in terms of travel industry revenues in the US (Curley et al., 2020) and also a heavy picture far beyond the sum of the 11/09 and 2008 crises (Molenaar et al, 2020). ICAO, on the other hand, estimated that the global airline industry, which has faced an unprecedented decline in passenger traffic since the Second World War in 1945, would lose up to approximately $418 billion (Sun et al., 2020).

Considering the other negative effects of Covid-19, as the most devastating global crisis, it is seen that aviation has lost its ability to connect the world for social, cultural, and economic profit. Thus, this period has been the period when aviation operations that connect people have decreased considerably, and the travel connection has been cut the most since the Second World War. Airlines had more robust balance sheets compared to during previous crises than COVID-19. However, a slowdown of this magnitude has left even the strongest players vulnerable (Curley et al., 2020). In this sense, remarkable development has been experienced to eliminate or minimize the negative effects of the pandemic on air transport. The pandemic period has taken its place in history as a crisis period in which substantial state support. Accordingly, some states have financially supported airlines and other key aviation stakeholders in the aviation value chain. In particular, the 173 billion dollars support provided to the flag carriers, the locomotives of the industry, by the states in which they are registered, has been the lifeblood of the airline industry. The main purpose behind government subsidies is to eradicate disconnection during the pandemic and protect the millions of jobs and workers supported by the industry. Thus, mass bankruptcies and possible bigger problems in employment have been prevented in many sectors such as primarily aviation and those which feed on aviation (Abate et al., 2020; IATA, 2021).

The COVID-19 pandemic crisis has delivered the biggest shock to air transportation and the whole aviation industry since the Second World War. Previously, the negative effects of the 9-11 terrorist attacks and the 2008 global financial crisis on aviation have been thought to be dramatic. After the pandemic, many studies comparing the impact of COVID-19 on the airline with the 11/09 attacks and the 2008 global financial crisis have revealed that the COVID-19 pandemic has affected the industry much more severely (Tuchen et al., 2020; Smith, 2020; Taylor, 2021). Compared to the 66% decline in revenue passenger kilometers (RPK) in 2020 globally, neither the 09/11 nor the 2008 crisis had such a negative impact (Lamb et al., 2020; Smidt, 2020) Figure 1 compares the results of past outbreaks experienced by the air transportation industry. The adverse effects of the COVID-19 pandemic have far exceeded the extent of the 2003 SARS epidemic, the most severe of the past epidemics, which caused 8 percent and 6 billion in total loss in terms of annual RPKs of Asian airlines (ICAO, 2022).

Figure 1. Past pandemics/outbreaks and their impact on air transport (based on RPK)
Source: (ICAO, 2022)

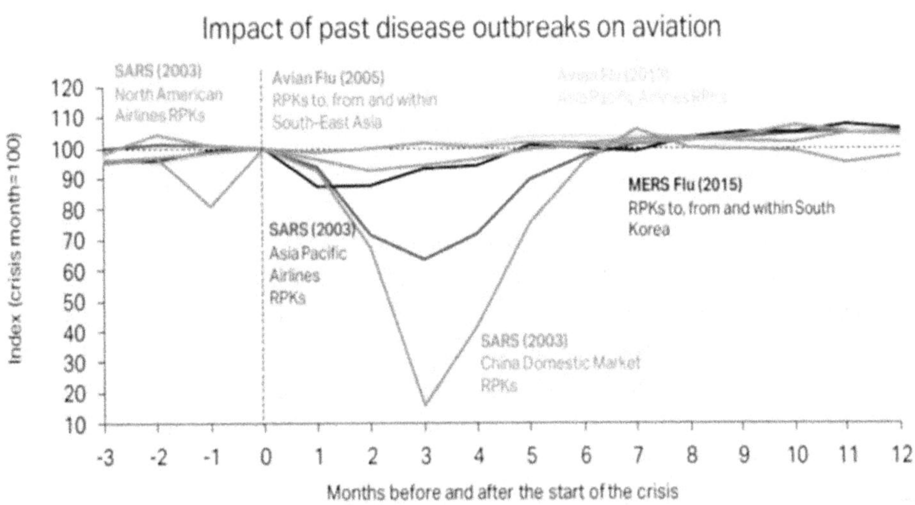

Figure 2, on the other hand, presents another perspective by comparing the COVID-19 pandemic crisis with other types of crises except for outbreaks experienced between 1945-2022. Accordingly, while other crises in the development process of air transportation only hindered the growth in the number of passengers traveling by air transport and caused it to remain at the same level, the Covid-19 pandemic crisis caused international airspace restrictions due to the increased contagion of air transportation. Thus, airlines grounded most of their aircraft. The result is extremely low passenger demand. In other words, although the global aviation industry has experienced steady growth even after previous global disasters such as the 11/09 terrorist attacks in 2001 and the 2008 global financial crisis (Lamb et al., 2020) the COVID-19 pandemic has brought air transportation down from the peak. In this respect, when expressing the situation of air transportation, it is compared with 2019, when it has reached the best figures. When the pandemic that occurred at the end of 2019 is compared with the following three years separately in terms of the number of passengers worldwide, we can express that there is a serious recovery in 2022 compared to 2019. In 2020, the effects of the pandemic were felt seriously and there was a 60% loss compared to 2019. Looking at 2021, the strategic responses and actions taken by the airlines ensured that the loss decreased to 49%, and by 2022, the 2019 figures were closed more and the 2019 figures were achieved by approximately 75%.

Figure 2. The crises between 1945-2022 and their effects on the number of airline passengers
Source: (ICAO, 2022)

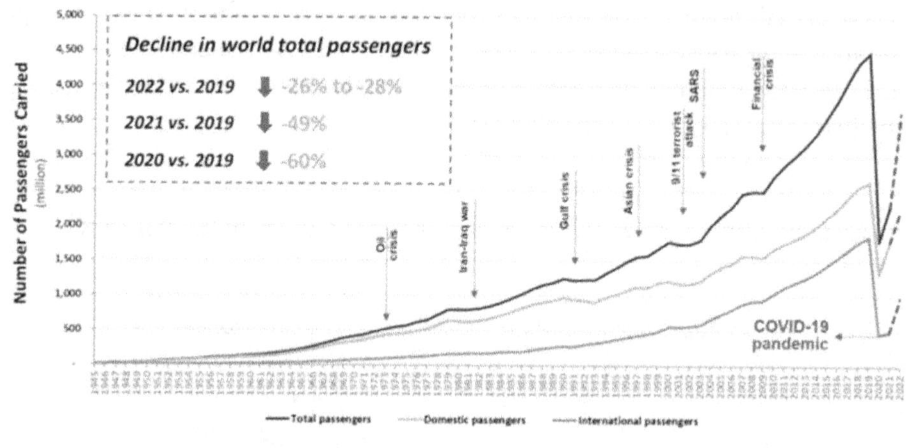

Historically, aviation has survived major shocks such as 9/11 or SARS within five to seven months. For the mid-long term impact of COVID-19 on aviation, certainly, the duration of the lockdowns and how quickly economic activity recovers in the post-pandemic period will be effective. Accordingly, industry experts have predicted a much slower recovery for air transportation post-COVID-19 than previous shocks such as 9/11 or SARS. This prediction has been made due to travel bans and closures (IFC, 2021). However, looking at Figure 3, which reveals the recovery in air transportation through the change in total seat capacity by continent post-pandemic, we can see that air transportation has almost recovered and caught the 2019 figures, especially in Latin America and North America. This shows that air transportation goes beyond forecasts and gives earlier turn signals.

METHODOLOGY

The main purpose of this study is to reveal the strategic response repertoire implemented by airlines against various crises. To do this, we reviewed articles on airlines' strategic responses in the literature. Accordingly, the relevant literature was limited in line with the following criteria before reviewing process, and then the studies that remained were examined: 1) Science Citation Index-Expended (SCI-E), Social Science Citation Index (SSCI), Emerging Sources Citation Index (ESCI), and Scopus indexes were scanned to reach related studies. 2) A broad search was conducted with the combination of keywords "strategic response" and "airline" by using sections of "topic" (For SCI-E, SSCI, and ESCI in the WoS database) and "titles, abstracts, keywords" (in Scopus) respectively. 3) The search was only performed on articles without using a time span.

Figure 3. Total seat capacity by regions (regional variations in resiliency and recovery time)
Source: (ICAO, 2022)

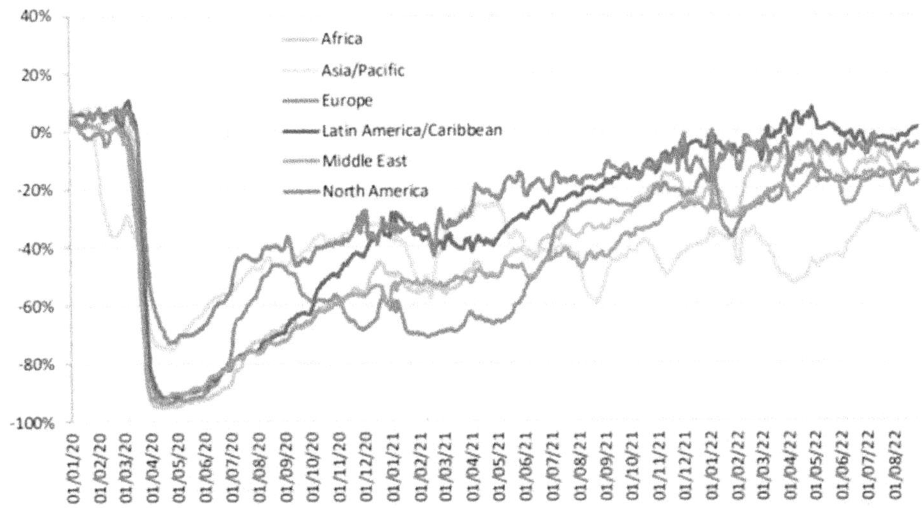

Titles and abstracts of 133 studies obtained in line with these criteria were independently examined by the authors in terms of suitability. Studies that were not conducted on responses to unexpected catastrophes, including strategic responses given by airlines in a competitive environment, were not included in this study. In addition, the article language was determined as English, and this criterion was also considered when excluding. As a result of the examination of all studies, a compromise was achieved on 17 studies by the authors, and these studies were included in the study and examined in detail.

RESPONSES EMPLOYED BY AIRLINE COMPANIES AGAINST UNEXPECTED CRISES

During periods of crisis or disaster, serious reductions in air traffic can be observed. Depending on the nature of the crisis experienced during these periods, some actions can be taken by carriers and countries. These actions, such as flight cancellations, travel bans, border closures, and aircraft groundings, taken to protect against greater negative effects, cause a decrease in flight traffic and occupancy rates. And thus, it causes a decrease in aviation and non-aviation unit and total revenues (ICAO, 2020). On the other hand, if we talk about all crises, it can be stated that the effects of these crises are not the same for every continent, country, or carrier, since the areas where the crises arise and the areas, they affect are different.

Despite the increasing number of strategic response studies against unexpected situations, how airline companies respond to these situations has been the subject of limited studies. Airline companies must develop a response to unexpected situations to ensure sustainability. In fact, it is a necessity to develop response strategies against unexpected situations to be sustainable (Volker et al., 2011), and thus airline companies need to know how to develop response strategies against unexpected situations. It is clearly seen that there is an increase in the number of response studies on the airline industry, especially for the term post-COVID-19. In this section, literature on strategic responses of airlines to unexpected events such as Covid-19, Ebola, the Economic Crisis, and September 11 have been examined. Findings obtained from these studies are evaluated under specific themes for possible crises in the future. These themes are preparedness/resilience efforts, resume, exit, adaptation, and action. Airlines' strategic responses to unexpected different events are addressed in Table 1.

Table 1. Airlines' strategic response repertoire

Title	Author	Unexpected Situation	Response strategies
Passenger airline and policy responses to the COVID 19 crisis: The case of South Korea (COVID-19)	Kim & Sohn (2022)	Covid – 19	• Retrenchment • Perseverance • Innovation • Exit
Strategic responses by European airlines to the Covid-19 pandemic: A soft landing or a turbulent ride?	Scheiwiller & Zizka (2021)	Covid-19	• Situational crisis communication
Strategic crisis response through changing message frames: a case of airline corporations	Ou & Wong (2021)	Covid-19	• Communication strategies • Verbal remedial strategies • Argumentation strategy to legitimize their positive image financial aid
European airlines' strategic responses to the COVID-19 pandemic (January-May, 2020)	Albers & Rundshagen (2020)	Covid - 19	• Retrenchment • Persevering • Innovating • Exit • Resume
European airline response to the COVID-19 pandemic – Contraction, consolidation, and future considerations for airline business and management	Budd et al. (2020)	Covid -19	• Changes to flight operations • Rationalizing the fleet • Reducing staff numbers and • Reconfiguring their networks and capacity
To bargain or not to bargain: Airlines, legitimacy, and nonmarket strategy in a COVID-19 world	Akbar & Kisilowski (2020)	Covid-19	• Strategic actions directed at influencing the governmental, legal regulatory, and societal environment of business
Note: Mayday, Mayday, Mayday! Responding to environmental shocks: Insights on global airlines ' responses to COVID-19	Amankwah-Amoah (2020)	Covid-19	• Internally generated short-term responses • Externally imposed short-term responses • Internally generated long-term responses • Externally imposed long-term responses
COVID-19 and the airline industry: crisis management and resilience	Suk & Kim (2021)	Covid-19	• Resilience • Cargo transportation • Request government relief
Moderating effects of orientation and innovation strategy on the effect of uncertainty on the performance of the business environment	Fernandes & Solimun (2017)	Turbulent environment	• Orientation • İnnovation Strategy
Ebola and Global Airline Business: An Integrated Framework of Companies' Responses to Adverse Environmental Shock	Amankwah-Amoah (2016)	Ebola	• Environmental Scanning—Monitoring and Assessment • Retrenchment (Downscaling) Strategy • Recovery Stage
A framework for effective crisis response	Pangarkar (2016)	Air crash	• Communication strategies • The high degree of readiness • Customer concern and responsiveness • Stakeholder orientation

continues on following page

Table 1. Continued

Title	Author	Unexpected Situation	Response strategies
Airlines' Flexibility in Facing Regulatory Uncertainty: To Anticipate or Adapt?	Engau et al. (2011)	Turbulent environment	• Anticipate • Adapt
Strategizing in an Unpredictable Climate: Exploring Corporate Strategies to Cope with Regulatory Uncertainty	Engau & Hoffmann (2011)	Turbulent environment	• Offensive • Defensive • Passive
The effects of 9/11/2001 on business strategy variability in the US air carrier industry	Goll & Rasheed (2011)	9/11 attacks	• Differentiation variability • Low-cost variability • Service scope variability
Adaptation of allocation of resources and attention in response to external shocks: The case of Southwest Airlines (11/09)	Blettner (2011)	9/11 attacks	• Allocation of resources
Relationships, layoffs, and organizational resilience airline industry responses to September 11	Gittell et al	9/11 attacks	• Relationships, • Layoffs, and • Organizational Resilience
Public relations in the airline industry: The crisis response to the September 11th attacks	Massey (2005)	9/11 attacks	• Strategic communication
The Asian economic crisis and the aviation industry: and impacts and response strategies	Henderson & Sadi (2000)	Economic Crisis	• Strategic alliances • Structural and operational reorganization • The application of new technologies

Scholars have conducted many studies on the strategic responses of airlines to unexpected events. When these studies are examined, it is possible to collect strategic responses under five headings. These titles are as follows:

- Preparedness/Resilience Efforts
- Resume
- Exit
- Adaption
- Action

When airlines encounter crises, they try to survive by resorting to these strategies. It is inevitable for airlines operating in a dynamic environment to encounter crises. Therefore, they are developing various strategies listed below.

Preparedness/ Resilience Efforts

It is a set of efforts to be prepared / to provide resilience before the crisis begins. Various studies show that the best response to unexpected events is to prepare in advance (Volker et al., 2011; Pangarkar, 2016). This is possible with organizational resilience. In the literature, organizational resilience is examined from two different perspectives. A group of researchers argues that organizational resilience is the ability to recover from unexpected, complicated, and negative situations and to continue their activities. In the second perspective, organizational resilience is used to express an ability to develop new skills, take advantage of new opportunities and create new opportunities. From this point of view, it is seen that the resources and capabilities of a business can be used not only to cope with emerging problems, but also to take advantage of opportunities (Lengnick-Hall et al., 2011).

Gittell et al. (2006) examined the responses of airlines after the terrorist attacks on September 11, 2001. According to their research, airlines with high resilience efforts have survived in the U.S. airline industry, while those with low resilience flagged bankruptcy. On the other hand, Suk and Kim (2021), argued that the Covid 19 pandemic put airlines' resilience efforts in serious trouble. In the study, it is stated that airline companies with high resilience efforts can continue their lives. Moreover, airline companies with low resilience efforts have faced serious problems and some have lost their sustainability. As a result, businesses with high resilience efforts can be successful despite experiencing unusual, uncertain, often negative, and often unstable conditions (Lengnick-Hall et al., 2011).

Resume

It is a fact that players who have made themselves resistant to crises before, get over the shock in a short time and continue their lives from where they left off. This strategy is a continuation of the previous section. Airlines with high resilience efforts can overcome crises in a short time and continue their lives from where they left off. In this context, Kim and Sohn (2022) made some suggestions in their study for airlines to continue their activities from where they left off. They investigated the responses of South Korean passengers, airlines, and governments to Covid 19. According to the research findings, they suggested policy directions for the pandemic and post-pandemic period. It is seen that passengers react by increasing their travel demands with the decrease in Covid 19 cases. Airlines used a variety of tools to cope with declining passenger numbers, such as domestic transits, expansion of freight transport, and mergers and acquisitions. Albers & Rundshagen (2020) states that the crises encountered will continue within a limited time frame. For this reason, they argued that the important thing is to get out of the crisis as soon as possible and to continue their activities from where they left off. In this context, they tried to categorize the responses of airline companies to the crisis during the Covid 19 period. They attributed the ability of airline operators to continue their activities from where they left off to four strategies. These are listed as typical crisis response strategies such as dismissal, perseverance, innovation, and exit strategies.

Exit

It is the complete withdrawal and disappearance of players with low resilience and endurance to act for the future, with or after the formation of the crisis. In short, it means that a business ceases its operations completely. Many studies show that airline companies withdraw from the sector on account of unexpected negative events (Kim & Sohn, 2022; Albers & Rundshagen, 2020). It emerges as a necessity or as a strategic response to crises in this crisis environment. So exit is not a failure in itself. It can also be expressed as the ability to use resources efficiently for future job opportunities (Wenzel et al., 2020). Today, the crisis periods of businesses operating in a dynamic environment take longer. It is not enough to simply ensure the survival of businesses in this environment. It is also necessary to consider the development and growth of the firm (Jin et al., 2022). In this context, exiting the sector can be considered an appropriate response if the development or growth of businesses does not seem possible after the crisis. While businesses can exit voluntarily due to their disappointing performance against crises (Wenzel et al, 2021), they can exit the industry to gain new opportunities and renew themselves (Dai et al., 2016). In this way, the scarce and valuable resources of the enterprises will be protected.

Adaption

Adaptation is one of the most frequently used strategic responses by businesses. It's an effort of businesses with even a little bit of resistance to get used to the unexpected. Adaptation is a continuous notion and exists at different levels. These are expressed as abilities that enable one to adapt to a certain market product area (Apaydın & Torlak, 2011). To sustain their activities and ensure their development, businesses have norms and values that will sense, understand, and activate the events around them. In this way, it allows businesses to adapt to unexpected events (Dension, 1990). Adaption has also been the subject of many studies in airlines (Budd et al., 2020; Massey, 2005; Volker et al., 2011; Kim & Sohn, 2022; Scheiweller & Zizka, 2021; Amankwah – Amoah, 2020; Fernandes & Solimun, 2017; Blettner, 2011).

Engau et al. (2011) emphasized that coping with crises is a fundamental problem for airline companies and the way to respond to crises is adaptation. They conducted research on airlines in crisis with the European Union emission plan. Two conclusions were reached for airline companies in their research. Either the airlines will anticipate and try to prepare for the potential consequences of the European Union Emissions Trading Plan conditions, or they will adapt to them quickly and effectively once the situation has become clear. In the study, which examined nine airline companies in the European airline sector, the organizational capabilities required to follow a predictive or adaptive strategy were defined and it was stated that each type of response required a certain combination of skills. Similarly, Fernandes and Solimun (2017) argued that the best strategic response for airlines to increase their performance in high uncertainty environments is "adaption".

Action

It is the efforts made to be in an important place in the new order as the environment begins to become clear and the strategies followed to plan the future. It expresses the need to take immediate action against unexpected events and the steps taken for the future. Many studies on airline companies have investigated how to take precautions against crises. Budd et al. suggested that airlines are driven by "changes to flight operations, rationalizing the fleet, reducing staff numbers and configuring their networks and capacity". In addition, Akbar & Kisilowski (2020) argued that it is necessary to take action as "strategic actions directed at influencing the governmental, legal regulatory, and social environment of business". Budd et al. have shown in their study how European airlines have taken action with the Covid 19 pandemic. In the study, airlines took action by trying to avoid direct operating costs and reduce indirect operating costs as much as possible to save cash against the sudden drop in demand and the resulting decrease in revenue. In addition to grounding planes as a first move, many airlines sought financial support from national governments, put their employees on unpaid leave or fired, and completely removed flights from some airports. As can be seen from the research, the way each airline takes action is different. It can be said that airlines take different actions according to their size, financial situation, and flight network.

CONCLUSION

In today's constantly changing and unexpected global business environment, businesses are faced with numerous unknowns and challenges that force them to respond for sustaining their operations (Amankwah-Amoah, 2020). In this respect, events that deeply affect the world and the airline industry, such as aircraft crashes, terrorist attacks, and outbreaks, are expressed as unexpected events since they are not predicted or desired events. The responses employed by the airlines during the face of these catastrophic crises are of great importance for the future of airline companies. In this context, this study aims to reveal strategic responses implemented by airline companies in case of facing unexpected events during the period post-Covid-19. Studies on the unexpected events were collected by scanning the Science Citation Index-Expended (SCI-E), Social Science Citation Index (SSCI), Emerging Sources Citation Index (ESCI), and Scopus index. The studies were comprehensively examined, and the strategic response repertoire employed by the airlines during past critical crises was determined.

As a result of the systematic review, the conceptual framework of the strategic responses developed by airline companies against unexpected events has been created and discussed under five themes preparedness/resilience efforts, resume, exit, adaption, and action. Strategic and tactical responses of airline companies are described in seven different conditions (COVID-19, Turbulent environment, Ebola, Air crash, 9/11 attacks, External shocks, Economic Crises) in the study. To summarize, business models, ownership structures, the financial strength of airlines, the severity of unexpected events, and where events occur have a decisive role in the responses implemented by airlines. According to another remarkable result, most of the studies reviewed have examined the Covid-19 pandemic crisis. This is a significant indicator clarifying the severity of the Covid-19 pandemic for the airline industry compared to previous crises. This study provides significant contributions to the airline management literature in terms of bringing all strategic responses employed by airlines by now together and forming a response repertoire for the unexpected crisis post-COVID-19 era. The study is expected to guide airline companies on how to respond to different unexpected events during the post-Covid-19 period. This will be a strategic tool that airline companies can use against unexpected situations that threaten their sustainability.

REFERENCES

Abate, M., Christidis, P., & Purwanto, A. J. (2020). Government support to airlines in the aftermath of the COVID-19 pandemic. *Journal of Air Transport Management, 89*, 1–15. doi:10.1016/j.jairtraman.2020.101931 PMID:32952317

Akbar, Y. H., & Kisilowski, M. (2020). To bargain or not to bargain: Airlines, legitimacy and nonmarket strategy in a COVID-19 world. *Journal of Air Transport Management, 88*, 101867. doi:10.1016/j.jairtraman.2020.101867 PMID:32834693

Albers, S., & Rundshagen, V. (2020). European airlines¢ strategic responses to the COVID-19 pandemic (January-May, 2020). *Journal of Air Transport Management, 87*, 101863. doi:10.1016/j.jairtraman.2020.101863 PMID:32834690

Amankwah-Amoah, J. (2016). Ebola and global airline business: An integrated framework of companies' responses to adverse environmental shock. *Thunderbird International Business Review, 58*(5), 385–397. doi:10.1002/tie.21789

Amankwah-Amoah, J. (2020). Note: Mayday, Mayday, Mayday! Responding to environmental shocks: Insights on global airlines' responses to COVID-19. *Transportation Research Part E: Logistics and Transportation Review, 143*. doi:10.1016/j.tre.2020.102098

Apaydin, F., & Torlak, N. G. (2011, January). Denison'un Önerdiği Uyum Sağlama Yetenekleriyle Miles Ve Snow'un Strateji Tipolojisinin Desteklenmesi Ve Açiklanmasi Ve Bunlarin Işletmelerin Performans Çiktilari Üzerine Etkilerinin Incelenmesi [Supporting and Explaining Miles and Snow's Strategy Typology with Denison's Suggested Adaptability Capabilities and Examining Their Effects on Business Performance Outputs]. In Journal of Social Policy Conferences. doi:10.1016/j.tre.2020.102098

Blettner, D. P. (2011). Adaptation of allocation of resources and attention in response to external shocks: The case of Southwest Airlines. *Management Research Review, 34*(3), 284–293. doi:10.1108/01409171111116303

Bouwer, J., Saxon, S., & Wittkamp, N. (2021). Back to the future? Airline sector poised for change post-COVID-19. In McKinsey & Company.

Budd, L., Ison, S., & Adrienne, N. (2020). European airline response to the COVID-19 pandemic – Contraction, consolidation and future considerations for airline business and management. *Research in Transportation Business & Management, 37*, 100578. Advance online publication. doi:10.1016/j.rtbm.2020.100578

Curley, A., Dichter, A., Krishnan, V., Riedel, R., & Saxon, S. (2020). *Coronavirus: Airlines brace for severe turbulence*. McKinsey & Company.

Dai, L., Eden, L., & Beamish, P. W. (2017). Caught in the crossfire: Dimensions of vulnerability and foreign multinationals' exit from war-afflicted countries. *Strategic Management Journal, 38*(7), 1478–1498. doi:10.1002mj.2599

Denison, D. R. (1990). *Corporate Culture and Organizational Effectiveness*. Wiley.

Engau, C., & Hoffmann, V. H. (2011). Strategizing in an unpredictable climate: Exploring corporate strategies to cope with regulatory uncertainty. *Long Range Planning, 44*(1), 42–63. doi:10.1016/j.lrp.2010.11.003

Engau, C., Hoffmann, V. H., & Busch, T. (2011). Airlines' flexibility in facing regulatory uncertainty: To anticipate or adapt? *California Management Review, 54*(1), 107–125. doi:10.1525/cmr.2011.54.1.107

Eurocontrol. (2020). *ANS Performance dashboard*. Eurocontrol. www.ansperformance.eu

European Commission, (2020). *Future of Transport: Update on the economic impacts of COVID-19. The* European Commission's science and knowledge service.

Fernandes, A. A. R., & Solimun. (2017). Moderating effects orientation and innovation strategy on the effect of uncertainty on the performance of business environment. *International Journal of Law and Management, 59*(6), 1211–1219. doi:10.1108/IJLMA-10-2016-0088

Gittell, J. H., Cameron, K., Lim, S., & Rivas, V. (2006). Relationships, layoffs, and organizational resilience: Airline industry responses to September 11. *The Journal of Applied Behavioral Science*, *42*(3), 300–329. doi:10.1177/0021886306286466

Goll, I., & Rasheed, A. A. (2011). The effects of 9/11/2001 on business strategy variability in the US air carrier industry. *Management Decision*, *49*(6), 948–961. doi:10.1108/00251741111143621

Henderson, J. C., & Sadi, M. A. (2000). The asian economic crisis and the aviation industry: Impacts and response strategies. *Transport Reviews*, *20*(3), 347–367. doi:10.1080/014416400412841

Johns Hopkins University. (2022). COVID-19 Dashboard by the Center for Systems Science and Engineering (CSSE). JHU. https://coronavirus.jhu. edu/map.html.

IATA. (2020). *Passenger air traffic dropped 90% at COVID-19 peak.* Airlines IATA. https://airlines.iata.org/analysis/passenger-air-traffic-dropped-90-at-covid-19-peak

IATA. (2021, 04 05). COVID-19 Travel Regulations Map (powered by Timatic). *Travel Center*. https://www.iatatravelcentre.com/world.php

ICAO. (2020). Effects of Novel Coronavirus (COVID-19) on Civil Aviation: Economic Impact Analysis. In International Civil Aviation Organization (ICAO), Montréal, Canada. IACO. https://www.icao.int/sustainability/Documents/COVID-19/ICAO Coronavirus 2020 04 02 Econ Impact.pdf

ICAO. (2022). Effects of Novel Coronavirus (COVID-19) on Civil Aviation: Economic Impact Analysis. IACO. https://www.icao.int/sustainability/Documents/COVID-19/ICAO_Coronavirus_Econ_Impact.pd

IFC. (2021). *The impact of COVID-19 on Airports: An Analysis*. IFC.

Jin, J., Li, S., Chen, Z., & Wang, L. (2022). Innovating, exit or both? Strategic responses to crisis revisited from resource redeployment perspective: evidence from China. *International Journal of Emerging Markets*.

Kim, M., & Sohn, J. (2022). Passenger, airline, and policy responses to the COVID-19 crisis: The case of South Korea. *Journal of Air Transport Management*, *98*, 102144. doi:10.1016/j.jairtraman.2021.102144 PMID:34539103

Lamb, T. L., Winter, S. R., Rice, S., Ruskin, K. J., & Vaughn, A. (2020). Factors that predict passengers willingness to fly during and after the COVID-19 pandemic. *Journal of Air Transport Management*, *89*, 101897. doi:10.1016/j.jairtraman.2020.101897 PMID:32837029

Lengnick-Hall, C., Beck, T., & Lengnick-Hall, M. (2011). Developing a capacity for organizational resilience through strategic human resource management. *Human Resource Management Review*, *21*(3), 243–255. doi:10.1016/j.hrmr.2010.07.001

Massey, J. E. (2005). Public relations in the airline industry: The crisis response to the september 11th attacks. *Journal of Hospitality & Leisure Marketing*, *12*(1–2), 97–114. doi:10.1300/J150v12n01_07

Mathurin, P., Aliaj, O., & Fontanella-Khan, J. (2020). Pandemic triggers wave of billion-dollar US bankruptcies. *Financial Times*.

Meyer, A. D. (1982). Adapting to environmental jolts. *Administrative Science Quarterly*, *27*(4), 515–537. doi:10.2307/2392528 PMID:10257768

Molenaar, D.-M., Bosch, F., Guggenheim, J., & Jhunjhunwala, P. Hean-Ho Loh, & Wade, B. (2021, March 29). *The Post-COVID-19 Flight Plan for Airlines.* Boston Consulting Group. https://www.bcg.com/publications/2020/post-covid-airline-industry-strategy

Ou, J., & Wong, I. K. A. (2021). Strategic crisis response through changing message frames: A case of airline corporations. *Current Issues in Tourism*, *24*(20), 2890–2904. doi:10.1080/13683500.2020.1849051

Pangarkar, N. (2016). A framework for effective crisis response. *Journal of Organizational Change Management*, *29*(4), 464–483. doi:10.1108/JOCM-04-2015-0063

Scheiwiller, S., & Zizka, L. (2021). Strategic responses by European airlines to the Covid-19 pandemic: A soft landing or a turbulent ride? *Journal of Air Transport Management*, *95*, 102103. doi:10.1016/j.jairtraman.2021.102103 PMID:36568570

Serrano, F., & Kazda, A. (2020). The future of airports post COVID-19. *Journal of Air Transport Management*, *89*, 101900. doi:10.1016/j.jairtraman.2020.101900 PMID:32834696

Sheng, J., Amankwah-Amoah, J., Khan, Z., & Wang, X. (2020). COVID-19 pandemic in the new era of big data analytics: Methodological innovations and future research directions. *British Journal of Management*, *32*(4), 1164–1183. doi:10.1111/1467-8551.12441

Suk, M., & Kim, W. (2021). COVID-19 and the airline industry: Crisis management and resilience. *Tourism Review*, *76*(4), 984–998. doi:10.1108/TR-07-2020-0348

Sun, X., Wandelt, S., & Zhang, A. (2020). How did COVID-19 impact air transportation? A first peek through the lens of complex networks. *Journal of Air Transport Management*, *89*, 101928. doi:10.1016/j.jairtraman.2020.101928 PMID:32952320

The World Bank. (2021). *Air Transport: Annual Report 2021.* World Bank. https://documents.worldbank.org/en/publication/documents-reports/documentdetail/099346308042217146/idu0a0f612fd04dd504b880866e0b851ba17ff30

Tuchen, S., Arora, M., & Blessing, L. (2020). Airport user experience unpacked: Conceptualizing its potential in the face of COVID-19. *Journal of Air Transport Management*, *89*, 101919. doi:10.1016/j.jairtraman.2020.101919 PMID:32874022

Volker, C. E., Hoffmann, H., & Busch, T. (n.d.). *Airlines' Flexibility in Facing Regulatory Uncertainty: To Anticipate or Adapt?* WHO. https://www.who.int/emergencies/diseases/novel-coronavirus-2

Wenzel, M., Stanske, S., & Lieberman, M. B. (2020). Strategic responses to crises. *Strategic Management Journal*, *41*, V7–V18.

Chapter 8
Portugália Airlines:
The Case of the Turin Route

Rui Quadros
https://orcid.org/0000-0003-0685-259X
Instituto Superior de Educação e Ciências, Portugal

Jorge Abrantes
https://orcid.org/0000-0003-4692-907X
Estoril Higher Institute for Tourism and Hotel, Portugal & Hotel Studies, Universidade Aberta, Portugal

Ana Barqueira
Instituto Superior de Educação e Ciências, Portugal

Antonio Rodrigues
https://orcid.org/0000-0001-5550-5581
Instituto Superior de Gestão, Portugal

ABSTRACT

Of all the Portuguese private airlines that for many years competed with TAP Air Portugal (the Portuguese owned state airline) only Portugália Airlines (PGA) resisted the hegemony of the Portuguese flag carrier until it was bought by TAP itself. Portugália Airlines, a private airline, managed to maintain its operation, even against the power of the state company. This case study shows how PGA operated certain routes (here the Turin example, in Italy) as a survival strategy. The approach is exploratory and based on an observational description of facts. In view of the recent events caused by COVID 19, the pandemic has brought about a change in business models, also providing the warning that anyone who does not have a clear strategy, dies at birth. The case study leaves some interesting clues to explore.

DOI: 10.4018/978-1-6684-6835-7.ch008

INTRODUCTION

The aviation industry community are aware the differences between companies that operate point-to-point (easyJet, for example) and those whose business model uses centrality (hub) to be able to distribute passengers to different routes (spokes).

We can, for example, reflect about Iberia Airlines of the IAG Group, as a Full-Service Carrier company (which converges its routes to the Madrid hub), Air Nostrum (an Iberia Airlines franchisee) as a regional company that supplies flights from Iberia and consider Iberia Express as a low-cost company (which operates point-to-point) that helps Iberia to grow.

It is clear that a point-to-point operation added to a network must be supported by a hub, that is, smaller-capacity planes converge to a hub to feed larger aircraft that can fly to more distant destinations and between continents.

If we mention the "point-to-point" business model and because there is no hub, but bases, the competitive advantages are always linked to cost leadership, where quality is not always valued.

Portugália (PGA), a regional airline company, since its first flight, operated with the characteristics of a full-service company, flying point to point, not having any network to distribute its passengers (a paradox). Clearly, the private Portuguese company operated with full-service characteristics but operated a point-to-point network.

After being sold to the Portuguese state company (TAP Portugal) in 2007, PGA finally began to fulfill its raison d'être. Until 2007, it competed with TAP and with the emergence of low-cost airlines.

Portugália, with its recognized quality attributes, became a money loser because it operated point-to-point and without any possibility of being a feeder of any other airline. Either become a Low Cost or join a hub company. It was purchased by TAP.

The general objective of this case study is to evaluate the catchment area of Piedmont (Turin) and not the regions of Lombardia (Milan) or Emilia Romagna (Bologna). The analysis focuses on the flight from Turin (which sells two destinations: Madrid and Lisbon) and its strategic options.

The specific objective of this case study is to demonstrate how a funnel flight type route managed to survive for many years, based on strategic customers and capacity management? It is intended to answer the following question, considering that the Turin route was one of the ones with the highest Yield. Was Portugalia as an airline company a project that was dead at birth?

The case study of a "funnel flight" route (Turin - Madrid – Lisbon), which will be discussed in this paper, was a strategy that the Portugália Airlines used in its network to survive, hoping that one day, it could entice a *shark* to its partner, and that could have been, for example, Air France.

We will explain, through the first person of one of the authors, how these strategies were that allowed a point-to-point company, with a premium service, to survive for so many years.

Today, and after we have overcome COVID 19, the strategy used by Portugália for years, would not have the slightest chance of working. While the so-called flag companies lose revenue on certain routes to gain on others, the strategic PGA model, would not work today.

It is also clear that it is not possible to use a point-to-point business model, and at the same time, practice premium fares in a high-cost structure, very similar to full-service airline business model.

Today, the industry is aware that TAP is still in great difficulties, aggravated by the pandemic, and that the help of the regional company (PGA) has been very appropriate in the strategy of TAP group, precisely the role of feeder.

Portugália Airlines (based in Lisbon), also known as Portugália or PGA, is a regional airline that belongs to the TAP Group since 2007. Founded in 1988 as a private airline, it operates domestic and international flights (short and medium haul) from Lisbon and Porto, under the TAP Express brand using Embraer 190/5 aircraft.

On January 14, 2016, TAP Express brand was announced, which thus replaced the PGA - Portugália Airlines brand, as well as the total renewal of the fleet (Lusa, 2016).

As a private company and a competitor for the same market as TAP, Portugália has never been a hub company, hence the impossibility of being a feeder for any company (Portugália Airlines, 2022).

The role of regional airlines as air service providers to small communities (smaller airports) dates to the regulated era (Forbes & Lederman, 2007). The role of feeder airlines is to transport passengers from destinations not served by large airlines to major hub airports.

The private Portuguese company of Grupo Espirito Santo never managed to be a feeder, worsening its financial situation whenever it increased its routes. Until 1999, and without competition, PGA already operated a daily flight between Turin and Madrid (Touring Club Italiano, 2007).

In the period between 1999 and 2003, Air Nostrum grew to operate three flights a day, helping to put PGA "out of the game", which ended up ceasing its operations in Turin (MD80.it, 2007).

The lack of a strategy, coupled with a "point-to-point" operation, operating only one daily flight, led the Portuguese regional airline to an announced death.

The business model of low-cost airlines is point-to-point flights, since connecting passengers involves many ramifications and affects the competitiveness of the business (Fageda et al., 2015).

It is intended to study the causes that led to the closure of the Italian route from Turin, specifically in the segments that connect TURIN (TRN) to MADRID (MAD), and TURIN to LISBON (LIS) in 2005, and under the acronym NI711 (*Nuovo volo Torino-Lisbona operato dalla TAP Portugal/New Turin-Lisbon Service Operated by TAP Portugal*, 2012).

The route had its origin in Lisbon, NI710 (NI – International Air Transport Association, IATA, two letter codes for Portugália Airlines), ending the flight at the destination of the capital of Piedmont (Turin) (*IATA - Codes - Airlines*, 2022).

Passengers, from Spain and Portugal, will not be considered in the objectives of this study.

As the *most important* players are in Italy, it is assumed that the largest share of revenue comes from passengers from Italy, whether to Madrid or Lisbon.

LITERATURE REVIEW

This brief literature review serves to support the case study in which the main players are Portugália Airlines and Air Nostrum.

The regional airlines of the Iberian Peninsula never received much attention from researchers, however, the liberalization of air transport in Europe contributed to these companies trying to take their place in the regional market around southern Europe.

Despite deregulation, competition in the Spanish domestic market between 1993 and 1997 remained limited because of three idiosyncratic characteristics: a) restrictions on the entry of foreign companies until 1997; b) Iberia's monopoly on access to slots; c) the management monopoly of the state-owned company AENA (responsible for all airports) until 1996 (Ruiz-Moreno et al., 2021).

From 1987 until liberalization in 1997, some newly arrived carriers (Air Europa and Spanair) competed on routes such as Madrid to Barcelona. Other Spanish carriers such as Air Nostrum, Panair and Air Truck also appeared, offering flights on regional routes.

In air transport and on business models, airlines can be divided into two main categories: Full-Service Carriers (FSC) and Low Cost Carriers (LCC) (*Impact of COVID-19 on different business models of European airlines*, 2021).

The deregulation policy has skyrocketed competition in the EU since 2000 (Gillen & Lall, 2004), with the consequence of lowering costs, fees and prices.

FSCs and LCCs differ markedly in terms of their business model: FSCs are sought after by travelers with organized trips who book through travel agencies, while LCCs are the preferred companies for those who are very price sensitive and therefore LCCs follow a very competitive pricing and cost minimization strategy with very simplified services (*Impact of COVID-19 on different business models of European airlines*, 2021).

The inability of airlines to sell their seats in an economical way has forced a very strict relationship with travel agents (Alamdari, 2002). In recent years, a lot has changed, first in the US and later in Europe (Alamdari, 2002). For what reason? The causes of such a change were as follows: a) Airlines with a tendency to reduce costs; b) the most direct contact with its customers; c) creation of direct channels with companies; d) the ability of large buyers to look for airlines.

Regional airlines are the major "allies" of major airlines to add capacity in regional markets, using regional carrier labor, but flying under the major carrier's code and livery (Reynolds-Feighan, 2018).

Major carriers use multiple regional carriers as regional feeders from the same airport, suggesting that major carriers "*do not choose a single regional carrier to serve all routes from a given airport*", and the analysis is by airport rather than by route (Reynolds-Feighan, 2018).

Regional carriers play an extremely important role in the air transport systems of the countries where they carry out short-haul and/or medium-haul operations, normally offering connections in regular traffic between smaller cities and larger cities (Tomová & Materna, 2018).

Regional airlines generally have load factors in the range of 75% to 79%, while FSCs are around 81% to 85% (Reynolds-Feighan, 2018).

Airlines may charge less economic fares each time they increase their flight frequency, which for a given demand, regional airlines may, as frequencies increase, also increase fares (*Air Services on Thin Routes*, 2012). On the other hand, low-cost airlines take advantage of their economies of scale to lower their prices by increasing their occupancy capacities. Low-cost airlines can try to take advantage of traffic density savings by using larger aircraft with higher load factors.

There are various types of aircraft operating in hub-and-spoke systems. Operationally, in a hub and spoke system, it is common for the aircraft type with higher seating capacity to operate routes with higher passenger flow and aircraft with fewer seats operate on routes with low passenger flow, to mitigate the unit costs of these routes (Integrated Aviation Model and Metaheuristic Algorithm for Hub-and-Spoke Network Design and Airline Fleet Planning | Elsevier Enhanced Reader, 2022).

The large-scale hub-and-spoke network can help increase airlines' market shares and an example of this are the regional feeder companies that operate in Europe and that serve to feed the main transport hubs.

In May 1997, Air Nostrum became a franchised member with Grupo Iberia for an initial period of 5 years, however it has been renewed since then. It became an active part of the Iberia Group (Iberia feeder), its flights being marketed by Iberia, Spain's national and international airline.

Regarding the way Lufthansa operates in its hubs, there is "Four bold waves hub pattern". The first wave at 7 am and lasting until 11 am, followed by smaller waves at midday and in the afternoon, and surrounded by high volume night waves lasting almost 4 hours (Števárová & Badánik, 2018). For Lufthansa, an FSC, focused on the morning and evening peaks of business travel, they are of great value (so these peaks are larger).

At night there is a large volume of aircraft arriving at remote stations during the night (night stop) to feed the hub with incoming traffic.

Figure 1. Banked structure of Lufthansa at Frankfurt [Compiled by authors, source of the data CAPA] Article: Performance of Hub and Spoke Networks of Selected Airlines

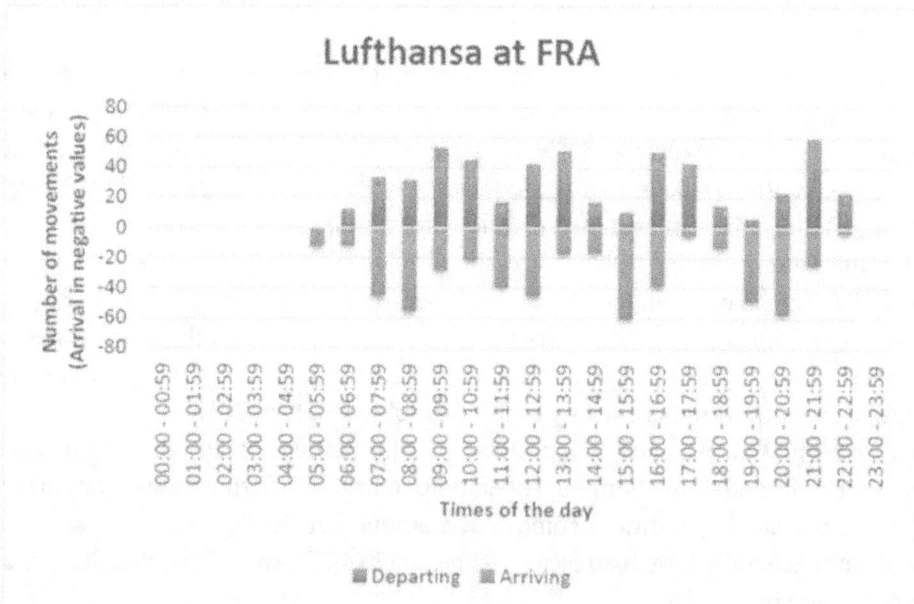

METHOD: CASE STUDY

Nothing relevant has been written about this airline. On the relationship between "funnel flights" and how they can generate revenue, studies are practically non-existent. For this reason, the exploratory approach of theoretical construction case study was adopted. The case study approach, which is based on a lived experience, can and should contribute to the development of theories around, for example, a case study of a route.

The Genesis of Portugália Airlines

The idea of creating a new company in the commercial aviation market was born in 1987, at the initiative of Coopav - Cooperativa de Pilotos, Transconsult - Office of Studies and Projects in the Transport sector and the Espírito Santo Group (*PGA,* 2022). The project came a year later with the creation, on July 25, 1988, of a limited liability company with exclusively Portuguese and private capital.

Although created in 1988, the Company only started operating in 1990. The city of Porto was the destination of the first official flight, operated by a Fokker 100, which departed Lisbon on the 7th of July of that year. That same day, the Lisbon/Faro route began.

Internationalization began in June 1992, with regular flights to Cologne, Turin, and Strasbourg. The subsequent acquisition of two new Fokker 100 allowed the conquest of niche markets in Europe. Flying to European destinations has allowed to conquer some interesting market niches in Europe, such as the city of FIAT (Turin).

With the full liberalization of European airspace in January 1993 (*EU Air Transport Liberalisation Process, Impacts and Future Considerations*, 2015), Portugália Airlines was able to continue its expansion projects, adding, in the years that followed, the cities of Madrid, Brussels, Mulhouse/Basel, Hannover and Stuttgart to its map of routes. To meet the new operational needs, the fleet was increased with two more Fokker 100 aircraft.

In April 1997, the Company opened eight new destinations in the Iberian Peninsula and the United Kingdom: Barcelona, Bilbao, Las Palmas, Gran Canaria, Tenerife, Valencia, Palma de Mallorca and Manchester (*PGA Sobre Nós*, 2022). To satisfy the growing operational need and make the fleet more flexible, six Embraer 145 aircraft (49 seats) are acquired.

The expansion in the French market took place between 1998 and 1999, with the opening of five important destinations: Nice, Lyon, Marseille, Bordeaux, and Toulouse. Also in 1999, a code-share agreement with the Moroccan company Regional Airlines added the route to Casablanca, the first non-European destination for Portugália Airlines (*PGA*, 2022). The Company's fleet stabilized in 2000, with 14 aircraft (6 Fokker 100 and 8 Embraer 145), a fleet that remained until 2016.

Portugália Airlines established itself in the European market as a private and innovative air carrier, guided by high standards of service. The efficient and rigorous management, the commitment of its team and the commitment to quality with its passengers have earned it international recognition, as evidenced by the awarding of the title Best Regional Airline in Europe in 2001, 2002, 2003, 2004, 2005 and 2006 and the Skytrax Best Cabin Staff Europe Award in 2005.

In 2004, Portugália Airlines made the Spin-off of its handling area (ground assistance), associating itself with the TAP Group. Thus was born SPdH – Serviços Portugueses de Handling, SA, in which the Company holds 6% of the share capital and which operates under the Groundforce brand (*PGA*, 2022).

In mid-2007, Portugália Airlines joined the TAP Group and assumed a new business model. The Company ceases to work autonomously in the regional market, starting to work within a Group logic (*TAP paga 140 milhões ao Grupo Espírito Santo pela Portugália*, 2006) integrates the TAP Group and starts to operate under a Wet Leasing (ACMI) regime for TAP Portugal, using its fleet, but without the NI code (PGA).

Portugália was acquired by TAP Portugal and started to include its operation, with some adjustments, in the national airline's network, using the TP (IATA two letter code for TAP Portugal) call sign on all flights operated by PGA.

Despite the existing contractual relationship, Portugália Airlines and TAP are independent companies that maintain their legal identity within the same Group.

In 2016, the Company renewed its fleet, starting to operate with a fleet consisting of 9 Embraer 190 aircraft and 4 Embraer 195 aircraft.

This renovation brought to Portugália Airlines a new image, an increase in capacity and greater flexibility while maintaining the same rigor and quality in the service provided to passengers.

The Company's aircraft currently fly under the TAP express brand, confirming the Group's commitment to the regional market (*PGA*, 2022).

Italy's Portugália Airlines Operations

In 1993, Portugália Airlines began operating between Lisbon and Turin (stopover in Madrid) using the Fokker 100 aircraft as its equipment. In seven days, it operates six flights a week offering 100 seats per flight (for the route = 200 seats). After Turin, the company also decided to operate Milan and Bologna departing from Lisbon.

Between 1993 and November 1999, the flight under study never had any type of competition.

The company was granted traffic rights and, therefore, the possibility to sell their seats from Turin to Madrid and from Madrid to Turin.

The first competition begins in the Winter IATA (Hayward, 2022), in November 1999, where Air Nostrum (subsidiary of the Spanish flag company (YW), Iberia) starts the operation with a daily flight from Turin to Madrid offering 100 seats daily. The aircraft chosen by YW was the CRJ Canadair with a capacity of 50 seats.

Portugália operated 6 flights per week on the Fokker 100 aircraft (and on Fridays the Embraer 145 operated with capacity to carry 49 passengers).

The Portuguese carrier offered 200 seats, roundtrip (daily) with the following schedule:

- Departure of PGA - NI711 - 7h30: Turin to Madrid and Madrid to Lisbon (same flight, same aircraft), offering 200 seats per spin.
- Departure from Air Nostrum / 7h15 (15 minutes before): Turin to Madrid, offering 100 seats per spin.

In June 2002, Air Nostrum placed a second daily flight using the same aircraft and capacity, offering more frequencies in a second schedule (*Air Nostrum_IBERIA Regional*, 2022):

- Departure, Air Nostrum from Turin/ 12h15

From 1999 to 2003, Air Nostrum tripled its capacity, increasing the number of seats daily and spread over three "slots":

1. A morning flight
2. A flight at noon
3. A late afternoon flight

With three daily flights, Iberia's subsidiary began to offer 300 seats daily, more 100 seats than PGA, and added two more frequencies than the Portuguese private company.

Between 1993 and 1999, PGA operated the route without any competition, offering an average of 200 seats per day, with only one frequency.

Table 1. Portugália Airlines annual seat capacity

PGA (1 Daily Flight)	Number of seats on Fokker 100 aircraft (100 seats)								
Lisbon-Madrid-Turin	Monday	Tuesday	Wednesday	Thursday	Friday	Saturday	Sunday	*Total Seat Capacity*	Annual 52 weeks
	100	100	100	100	50	Não opera	100	550	28 600
	Monday	Tuesday	Wednesday	Thursday	Friday	Saturday	Sunday		
Turin-Madrid-Lisbon	100	100	100	100	100	50	Não opera	550	28 600
								Annual Seat Capacity	57 200

Source: Portugal Airlines (2006)

PGA flight with the acronym NI and the number 711, offered two destinations (Madrid and Lisbon). In terms of revenue optimization and being only one flight (NI711), the revenue management analyst should allocate, in theory, and according to history, 50 seats for each route.

From Turin to Madrid assigns a quota of 50% (50 seats) and on the remaining route, the 50 seats (remaining 50%). In this way, flight NI711 was allocated 100 seats, corresponding to 50 seats per route.

Clearly, a sales inhibition strategy for any route, the only possible variable being the attempt to sell higher fares where there is "corporate" traffic, inhibiting the lower classes, as was the case with passengers who flew between Turin and Madrid.

If demand increases (by premium or leisure passengers) and exceeds 50% (Table 2) of the allocated quota, the destination of Lisbon will have fewer seats to sell and will be negatively affected.

The flight between Madrid and Lisbon will be affected equally.

In a situation of full competition, such as the one suffered by Air Nostrum the PGA flight would be doomed.

Table 2. Seat allocation capacity for Portugália

		Number of seats allocated by flight segment (50%)							
Flight Segments	Monday	Tuesday	Wednesday	Thursday	Friday	Saturday	Sunday	Sunday	
Turin Madrid	50	50	50	50	50	25	No oper	275	14 300
Madrid Lisbon	50	50	50	50	50	25	No oper	275	14 300
Lisbon Madrid	50	50	50	50	25	nop	50	275	14 300
Madrid Turin	50	50	50	50	25	nop	50	275	14 300
Half capacity	57 200	2	28 600				Annual Seat Capacity in Portugal and Italy		57 200

Source: Portugália Airlines (2006)

Despite the excellent service on board and the very convenient timetable for the businessmen segment, the existence of only one flight, in competition with another company, in which it operated three daily turn arounds, PGA was doomed to continue.

Table 3. Air Nostum´s capacity offer

Air Nostrum	Monday	Tuesday	Wednesday	Thursday	Friday	Saturday	Sunday	Total seat Capacity
Turin - Madrid	150	150	150	150	150	150	150	54 600
Madrid - Turin	150	150	150	150	150	150	150	54 600
						Annual capacity offer		109 200

Source: Statistical Yearbook ENAC (2006)

At the end of 2003 and due to Air Nostrum's investment (Table 3) in Piedmont catchment area, the total annual capacity for Madrid increased by 91%. PGA capacity decreased to 34% (from 100% in early 2000) and YW from 1999 to 2003 stabilized capacity participation by 66%.

The Load Factor may even have stabilized, but for sure the average fare ended up "falling", because if the competition offers more availability in terms of seats and frequency, it obviously puts PGA in great difficulty.

PGA used a business model like that of Air Nostrum but failing to meet the objectives of that same model, it ends up once again jeopardizing its operation because it has no distribution capacity, much less the hub effect in Lisbon (Lisbon is TAP's home market).

Wrongly, PGA operates in the traditional business model of Full-Service Carriers, but with an operation with characteristics like a point-to-point company, characteristic of low-cost companies.

The Spanish company operated on a "feeder" basis, capturing traffic from all over Europe converging to the Madrid HUB, where from the Spanish capital, Iberia transports it to the rest of the world (Buyck, 2009).

Portugália Airlines abandoned the Turin route in the first quarter of 2005 (*Nuovo volo Torino-Lisbona operato dalla TAP Portugal/New Turin-Lisbon Service Operated by TAP Portugal*, 2012).

PORTUGÁLIA AIRLINES OPERATIONS

The Table 4 shows the evolution of the number of passengers between Turin (TRN) and Lisbon (LIS). The airline Air Nostrum began operating in 1999, not creating difficulties for PGA's operation to Lisbon.

In fact, the number of passengers boarded (PGA) has increased significantly until 2004, and 2003 was the year the most passengers carried.

As YW flights grew, the number of passengers on the PGA-operated "leg" between Turin and Madrid was decreasing, which is why the number of passengers was increasing to Lisbon.

Being a flight that provided the revenue for two destinations (Madrid and Lisbon), and with great anticipation to Lisbon, the YW effect, generated an increase in passengers to Lisbon.

All numerical data presented are taken from reports by ENAC (Ente Nazionale Aviazione Civile), SAGAT (Turin Airport) and the municipal institution, Torino Citta.

Table 4. Portugálias´s passengers transported between 2000 and 2005

Passengers transported - PGA (ENAC SAGAT TORINO CITTA)					
Turin -Lisbon - Turin					
2000	**2001**	**2002**	**2003**	**2004**	**2005**
11 055	*11 292*	*11 555*	*14 717*	*16 432*	*2 951*

Source: ANAC/SAGAT/Torino Citta (2006)

In 2005, the year with fewer passengers, because of the closure of the route (as can be seen in the table 5), Portugália Airlines stopped operating between Turin and Lisbon.

The Portuguese private company transferred the flight to Milano Malpensa Airport, where it operated in a more robust, but very unsustainable way.

Table 5. Comparison of passengers transported between PGA and Air Nostrum to Madrid

Turin - Madrid - Turin					
IBERIA Air Nostrum (passengers)					
2000	2001	2002	2003	2004	2005
35 270	35 627	50 852	51 087	101 691	62 410
Turin - Madrid - Turin PGA					
13 882	14 206	18 515	13 507	11 510	2 111
Total passengers PGA + IB/YW (TurinMadridTurin)					
49 152	49 833	69 367	64 594	113 201	64 521
Air Nostrum (Share)					
72%	**71%**	**73%**	**79%**	**90%**	**97%**
28%	29%	27%	21%	10%	3%
PGA (Share)					

Source: ANAC/SAGAT/Torino Citta (2006)

In five years, Iberia/Air Nostrum increased the number of passengers carried by 77% according to the investment in the capacity, materialized by offering three daily flights from Turin to Madrid.

Passengers to Lisbon decreased dramatically when the Spanish regional airline introduced the third flight.

Table 6. Passengers transported by Portugália from Turin to Madrid

TOTAL Turin - Madrid - Lisbon - Passengers PGA					
2000	**2001**	**2002**	**2003**	**2004**	**2005**
24 937	25 498	30 070	28 224	27 942	5 062
Share Turin Lisbon					
44%	44%	38%	52%	59%	58%
56%	56%	62%	48%	41%	42%
Share Turin Madrid					

Source: ANAC/SAGAT/Torino Citta (2006)

In 2003, Air Nostrum implements the third flight to close the cycle of three spin cycle per day. The load factors rate for the flight to Lisbon is starting to increase, however the occupancy of the "leg" between Turin and Madrid is starting to decrease.

The share corresponding to the occupancy of the number of passengers to Lisbon, from 44%, rose to almost 60% (58%), Table 6. The market share of the flight between Turin and Madrid had the exact opposite behavior: from almost 60% (56%) it fell to 42% (2005).

The PGA branch in Italy (2005) had only five people, operational and commercial costs were fixed charges of the company, however the handling of passenger and aircraft support was outsourced.

The number of passengers has substantially reduced in the year in which the PGA leaves the area of influence of the Piemonte region. It remains to be seen whether the PGA would be able to remain in the market or if it in fact had no more means to continue.

A review of the business model would have been the most appropriate strategy. The company operated a route with very controlled costs, without any investment, practicing high-value tariffs, which in turn would have to cover the costs of other loss-making routes on the network.

MARKET IMPLICATIONS

While Portugália presents a very fragile product, Iberia can have a distinctive offer capable of sustaining any advances in any competition. Air Nostrum offers the Madrid hub to transfer and "absorb" passengers in a feeder sixth freedoms logic.

While PGA was operating without competition, fares could not decrease either, as well as the strong tendency for passengers to fly beyond the Madrid hub (or Lisbon and then with TAP), for example Rio de Janeiro and São Paulo, using the PGA flight in an interline logic.

These operating conditions favor the increase in average fares, keeping passenger traffic at a premium rate.

As Iberia began to offer three daily frequencies, *interline passengers traveling with* PGA to Madrid, and to other destinations on the network that the PGA flight was accustomed to operating *as a feeder*, could no longer make it, creating a sharp decrease in revenue from the Portuguese company.

All passengers who flew "beyond Madrid", and who, before the entry of Air Nostrum, were flying with PGA, automatically began to book with the Spanish regional airline.

Traditional travel agents or those selling over the Internet (Online Travel agents, OTA) drastically reduce their sales at PGA as the offer improves substantially with the arrival of Air Nostrum, and sales to PGA become marginal.

Another consequence is the widespread lowering of fares, namely the "appetite" of online sales distributed by travel agents who place all their inventory on online platforms.

If sales increase for the Spanish airline and decrease for the Portuguese carrier, so does interest in working with PGA decrease substantially, as the travel agent, an intermediary retailer, lives off the commissions it receives and the diversity of flights it can offer its customers. This possibility also substantially reduced Portugália's operation.

The effects of Air Nostrum's entry into the market have left PGA (marketing *mix* variables) completely unbalanced and weakened.

Some customers (preferred travel agents) remained the only force in the operation, in a synergistic effort between flights from Milan and the Turin flight, however, and although the number of passengers had stabilized, average fares fell, and revenue swelled significantly.

DISCUSSION

In 1993, Portugália Airlines launched the first route to Italy (Turin), due to the growing interest in the business generated by Fiat, Pirelli, Ferrero, Iveco, Magneti Marelli and Alenia Aerospace, just to name a few of the premium customers.

The flight profile indicated departure from Lisbon at 17:15 (NI710) to Madrid (boarding and disembarkation), and proceeding to Turin, arriving at 22:30.

The next day, because of the nightstop, the flight begins its cycle at 7:30 am (NI711) with the first destination being Madrid, with arrival scheduled for Lisbon at 10:30 am.

It was possible to travel to Madrid or Lisbon and return to Turin on the same day. The Lisbon – Turin flight, despite being scheduled as direct and marketed in the same way, made a stopover in Madrid (funnel flight). Flight NI710 did not operate in a "normal" way in the sense that it was possible to sell seats: from Lisbon to Madrid, from Lisbon to Turin and from Madrid to Turin.

For the same reason, the same happens from Turin to Lisbon, from Turin to Madrid and from Madrid to Lisbon.

This flight profile clearly demonstrates the inability to allocate more seats to the higher value segment (TRN-MAD), since, selling the full capacity to MAD, it did not allow sales to LIS; selling the 100 seats to LIS inhibited sales to MAD and to top it off, seats between MAD and LIS could not be sold. Obviously, this flight could not last much longer if it were not divided into two segments: LIS and MAD.

The initial idea was to operate between Lisbon and Turin without ever involving the Madrid route. Also, during the first visits to Turin it was immediately understood that the strength of the route would always be in Turin and preferably towards Madrid. Corporate traffic was a reality between Turin and Madrid, due to the commercial relations between the two cities and the average values sold per seat were quite desirable. Thinking only from an economic and financial perspective, the route was launched with a stop in Madrid, taking advantage of the traffic rights between Madrid and Turin and vice versa.

At the time, it was clear that the strength of the route was in the segment between TRN and MAD, and that if one day there was competition, it would appear precisely in the segment where businessmen flew.

The purpose of the TRN-MAD flight was to transport premium passengers at high-yield fares. For the TRN-LIS segment, the intention was to optimize sales with full fares whenever possible, increase load factors rates with discounted fares and in hiring through travel agencies (Tour operators for individual and group sales).

This was the strategy chosen to balance the "offers" in the flight segments, and it worked perfectly, while the company operated without competition.

In November 1999, Air Nostrum Regional Airlines, Iberia's feeder company, took the first step to compete with PGA, placing the first flight from Turin to Madrid and with a timetable very close to the Portugália flight.

In 2003, Air Nostrum (Fig 3) had already launched the third flight, ending up annihilating PGA operation, which began to lose passengers very sharply between 2003 and 2005.

On the contrary, *Iberia's three-flight feeder* wave generated more traffic, and only stopped the trend with the cessation of the PGA flight in 2005 (see figure 2).

The strong competition of the Spanish airline was the main cause of the departure of Portugália Airlines from Sandro Pertini Airport. The lack of investment on the part of the Portuguese company was also an extremely negative point in the company's strategy.

Figure 2. Evolution of transported passengers (Portugália Airlines)
Source: Portugália Airlines

Figure 3. Evolution of transported passengers (Air Nostrum)
Source: ENAC (2006)

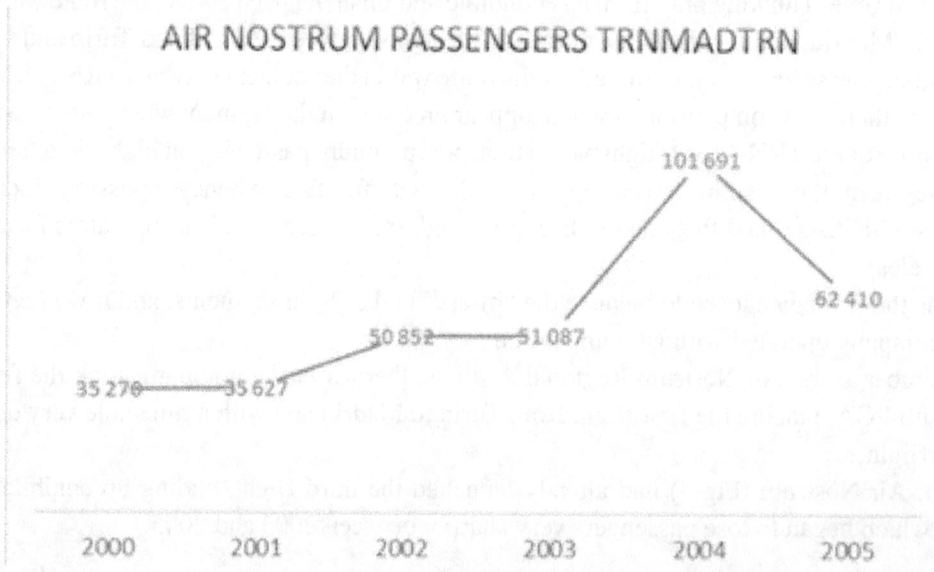

The Turin office was transferred to Milan (Malpensa) and all operations concentrated in Milan/ Malpensa, thus competing directly with TAP Portugal.

PGA now has four daily flights, but with very unattractive schedules for customers. PGA was forced to close a flight with more than 10 years of existence because of enormous pressure promoted by Iberia Regional Airlines, Air Nostrum at Turin/Piedmont airport.

The failure of Portugália?

Defeated by competition and forced to leave Turin, the airline decided not to lose market in Italy, increasing its capacity at Milan Malpensa airport with a fourth flight, leaving the Piedmont region served by Sandro Pertini airport in Turin.

The seven "deadly" PGA sins are described below.

1. PGA missed the opportunity to consolidate traffic, especially customers traveling on business generated by very important companies operating in Turin, Madrid, and Lisbon.
2. Missed the opportunity to establish a strategic positioning throughout northern Italy, as Milan and Turin could be the great "catchment area".
3. Missed the opportunity to have a valuable product, and to negotiate the best rates with the main players in the market.
4. Lost the ability to negotiate with the Italian BSP's most potential agencies.
5. Lost the commercial capacity to negotiate interline agreements, special prorate agreements, and in particular, the impossibility of entering code sharing agreements.
6. Lost the possibility of having competitive advantages to negotiate with Alitalia trying to fight Air Nostrum on the one hand and then build a solid relationship with Alitalia, increasing the offer to Spain and Lisbon.
7. As PGA was approaching the SKY Team with players like Air France and Alitalia, Turin could have been a very interesting ally to strengthen ambitions.

PGA revealed that it did not have a strategy for the future and, for that reason, never invested in a route that proved very important for Air Nostrum.

For six long years, the Turin-Madrid route was operated by Portugália with a daily flight offering 200 daily. The market was "asking" for a second flight, as it was necessary to end the cycle with at least two daily flights.

Given the results, Portugália has never considered Turin a strategic route, although it has had great results, both in terms of average fares and in terms of operating costs.

Most likely, and in a network logic, the gains always served to balance the accounts of loss-making routes of Portugália´s network.

When Iberia started the first flight, PGA did not respond, and lost the opportunity to establish a position in the market. TAP, in turn, was investing year after year.

PGA and Air Nostrum, despite operating with the same business model, the Portuguese company did not have sixth freedom traffic rights.

The strategy was (or could have been) to increase capacity and find partners to grow, such as Alitalia, as Air France and Alitalia, both in the Sky Team Alliance, could be natural allies of the PGA. One could imagine a triangle between Milan (possibly Madrid), Lisbon and Paris.

PGA never had a strategy for the Italian routes, on the contrary, TAP, every year put more capacity on its routes, also increasing the pressure of competition against Portugália.

TAP did not operate from Turin (after purchasing PGA in 2007, it started operating a few years later) and until 1999, Air Nostrum did not fly from Turin to Madrid (only Barcelona) and, for that reason, PGA had such a competitive advantage.

CONCLUSION

The first PGA flight impressed businesspeople traveling between Turin and Madrid. The entry of Air Nostrum and consequent investment in the route forced Portugalia to withdraw.

In 2003, when the third flight was launched by Air Nostrum, the business environment changed to Portugália, which abandoned the route and repositioned itself in Milan, launching another turn around between Milan and Lisbon.

It was evident that PGA's business model lacked a hub, as the possibility of distribution in the network was very weak. The company sustains its results based on a typical point-to-point operation, something unsustainable for its future.

The Turin route, despite being a funnel flight, had a huge shortcoming as PGA did not have a hub. The flight from Turin to Madrid, largely because of the business travel traffic, kept the route operating. Obviously, there was never a desire to invest much for lack of a strategy, however, it always served to pay for other loss-making routes for the company.

For years it was argued that PGA could have been a feeder for TAP, taking advantage of the Lisbon hub. The idea never gained great supporters, except that at the end of 2007 the Portuguese state company acquired the private carrier of Grupo Espírito Santo.

In 2007 the company was purchased by TAP (not merged) and transformed PGA into TAP Express, with strategic objectives like those declared in 2000 by the President of PGA himself.

Regarding the moment we are going through (post-pandemic), we have seen a decrease in business travel, an increase in the average fare, but cost containment should continue.

All this to say that nowadays it would be very difficult to see a company born with the same characteristics as PGA because the business environment changed the paradigm that was registered before Covid 19.

REFERENCES

Aeromedia. (2012). Nuovo volo Torino-Lisbona operato dalla TAP Portugal/New Turin-Lisbon Service Operated by TAP Portugal [New Turin-Lisbon flight operated by TAP Portugal/New Turin-Lisbon Service Operated by TAP Portugal.]. *Aero Media.* http://www.aeromedia.it/taptolis.html

Air Nostrum_IBERIA Regional. (2022). Air Nostrum. http://www.airnostrum.es/home.asp?idioma=2

Alamdari, F. (2002). Regional development in airlines and travel agents relationship. *Journal of Air Transport Management, 8*(5), 339–348. doi:10.1016/S0969-6997(02)00014-5

Buyck, C. (2009). Iberian Success: Air Nostrum Thrives as an Independent Regional With Strong Ties to Iberia. *ATW: Air Transport World, 46*(6). https://trid.trb.org/view/890798

Codes—Airline and Location Codes Search. (n.d.). IATA. https://www.iata.org/en/publications/directories/code-search/

EU Air Transport Liberalisation Process, Impacts and Future Considerations. (2015). International Transport Forum Discussion Papers N. 2015/04; International Transport Forum Discussion Papers, *vol. 2015/04*. doi:10.1787/2223439X

Fageda, X., Suau-Sanchez, P., & Mason, K. J. (2015). The evolving low-cost business model: Network implications of fare bundling and connecting flights in Europe. *Journal of Air Transport Management, 42*, 289–296. doi:10.1016/j.jairtraman.2014.12.002

Fageda, X. & Flores-Fillol, R. (2012). Air services on thin routes: Regional versus low-cost airlines. *Elsevier Enhanced Reader*. doi:10.1016/j.regsciurbeco.2012.03.005

Forbes, S., & Lederman, M. (2007). The Role of Regional Airlines in the U.S. Airline Industry. *Advances in Airline Economics, 2. Full article: Impact of COVID-19 on different business models of European airlines*. https://www.tandfonline.com/doi/full/10.1080/13683500.2021.1960284

Gillen, D., & Lall, A. (2004). Competitive advantage of low-cost carriers: Some implications for airports. *Journal of Air Transport Management, 10*(1), 41–50. doi:10.1016/j.jairtraman.2003.10.009

Hayward, J. (2022, january 13). *What Is The IATA Summer And Winter Schedule?* Simple Flying. https://simpleflying.com/iata-summer-winter-schedule/

ItalianoT. club. (2007). *Portogallo*. Touring Editore.

Lusa, A. S. Agência. (2016). Portugália passa a TAP Express e ganha novos aviões [Supporting and Explaining Miles and Snow's Strategy Typology with Denison's Suggested Adaptability Capabilities and Examining Their Effects on Business Performance Outputs]. *Observador*. https://observador.pt/2016/01/14/pinto-anuncia-fim-da-portugalia-lancamento-da-tap-express-novos-avioes/

MD80.it. R. (2008). Aeroporto di torino: 2007 da record, oltre 3 milioni e mezzo i passeggeri [turin airport: 2007 a record, over 3 and a half million passengers]. *MD80.it*. https://www.md80.it/2008/01/15/aeroporto-di-torino-2007-da-record-oltre-3-milioni-e-mezzo-i-passeggeri/

PGA Homepage. (n.d.). PGA. https://www.portugalia-airlines.pt/

Publico. (2006). TAP paga 140 milhões ao Grupo Espírito Santo pela Portugália [TAP pays 140 million to Grupo Espírito Santo for Portugália]. *PÚBLICO*. https://www.publico.pt/2006/11/07/jornal/tap-paga-140-milhoes-ao-grupo-espirito-santo-pela-portugalia-106011

Reynolds-Feighan, A. (2018). US feeder airlines: Industry structure, networks and performance. *Transportation Research Part A, Policy and Practice, 117*, 142–157. doi:10.1016/j.tra.2018.08.009

Ruiz-Moreno, F., Mas-Ruiz, F. J., & Sancho-Esper, F. M. (2021). Strategic groups and product differentiation: Evidence from the Spanish airline market deregulation. *Research in Transportation Economics*, *90*(C), 101030. https://ideas.repec.org/a/eee/retrec/v90y2021ics0739885921000020.html. doi:10.1016/j.retrec.2021.101030

Sobre Nós [About Us]. (n.d.). PGA. https://www.portugalia-airlines.pt/sobre-nos#historia

Števárová, L., & Badánik, B. (2018). Performance of Hub and Spoke Networks of Selected Airlines. *Transportation Research Procedia*, *35*, 240–249. doi:10.1016/j.trpro.2018.12.024

Tomová, A., & Materna, M. (2018). Miscellaneous "modi vivendi" of regional and network airlines: The tracks for future research. *Transportation Research Procedia*, *35*, 305–314. doi:10.1016/j.trpro.2018.12.007

Wu, J., Zhang, P., Wang, Y., & Shi, J. (2022). Integrated aviation model and metaheuristic algorithm for hub-and-spoke network design and airline fleet planning. *Elsevier Enhanced Reader*. doi:10.1016/j.tre.2022.102755

Chapter 9
A Case Study of Lufthansa, Vueling, and Wizz Air Aircraft Turnarounds at Lisbon Airport

João Pedro Costa
Instituto Superior de Educação e Ciências, Portugal

Rui Quadros
Instituto Superior de Educação e Ciências, Portugal

Ana Barqueira
Instituto Superior de Educação e Ciências, Portugal

ABSTRACT

Ground handling services must be provided in accordance with the airlines' strategy in order to add value and create competitive advantage for the contracting airline, and it is therefore critical to understand how airlines shape ground handling services based on their business models. This study addresses ground handling services by focusing on the aircraft turnarounds of full-service, hybrid and low-cost airlines. It aims to extend the knowledge of the procedures, activities, and resources of aircraft turnarounds by comparing, based on qualitative research using the case study method, the characteristics of Lufthansa, Vueling and Wizz Air aircraft turnarounds. The results point out that the tasks related to unloading/unloading the holds, cleaning the cabin, and disembarking/embarking passengers, are the turnaround tasks that show the most differences between the three business models, playing an important role in the value creation strategy and in increasing the competitive advantage of airlines in ground handling services.

INTRODUCTION

Airline business models can be classified into three types, Full-Service, Hybrid and Low-Cost, and it is through these that airlines create, deliver and capture value to remain profitable and competitive. It is therefore essential that at every stage of the flight the airline's services match the strategies of its busi-

DOI: 10.4018/978-1-6684-6835-7.ch009

ness model (Chesbrough, 2007). Ground handling services are no exception and must match the airlines' strategy and desired levels of quality and efficiency.

In order to analyze how different airlines create competitive advantage and capture value through their ground handling services, the author proposed to carry out, based on a qualitative line of research, a case study that analyzes the turnarounds of the three business models by selecting three airlines that operate the same type of aircraft at Lisbon Airport, namely, Lufthansa (Traditional), Vueling (Hybrid), Wizz Air (Low cost).

Although there is a vast literature on the differences in the airlines' business models in the various areas of their operations, the approach to ground handling services in the different business models is not so common and is usually carried out focusing only on the full-service and low-cost business models.

Ground handling services are subdivided into several areas and include the provision of various essential services for air transport operations at different time periods and locations from the airport, with a large part of these services occurring on the airside (with a focus on boarding and servicing the aircraft) and involving the performance of various tasks using teamwork as well as driving and approaching several vehicles simultaneously around the aircraft (Ashford, 2013).

İnan (2018) states that in order to cope with the inherent complexity of ground handling services, it is of utmost importance to implement systems and strategies aimed at increasing task efficiency in order to reduce delays caused by airside ground handling services, thereby improving ground handling service levels as well as airport and airline performance levels (Schmidberger et al., 2009).

To enhance the understanding of how airlines shape their aircraft turnarounds in order to create competitive advantage, in a last step, three schemes demonstrating the relationship between the operational characteristics of aircraft turnarounds and the accomplishment of the tasks performed by the ramp operator teams.

Objectives

The main objective of this work is to analyze the aircraft turnaround of full-service, hybrid and low-cost airlines, with the aim of presenting an empirical view of ground handling services and how these are processed and shaped by each airline's business model.

Its specific objectives are therefore to analyze the turnarounds of Lufthansa (full-service), Vueling (hybrid) and Wizzair (low cost) at Lisbon Airport, carrying out for the three types of airline a description, analysis and comparison of the ground handling services, equipment and staff used in their turnarounds.

Research Question

Ground handling services must be provided in accordance with the airlines' strategy and levels of quality and efficiency in order to add value and create competitive advantage for the contracting airline, and it is therefore critical to understand how airlines shape ground handling services based on their business models.

Although there is a vast literature on the differences in airline business models in different areas of operation, addressing ground handling services in different business models is not as common, and is usually carried out in a general way and focusing only on Full-Service Carriers (FSC) and Low-Cost Carriers (LCC). Gomez and Scholz (2007) suggest that the factors that most distinguish each airline's ground handling services are largely related to fleets, routes and service providers, most notably the difference in aircraft turnaround times, with LCC having lower turnaround times than FSC.

Thus, in order to increase the knowledge of the impact of airlines' business model on ground handling services, the author proposes a more in-depth and detailed analysis of airlines' ground handling services, with the research question "How are aircraft turnarounds influenced by full-service, hybrid and low-cost business models?".

LITERATURE REVIEW

Airlines Business Models

The airline business model dictates how companies create, deliver and capture value to remain profitable and competitive. The most important indicators representing the business model of a successful airline can be presented in terms of how it generates revenue (aeronautical/non-aeronautical revenue, sources of revenue, etc..), what type of product it offers, and what its target audience is (Chesbrough, 2007)

Although several authors define and characterize business models with different approaches, according to Klophaus et al. (2007) in Tomová & Ramajová (2014), airlines can be classified into the following models:

- Low-cost
- Hybrid
- Full-Service

Hybrid models are models where airlines adopt strategies from both models in order to increase their competitive advantage. Whereas in pure Low Cost or Full Service models the management and operation of the airline relies only on the strategies of the business model adopted by the company (Vidović et al., 2013).

The degree of flexibility and independence in decision making as well as the level of efficiency of an airline is reflected by the business model adopted, being pointed out by Munisamy et al. (2013) an important factor when an airline faces turbulent periods that may negatively affect its financial position. The same authors emphasize that a well-designed business model allows airlines to face periods of crisis more successfully by making critical decisions more quickly, which, by the brevity and effectiveness with which they are made, allow airlines to minimize losses from poor business performance.

The role of business models in airline efficiency has been addressed more frequently since the emergence of low-cost airlines following the deregulation of the airline market in 1978 in the United States and in 1997 in the European market (Inglada et al., 2006). However, the first studies conducted in this field focused their research on airlines with relatively similar cost structures such as Adler & Golany (2001), Fethi et al. (2001), Gillen & Lall (1997), Good et al. (1995), Land et al. (1993) and Semenick Alam & Sickles (1998).

In this field, Scheraga (2004) conducted one of the first studies that included the analysis of airline efficiency across the wide range of cost structures in the air transportation industry. Over the years, other authors such as Alves & Barbot (2007), Assaf & Josiassen (2012), Barbot et al., (2008), Lee and Worthington (2010) and Vlaar et al. (2005), have written studies showing that LCC, due to their greater ability to reduce operating costs, by practicing lower average fares that increase passenger load factor levels, have a higher level of operating efficiency compared to FSC (Chesbrough, 2007).

The comparison between the characteristics of each business model, as well as the advantages that each model offers, has been a subject of much research, especially since the beginning of the century, due to the existence of larger amounts of data available to be analyzed by researchers. These comparative studies range from operational issues Dobruszkes (2006), Hunter (2006), Lordan (2014) and Reynolds-Feighan (2010) to analyses of the position of each airline in the air transport market Barrett (2004) and Rozenberg et al. (2014) or issues related to passenger satisfaction in each airline Chiou and Chen (2010), O'Connell & Williams (2005) and Sezgen et al. (2019).

Shank & Govindarajan (1993) suggest that there are three factors where commercial airlines can provide value to their customers. These are the provision of reservation and ticketing services, the operation of aircraft from point A to point B, and the provision of efficient passenger services before the flight, during the flight, and after arrival. Ground handling services are then directly linked to these three factors, acting as a partner to airlines enabling their overall performance, in all phases of flight and passenger services, to exhibit higher levels of quality, and also playing a key role in the operational performance of airports (Kolukısa, 2011).

The evolution of airline business models is a topic widely covered in the literature Carmona, (2008), Diaconu (2012) and Vidović et al. (2013) and contributes to the increase of knowledge about airlines and how they react to external factors that affect the air transport, as for example the recent pandemic crisis that in the last two years has motivated several studies Akbar & Kisilowski (2020), Atems and Yimga (2021), Bauer et al. (2020) and Kökény et al. (2022).

Daft & Albers (2014) suggest that over the years there has been an approximation of the FSC and LCC business models, with fewer companies adopting purely Full Service or Low Cost strategies. According to the authors, this approximation is due to the airlines' attempt to expand their business by providing services for different types of passengers in order to increase their competitiveness. The homogenization of fleets and the reduction of cost structure by FSCs or the bet of LCCs on marketing more elaborate products such as loyalty programs, seat classes or routes to major airports, are some examples of the measures adopted in recent years by both types of airlines, which contribute to the increase of airlines with hybrid business models

In the study of the impact of covid-19 on airline business models, the results of research carried out by Mrázová and Kazda (2021) propose that there are more and more hybrid Airlines due to the crisis created by covid-19 and the need to adapt to market needs and capture new target audiences, predicting that this convergence, also pointed out by Daft and Albers (2014) will be increasingly noticeable in the coming years.

Diaconu (2012) suggests that in general, traditional airlines have higher levels of service quality in all phases of flight (pre-flight, flight and post-flight) than low-cost airlines, which have shorter turnaround times than traditional airlines. The author also highlights the tendency on LCC to have a homogeneous fleet with a high density of seats per aircraft, while FSC generally opt for a heterogeneous fleet with several aircraft models and lower seat density. Dobruszkes (2006) and Lordan (2014) also point to the routes of each type of business model as being quite different, with FSC flying mainly to main airports with a hub-and-spoke network while, by contrast, LCC opt mostly for secondary airports operating on a point-to-point route system.

Full-Service Network Airlines

Full-Service Network airlines, also known as Legacy airlines, are airlines that started their operations before the airline deregulation law in the U.S. market in 1978. and have as their main strategy to provide high quality differentiated services (Lawton et al., 2011).

According to Porter (1980) the differentiation strategy is based on the creation of a unique product and/or service, which can be achieved through the following aspects: design and brand image, implementation of innovative technologies, quality of customer service, among others. This differentiation in legacy airlines is made through an offer of attractive and unique services on board, as well as the offer of heterogeneous services and fares that aim to attract and serve the interests of different types of passengers (business, VFR, leisure, etc...) while also focusing on customer loyalty through frequent flyer reward programs (Vidović et al., 2013).

The impacts of the airline market deregulation law were so significant that legacy airlines that were previously state-owned companies with monopolistic and consolidated routes had to rethink their business models to remain competitive in the face of industry changes and growing competition, especially from low-cost carriers. Over the years, several prominent legacy airlines such as PanAm and SwissAir have been terminated, others, such as Aer Lingus, have reinvented themselves as hybrid carriers adopting low-cost operating characteristics or have chosen to merge, as is the case with Air France and KLM (Lawton et al., 2011).

Low-Cost Airlines

Low Cost Carrier's are airlines whose business model is to operate airline flights based on a cost leadership strategy (Hunter, 2006). These airlines offer a more "basic" standard service when compared to FSC which consequently allows them to have lower operating cost structures and therefore offer passengers more affordable fares (Savenkovs & Sztore, 2021).

Low-cost airlines emerged in the United States of America with the deregulation of the airline market in the late 1970s, and by the 1980s several low-cost airlines were operating flights between airports in the United States of America. In the European market, air market deregulation occurred later and in a lagged manner, with the first package of measures aimed at liberalizing European airspace being issued in 1987, the second in 1990, and the last in 1993 (Diaconu, 2012). These measures to deregulate the airline market have allowed low-cost airlines to grow in the European air market reaching a market share of 44.5% of the available seats of the European air market by the year 2020 (Mazareanu, 2021).

Ground Handling Services

The Europen Comission (1996) defines ground handling services in the Annex to Directive 96/67/EC by classifying them into 11 categories that are subdivided into a number of services essential for ground handling airline operations. These are:

- Category 1 - Ground administrative assistance and supervision, which includes four subcategories of services;
- Category 2 - Passenger handling
- Category 3 - Baggage handling;

- Category 4 - Cargo and mail handling, which includes two subcategories of services;
- Category 5 - Ramp handling, which includes seven sub-categories of services;
- Category 6 - Aircraft cleaning and service assistance, which includes three sub-categories of services;
- Category 7 - Fuel and oil handling, which includes two subcategories of services;
- Category 8 - Line maintenance assistance, which includes four subcategories of services;
- Category 9 - Air operations and crew management assistance, which includes four subcategories of services;
- Category 10 - Ground transportation handling, which includes two subcategories of services;
- Category 11 - Catering assistance, which includes four subcategories of services.

Ashford (2013) states that ground handling services are subdivided into several areas and include the provision of several essential services to air transport operations. These services are processed at different time periods and locations at the airport and, according to the same author, it is common, when addressing the subject of ground handling services, to divide activities that take place airside (with a greater focus on aircraft handling) with those performed landside (with a greater focus on passenger handling). However, the author suggests that despite this theoretical separation, in practice this division is not so linear, as the work teams and operations involved are mostly interconnected and therefore not necessarily restricted to one area. The relationship of ground handling services to airports and how they influence each other are studied in Andersson et al. (2000), Schmidberger et al. (2009) and Yager (1983).

Ground handling services that take place on the landside are mostly performed in the passenger terminal and correspond largely to the Category 2 (Passenger Handling) services of the ground handling services defined by the European Commission. These services are then directly related to passengers, whether they are arriving, departing, transferring or in transit, and according to Schultz & Fricke (2011) these services can be divided into five main activities, namely ticketing, check-in, baggage check-in, embarkation/disembarkation assistance and the provision of support/information services. In addition to these, passenger handling services also include the provision of special handling services that may be requested by passengers, such as VIP (Very Important Person), CIP (Commercially Important Person), escorting minors flying alone, and assistance to passengers with reduced mobility (Türeli et al., 2019).

With regard to ground handling services provided on the airside, these are mostly performed on the apron and focused on assisting the aircraft, and it is common to have a separation between activities performed over the wing and activities performed under the wing. The tasks performed above the wing are activities related to the passenger cabin, such as cleaning, preparing galleys, and managing the embarkation and disembarkation of passengers. The tasks performed below the aircraft wing, on the other hand, are related to the loading and unloading of cargo in the aircraft holds as well as to tasks supporting the aircraft on the apron, such as power supply, air conditioning, dumping services, drinking water, and fuel supply (Tabares and Mora-Camino, 2017).

In order to ensure that the services provided by handling agents correspond to the strategy and levels of quality and efficiency desired by the airlines that contract them, it is essential to have a contract that describes them in detail, so that there are guidelines in the provision and monitoring of ground handling services. These contracts are called Service Level Agreements (SLAs) and they are defined by IATA (2018) as a contract concluded between two or more parties where the terms of the services to be contracted and provided are defined.

The Airport Research Center (2009) highlights Service Level Agreements, along with airport infrastructure and service providers, as the main factor influencing the quality levels of ground handling services.

Aircraft Turnaround

Aircraft turnaround, which can be considered the core of the ground handling business, runs between the in-blocks and off-blocks of the aircraft and can be defined as the process of "getting the aircraft ready for the next flight" by performing several tasks performed simultaneously by different operational departments, and its efficiency is therefore related to how well the tasks performed by the different departments are managed and coordinated (Kolukısa, 2011).

Figure 1. Main tasks and services of an aircraft turnaround
Source: Padrón and Guimarans (2019)

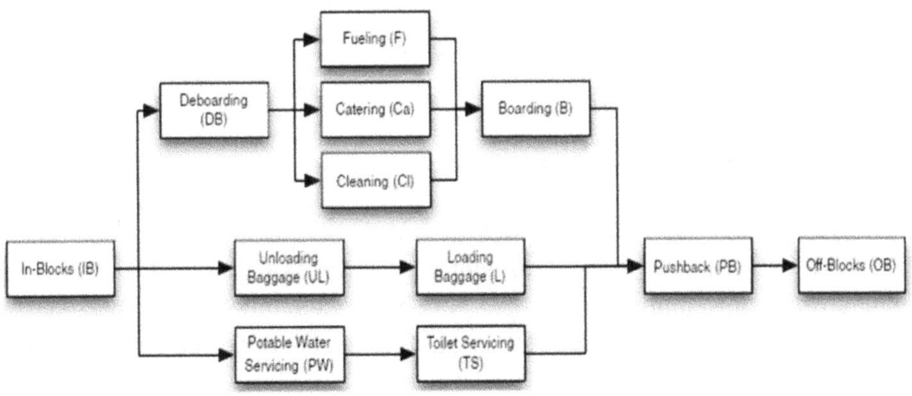

The duration and complexity of a turnaround depends on several factors mostly related to operational variables such as, the type of aircraft (family, size, number of seats, etc.), parking position (near the terminal or in a remote position), or the types of ground handling services provided based on the airline's business model and policies (Padrón & Guimarans, 2019). Security constraints, parking stand and airport space requirements, adaptability to traffic peaks and adverse weather conditions may be other variables that can affect the completion of the tasks inherent to a rotation, influencing its duration and complexity (Kierzkowski & Kisiel, 2017; Tabares & Mora-Camino, 2017)

In addition to these variables, there are other operational factors that are not predictable and can influence the duration and complexity of a turnaround. Ashford (2013) gives the example of a malfunction in a ULD loading equipment, showing the impact of this malfunction on the aircraft turnaround time. According to the author, when faced with a malfunction, ramp operators must proceed to assess and resolve the problem as quickly as possible, and the following actions must be taken:

- Assess the nature of the problem and understand how long the problem will take to be resolved.
- Take corrective action such as calling the staging teams and asking for mechanics at the booth or, if possible, continue the task using other equipment.

- Inform all other activities that will be affected by the breakdown giving instructions as necessary.

The author then stresses the importance of the previous actions in order to inform those responsible for operational management so that they can decide, depending on the importance of the interrupted task, to cancel that task or shorten it in order to avoid aircraft delay which can sometimes be more detrimental to the airline than the non-performance or non-finishing of the task.

Schultz et al. (2020) addressed the introduction of cabin disinfection procedures between each flight, as well as boarding/disembarkation with minimum physical distance due to the restrictions caused by COVID 19, concluding that the introduction of these restrictions increase the turnaround time and the associated workload even if process optimization mechanisms are adopted and the work teams allocated to cleaning and disinfection of the aircraft are reinforced. According to the results of the study, to match the pre-pandemic turnaround times of a given aircraft without having to reinforce the work teams allocated to cleaning and disinfection of the aircraft, the aircraft cannot have a load factor higher than 67% and it is necessary that the disembarkation procedure occurs simultaneously at both doors of the aircraft. In positions near the terminal where passengers board only via the telescopic bridge, turnaround times increase by 10% even with reinforced crews and 20% if crews for cleaning and disinfection of the aircraft cabin are not reinforced.

Ground Support Equipment (GSE)

Ground support equipment is used to help workers perform ground handling tasks for airline operations, allowing heavy loads to be transported and loaded/unloaded more quickly, efficiently, and safely.

This equipment can be powered or non-powered. Non-motorized equipment can be towed using special tractors or labor. Ground support equipment for aircraft ground handling can range from more conventional vehicles and equipment such as buses, trolleys and baggage carts, to less conventional vehicles and equipment such as loaders, pushbacks and dollies. GSE's can be owned by the airport, the airline or the ground handling service providers. GSE's typically provided by centralized airport infrastructure are the telescopic bridges for passenger boarding/disembarkation, the GPS (Ground Power System) for power supply to the aircraft, and the PCA (Pre-Conditioned Air) for air conditioning supply to the aircraft cabin. Despite this, not all aircraft parking positions have the above services, so it is up to the ground handling service provider to manage the allocation of the GSE's needed for each platform and parking positions taking into account their characteristics as well as the characteristics of the aircraft and the services needed (Tabares & Mora-Camino, 2017).

Despite the benefits that ground support equipment offers, the simultaneous driving of several vehicles next to the aircraft as well as their approach to its fuselage, turn the apron into a working environment that represents high levels of risk and danger for workers and aircraft. According to Balk (2007), 26.4% of all aviation incidents can be categorized as ground incidents, of which 84% of these incidents occur when the aircraft is stationary in the parking position. Graph 1 provides an overview of the distribution of accidents in each phase of flight on the ground, and it is easily perceptible the predominance of incidents in the phase when the aircraft is parked and that are directly linked to ground handling services.

The Flight Safety Foundation (n.d.) estimates that 27,000 accidents and incidents occur annually on airport tarmacs, a figure that corresponds to 1 accident/incident for every 1,000 departures. The same foundation, which has been dedicated since 1947 to providing impartial, independent and expert information to the aviation industry, further estimates that accidents/incidents on the ground result in injuries to 243,000 people annually, which corresponds to 9 injuries for every 1,000 departures.

In these types of operations, such as those that take place on the airport apron and which largely involve moving close to the aircraft, the economic situation of the airline and ground handling service provider can be substantially influenced by the effectiveness of operational safety management, since damage to the aircraft represents high repair costs right away, as well as high indirect costs related to flight delays and cancellations due to the accident (Rose et al., 1990; Ek & Akselsson, 2007). Forecasts in the IATA Ground Damage Data Base (GDDB), suggest that by 2030 aircraft accidents on the ground will cost the aviation industry $8 billion (Suidan et al., n.d.).

Figure 2. Example of GSE's in an aircraft turnaround
Source: Ashford (2013)

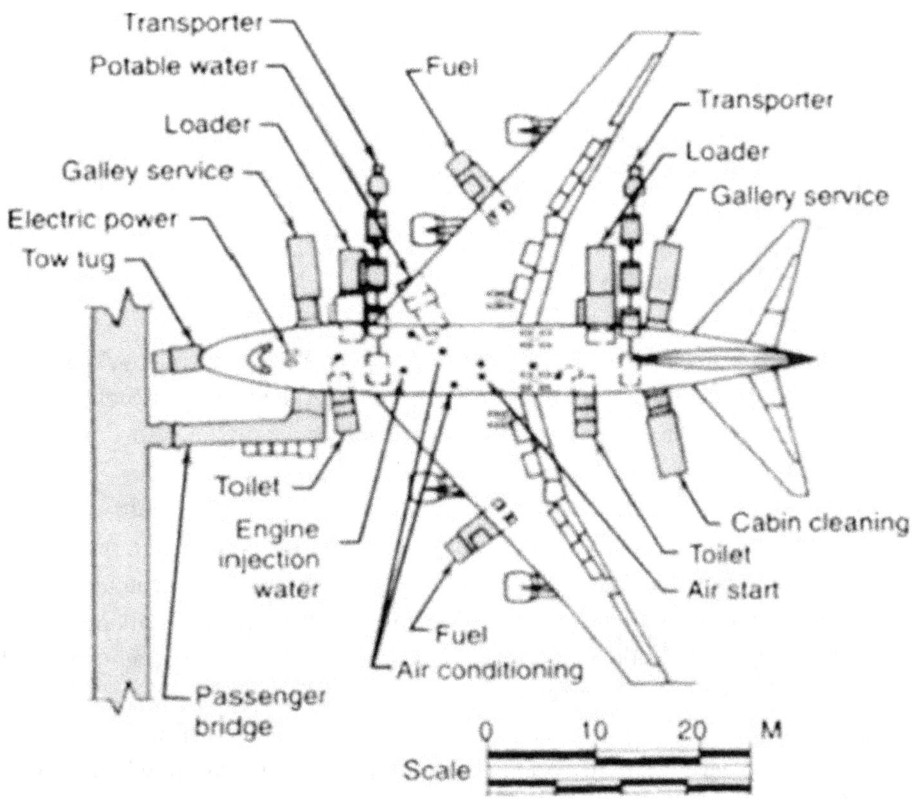

Key Performance Indicators

Schmidberger et al. (2009) point out that given the fact that most ground handling services occur on the airside of the airport, they represent an important variable in the performance levels not only of airlines, but also of airports and the ground handling service providers themselves. That said, due to the overall complexity of the services practiced near the aircraft and their importance to air transport performance levels, İnan (2018) points to the importance of implementing control systems and strategies that monitor the service levels agreed in SLAs, with the aim of increasing efficiency and reducing delays caused by airside ground handling services.

According to Tabares & Mora-Camino (2017) in ground handling services provided on the apron, the main KPI's that can be used to measure performance levels of ground handling service providers are, punctuality of services, meeting Airport Collaborative Decision Making (A-CDM) milestones, the number of accidents, incidents and aircraft damage arising from service provision (Safety), as well as minimizing the number of resources (GSE's, Infrastructure, etc...) required to perform the services. Some performance indicators are somewhat contradictory or at least point in different directions, as is the case of meeting the punctuality of services with the minimization of resources needed to perform the tasks inherent to them.

The management of the work teams allocated to each service is an important variable in increasing performance levels and consequently reducing delays caused by ground handling services. According to Clausen (2010) the providers of these services should take into account the general needs of the tasks to be performed in a given period of time, so that they can allocate the correct number of workers to each task. As not all workers have the same productivity levels when performing their tasks, it is also important that the management and allocation of work teams to each service take into account the experience, the level of training and the set of skills of the workers that will be present in each service, being fundamental to create multidisciplinary work teams to cope with the inherent complexity and volatility of ground handling services to aircraft (Baggia & Rodič, 2010).

Staff Schedule and Allocation

An organization's employee management and planning practices are pointed out by Amaeshi et al. (2013), Ernst et al. (2004), Katou (2008), Rathnaweera (2010), and Savaneviciene & Stankeviciute (2011) as predictive factors of employee satisfaction and retention levels in the organization, directly affecting the organization's performance at various levels.

As indicated by Ashford (2013) ground handling services are subdivided into several areas and are processed at different time periods and airport locations. Due to this complexity, resource management for the performance of ground handling services represents a challenge for ground handling providers (Clausen, 2010). The problem of human resource allocation, work team management and support equipment is addressed by Baggia & Rodič (2010), Cowling et al. (2006), Ernst et al. (2004), Padrón et al. (2015), Saggar et al. (2021) and Ter Mors et al. (2007).

Ernst et al. (2004) stresses the importance of allocating to each task the sufficient number of employees with adequate skills and experience to respond to the complexity of the task to which they are allocated. Ter Mors et al. (2007) solved the problem of equipment allocation in situations where equipment is scarce by using a "First Come First Served" based resource allocation, prioritizing the allocation of equipment to aircraft according to their arrival, in order to minimize flight delay times.

Planning for scaled care teams has implied a number of complex optimization problems (Clausen, 2010). The workload is represented by a list of tasks that must be performed by employees of appropriate qualification to the task requirements. It is common to classify the staffing problem according to time dependencies into strategic, tactical, operational, and real-time planning. Each of these stages are strongly interrelated and different optimization problems must be solved at each of them (Popova, 2015).

Task scheduling assigns tasks to shifts (which can broadly include ramp operators, equipment or resources) for execution on the day of the operation. Task scheduling includes analyzing and responding to operational problems in real time based on flight schedules, aircraft types to be serviced, and also expected airport resources (Clausen, 2010).

Cowling et al. (2006) address the problem of scheduling and allocating work crews to the tasks of the day by dividing the problem into 4 main components:

- Tasks
- Resources
- Skills
- Site requirements

The study by Mahmood et al. (2016) concluded that implementing a system for allocating work crews to tasks based on linear programming systems enables ground handling service providers to significantly reduce total work time, reducing equipment operating costs, personnel costs, gas emissions and noise pollution, as well as reducing traffic on the ramp.

According to Chang and Chen, (2007) integer linear programming is an excellent tool to use in the planning and management of ground handling equipment because it allows to accurately represent the costs, times or distances during operations on the apron.

According to Ernst et al. (2004) the development of the mathematical models involves:

1. A preliminary study aimed at collecting and using historical data to predict the demand for the services subsequently converting this data to the staffing levels required to meet the service standards;
2. Considering the necessary solutions to be implemented in the tool in order to satisfy the existing constraints in the workplace, meeting the organization's objectives and responding to the needs of the employees, also seeking to minimize costs by satisfying the employees' preferences as much as possible;
3. Develop a reporting tool that displays solutions and provides performance reports.

Furthermore, it is essential that in formulating the mathematical models include real constraints that occur in the operations of a rotation (Mahmood et al., 2016). According to Cowling et al., (2006) there are three main groups of constraints: task constraints, resource constraints, and constraints concerning location.

In Mahmood et al. (2016) the authors resort to integer linear programming to simulate and find the best scheme for allocating tasks to scaled care workers including, for example, constraints such as:

- A task must be performed only once;
- Workers move between aircraft until they are back at the operational base where all transportation equipment is parked;
- Workers cannot assist more than one aircraft at the same time;
- An aircraft cannot take off until the ground handling operations for that aircraft are completed. Therefore, all necessary ground handling operations must be performed and completed before the scheduled departure time;
- To ensure that the model calculates the minimum time required to serve aircraft rather than the overall minimum time (i.e., zero), non-negativity constraints must be added.

METHODOLOGY

In the development of this study, a qualitative research line was used, opting for the case study method as a research strategy.

Qualitative research is indicated when the intention is to study "how" questions. This type of research assumes direct, thorough, and prolonged contact of the researcher with real-life situations relevant to the study, describing the data collected in these situations through words, tables, and graphs (Bogdan & Biklen, 1994). This approach places the researcher as the primary instrument and allows for an intense, in-depth, and inductive analysis of the characteristics of the object under analysis, increasing the level of understanding of the meanings that emerge from the observed reality (Quivy & Campenhoudt, 1998).

The case study is one of the most used methods in qualitative research Aires (2015) and consists of the detailed observation of the context of a specific event, and should be considered when the aim is to understand the situation in depth, as well as to understand the meaning of the situation for those involved in it (Quivy & Campenhoudt, 1998).

Data Collection

The collection of information and data should cover the whole situation under study, and, therefore, it is performed through the selection of a set of appropriate methods and techniques, which allow obtaining numerous and detailed data for later processing and analysis (De Brune et al., 1975). In this study, data were mostly collected by the researcher, mainly using the direct observation method, which puts the researcher in contact with the objects that will allow for the immediate perception and recording of the observed data (facts, procedures, behaviors, actions) at a given moment in a given situation (Deshaies, 1997; Roegiers & Ketele, 1993).

In this sense, at least five short turnarounds of Lufthansa, Vueling and Wizz Air were observed, allowing to collect data related to operational procedures, services provided and timings to understand in more detail the aircraft turnarounds of full-service, hybrid and low-cost airlines. Roegiers & Ketele (1993) state that both direct observation and the analysis of certain significant documents provide valuable complementary information for the study of a theme. Thus, the author resorted to document analysis in order to contextualize the case, add information, and validate evidence from other sources. The documentary collection was carried out by reading a set of scientific documents as well as ground handling manuals, especially those written by organizations relevant to air transport, such as the AHM and IGOM written by IATA.

Taking into account the objectives of the study, the site selected for data collection was Lisbon Airport. This choice is directly linked to logistical issues, namely due to the fact that the author, working at the airport, has access to the reserved areas of the airport, thus enabling direct observation of aircraft rotations. However, Quivy & Campenhoud (1998) mention that when using direct observation as a data collection tool, the researcher must, among many factors, be aware of and take into account the context in which the data are observed. In this sense, it is important to highlight that data collection and the results obtained are irremediably linked to the operational context of the Lisbon Airport. An analysis carried out in other airports may present different results, given that the services contracted by airlines are influenced by several factors, such as the operational characteristics of airports and the airlines' own objectives and target audience for each market.

Sample Selection

To correctly investigate the problematic of the study, the process for selecting the sample for data collection had as its main parameter the business model of the airlines, seeking to obtain a sample that had a Full-Service, Hybrid and Low-cost carrier. The selection of the sample also took into account other parameters such as the type of operation (scheduled, charter, etc...) and the type of aircraft operated, seeking in these parameters, unlike the first, to approximate the characteristics of the selected airlines in order to highlight the problem of this study. In this sense, based on the parameters presented above, the following European airlines were selected for the study sample: Lufthansa (Full-Service), Vueling (Hybrid) and Wizzair (Low-Cost). All three airlines operate on a regular basis to Lisbon Airport with the same type of narrow-body aircraft of the Airbus A320 Family, thus meeting the "requirements" selected by the author.

Sample characterization

Lufthansa

At the end of 2021, the German flag carrier operated flights to 183 airports in 71 countries (Star Alliance, 2021), and currently has 35,738 employees and a diversified fleet of 389 aircraft divided into 27 Boeing 747 aircraft, 31 Bombardier CRJ, 26 Embraer and the remaining aircraft from the manufacturer Airbus (A320 Family, A330, A340, A350, A380), offering fares to passengers with very complete options in first class, business class, premium economy and in economy class (Lufthansa Group, 2022).

In terms of ground handling services, Lufthansa has a subsidiary in Portugal, Lufthansa Ground Services Portugal (LGSP), a company which is 100% owned by the airline and which self assists its operations and the operations of the companies of the Lufthansa Group in the services of Station Administration, Passenger Check-in, Ground Handling Supervision and Ticket Sales.

Vueling

Currently the airline operates scheduled passenger air transport flights (and exceptionally charter flights), to almost 130 destinations with a fleet of 117 young and constantly renewed Airbus A319, A320 and A321 aircraft that ensure the safety of its operations, comfort and passenger experience, while minimizing fuel consumption, CO2 emissions and personnel training and aircraft maintenance costs due to the standardization of the fleet (IAG, 2022).

Its business model is based on low cost combined with high passenger service, offering a hybrid operation between the pure low cost model and the full service model of legacy airlines, providing differentiated fares within the economy class spectrum, namely the Basic fare, Optimum fare and TimeFlex fare, which include several services such as the possibility of seat choice, priority boarding, free cancellation, checked luggage, exclusive counters (Díez, 2021).

Wizz Air

The airline currently employs around 5500 staff and operates scheduled commercial flights to 190 airports in 51 countries including Europe, North Africa, the Middle East and Asia. Its fleet consisting of

153 A320 and A321 aircraft (equipped with new generation engines) with an average age of 5.2 years, has enabled Wizz Air to rank among the top ten safest low-cost airlines of 2019 (WizzAir, n.d.).

Wizz Air's mission is to offer affordable travel opportunities for the lowest possible price by creating, through an operation consisting mostly of essential services, equal value for all passengers, while also adopting a conscious stance on its environmental responsibility. The company continuously seeks opportunities to expand its destination network and provide friendly and fast customer service at low costs, and is committed to achieving the lowest cost base in the Central and Eastern Europe region with a mix of primary, secondary and regional airports (WizzAir, n.d.).

RESULTS AND INTERPRETATION

Case Study

To facilitate the characterization of the tasks of the Lufthansa, Vueling and WizzAir turnarounds, the analysis of the turnarounds was divided based on the main activities of a turnaround proposed by Padrón and Guimarans (2019) namely, In-blocks, disembarkation, unloading of holds, potable water and toilets service, fuel, cabin cleaning, loading of holds, boarding, pushback and off-blocks.

Although Padrón and Guimarans (2019) include Catering service in the main activities of a turnaround, in the analysis performed on the ground handling services of the three airlines at Lisbon Airport, no Catering service was provided. For this reason the Catering service is not addressed in the present analysis.

General Characteristics

The table below aims to present in a summarized way a set of relevant characteristics, which will serve as a basis for the more detailed analysis of the turnaround operational characteristics observed at each airline. Factors such as seat density and type of aircraft holds, cargo transport, the ground handling providers with whom they have a contract for aircraft handling and the terminal where they operate are addressed here in order to introduce the analysis of the differences between the three airlines' business models.

Aircraft Type

The A319 aircraft, the smallest of the A320 family, is operated by Lufthansa and Vueling but, in addition to being the smallest aircraft in their fleets, it is also the least used aircraft in their flights to Lisbon Airport. For example, in an analysis carried out by the author on the ANA Aeroportos "Flight4Staff" platform, which is intended to provide flight information to the Portuguese airports' employees, it was found that in August Lufthansa and Vueling did not operate any A319 to the Lisbon Airport. For this reason, and due to the fact that Wizzair does not operate this model, the A319 is not included in the analysis of this study.

With regard to aircraft characteristics, through the analysis of data provided in the same platform it was found a higher density of seats on Wizz Air and a lower density on Lufthansa. Vueling's seat density is between the two companies' seat densities.

Table 1. General characteristics of Lufthansa, Vueling and Wizz Air

		Lufhtansa	Vueling	Wizzair
General Characteristics	Aircraft Type	Family A320	Family A320	Family A320*
	A320 ceo	174 seats	180-186 seats	186 seats
	A320 neo	180 seats	186 seats	189 seats
	A321 ceo	200 seats	212 seats	230 seats
	A321 neo	215 seats	220 seats	239 seats
	Holds	Containerized	Bulk	Bulk
	Cargo/Mail Transport	Yes	No	No
	Ground Handler	Groundforce	Groundforce	Portway
	Terminal	Terminal 1	Terminal 1	Terminal 2

* Doesn't operate A319.

Type of Holds

When loading and unloading aircraft, one of the characteristics that most impacts the way these tasks are performed is the configuration of the aircraft holds. Despite operating the same models, Lufthansa operates aircraft with containerized holds, while Vueling and Wizz Air operate aircraft with bulk holds.

Cargo/Mail Transport

Of the three airlines, only Lufthansa transports (at least on a regular basis) cargo in its commercial flights to Lisbon Airport. This transport is mostly done in ULD's that, although they are loaded and unloaded in the aircraft by Groundforce, their processing outside the stand is done by another ground handling company, Portway.

Ground Handler

Ground handling services for Lufthansa and Vueling operations are provided by Groundforce. On the other hand, ground handling services for Wizz Air operations are provided by Portway.

At Lisbon Airport, Groundforce is licensed by ANAC to provide ground handling services to third parties in all subcategories of Category 1 "Ground administrative assistance and supervision"; Category 2 "Passenger assistance"; Category 3 "Baggage assistance"; Category 4 "Cargo and mail assistance"; Category 5 "Ramp operations assistance"; Category 9 "Air operations assistance and crew management"; Category 10 "Ground transportation assistance". In addition to these categories, the company is also licensed under subcategory 6.1 for aircraft exterior and interior cleaning, lavatory service and water service, as well as subcategory 8.4 which allows it to request or reserve on behalf of the airline a parking space and/or hangar for maintenance.

At Lisbon Airport, Portway is licensed by ANAC to provide ground handling services to third parties in all subcategories of Category 1 "Ground administrative assistance and supervision"; Category 2 "Passenger assistance"; Category 3 "Baggage assistance"; Category 4 "Cargo and mail handling"; Category 5 "Ramp operations assistance"; Category 9 "Air operations assistance and crew management"; Category 10 "Ground transportation assistance". Portway is also licensed in two subcategories of category 6, namely 6.1 and 6.3, which allows it to provide external and internal aircraft cleaning, lavatory and water service, as well as packing the cabin with its own equipment and storing the same.

Terminal

Lisbon Airport has two passenger terminals, Terminal 1 and Terminal 2.

Terminal 1 is the main terminal, and serves the departures and arrivals of most airlines operating in Lisbon including Lufthansa and Vueling. Terminal 2 is dedicated only to the departure flights (disembarkation is performed in Terminal 1) of low-cost airlines, including Wizz Air, and is a terminal that has a simplified service, which makes the whole boarding process faster and more convenient.

In Blocks

The term "In-blocks" refers to the time when the procedure of placing chocks on the aircraft wheels is completed. This procedure is intended to ensure that the aircraft is safe and do not move during the time that services are being provided. Usually 6 chocks are placed, two on the front gear and two on each side of the center gear. Nevertheless, the author includes to this activity the performance of two other tasks that take place immediately after chocking. These are, signaling the aircraft at its ends and engines and connecting the GPS/GPU to provide external power to the aircraft. The completion of these tasks, although there may be some differences depending on the airlines' safety requirements or operational issues such as equipment failure, are fairly standardized tasks, with these three airlines taking 2-4 minutes to complete this set of tasks.

Disembark

After the aircraft is secure, i.e. with the chocks on the wheels and properly marked, the disembarkation procedures can be started. In the case where aircraft are in positions with telescopic bridges, the operator of this equipment can begin his approach to the door 1 of the aircraft as soon as it is secure (chocks). At the Lisbon Airport, Lufthansa and Vueling position mostly in stands next to Terminal 1 that have telescopic bridges, thus the disembarkation of passengers is done only through door 1. In Lufthansa rotations, a stair to allow ground services access to the aircraft cabin is leaning against door 2.

Despite positioning mainly in stands with telescopic bridge, due to management reasons and restrictions related to airport issues, Lufthansa and Vueling sometimes park in remote stands where, as in all Wizz Air turnarounds, mobile stairs and buses are used to disembark and transport passengers to Terminal 1. In this case, all three airlines allocate the same resources, two stairs and three passenger buses for A320 and four passenger buses for A321.

Unloading

The unloading of aircraft holds is greatly influenced by their typology and can be started as soon as the tasks mentioned in the first activity are completed. However, in remote positions, before starting the procedures for unloading the holds, the ramp staff should give priority to the staircase pull-down so that the passenger disembarkation can be started.

Although Lufthansa's and Vueling's holds are different, their hold unloading times are quite similar at around 20 minutes. In the case of Wizz Air the unloading time is around 10 minutes.

The similarities between Lufthansa and Vueling in this parameter are greatly influenced by the resources used as well as the amount of baggage allowed per passenger on their flights. While on the basis of the fares charged by both airlines, Vueling carries less baggage compared to Lufthansa, the fact that baggage is unloaded manually makes unloading as time-consuming as unloading large amounts of baggage and cargo carried by Lufthansa. Unloading holds on Lufthansa are also influenced by a number of variables that can make unloading difficult or easy, increasing or decreasing the time it takes to unload. For example, space-constrained parking positions that make it difficult to drive the loader and transfer the ULDs to the dollies can increase unloading time. On the other hand, the allocation of more experienced ramp staff who have more training and knowledge in driving the equipment, can increase the speed in unloading the holds to containers.

Comparing Vueling and Wizz Air, both have the same type of holds, but unloading times differ by 10 minutes. This difference is mainly created by two factors, the number of human resources allocated to each turnaround and the amount of luggage carried.. Although, by direct observation, it is difficult to accurately determine the amount of baggage each one carries, the author verified, in all turnarounds observed, the amount of baggage for two baggage carts on Wizz Air and three on Vueling on the A321. In addition to carrying less baggage, in Wizz Air turnarounds, one more element is allocated to the team of ramp staff that assist the aircraft, and at Vueling 3 ramp operators are allocated while at Wizz Air 4 operators are allocated.

Loading

The loading of holds can be started as soon as their unloading is complete. In general, the time and manner of loading the holds is quite similar to the unloading, being influenced by the same variables, such as the type of holds, the amount of baggage, the resources allocated, among others.

However, unlike unloading, where the passenger disembarkation has little impact on how the aircraft is unloaded, the loading of holds is a procedure that depends heavily on the passenger boarding due to operational and security issues. One example that increases hold loading time is the fact that the holds have to remain open, or have to be opened after loading, if a passenger does not board or is not boarded on the aircraft, so that the baggage of the passenger who has not boarded can be removed for security reasons.

Baby strollers and wheelchairs delivered to the aircraft door also require the loading to be extended to the end of boarding. In the case of Lufthansa, the front and rear holds are usually closed once the loading of ULDs is complete, leaving only the bulk compartment open where loading is done via the conveyor belt.

Another example that can affect hold loading time is transfer baggage from connecting flights. This case applies to Lufthansa because of the code-shares the company has with other airlines that allow passengers of other airlines to transfer from a flight of one airline to a Lufthansa flight without having to check in and check their hold baggage. In this way baggage from flights with short connections can be delivered later in the rotation or even wait until the passenger boards the aircraft to be loaded onto it.

If on the one hand Lufthansa has a more complex loading structure due to its network and code-share operation, on the other hand, the company uses the BRS system that allows the ground handling service provider to locate the baggage at the airport as well as locate the baggage in the hold compartments, thus increasing the speed with which baggage is removed, should it need to be removed.

This period of time when the loading of the holds is in "stand-by" is represented by dashes in the schedule (see figure x) and as mentioned above, its time is highly dependent on the boarding of passengers. In this aspect Wizz Air presents, similarly to the unloading time, a much shorter loading time compared to Lufthansa and Vueling due to its way of boarding that allows earlier identification of the need to remove luggage and allows to decrease the time that the ramp staff have to wait for last minute luggage due to the earlier start and end of boarding.

Potable Water and Toilet Service

Of the three airlines, only Lufthansa performs potable water and toilet services at Lisbon Airport. Both are done by Iberlim, a company which at Lisbon Airport is licensed by ANAC to provide third parties with category 6 ground handling services for cleaning assistance and aircraft services, and is therefore able to perform aircraft exterior and interior cleaning, lavatory service, drinking water service, air conditioning, cabin heating, aircraft snow and ice removal, and the packaging of cabin equipment and its storage.

On Lufthansa turnarounds, these services are usually performed between the fifth and fifteenth minute of the turnarounds, with one car for the potable water service and another for the toilets service. Each service lasts about 5 minutes, from the time the equipment is reclined to the time it is released, and the duration may vary a little more or less depending on the needs of the aircraft. The advantage of these services is the fact that they are performed from under the tail of the aircraft, thus not affecting the other tasks and consequently the turnaround time.

Although in the turnarounds of Vueling and Wizz Air the potable water and toilets service are not performed on a regular basis, this does not imply that in occasional situations, if necessary, these services are not performed to their aircraft.

Cabin Cleaning

Similarly to the potable water and toilet services, of the three airlines, only Lufthansa performs regular cabin cleaning in its turnarounds. This service is also performed by Iberlim that starts the cleaning tasks as soon as the passengers disembark, taking about 20 minutes to complete the cabin cleaning, which also includes the collection of garbage from the arrival flight.

On Vueling and Wizz Air the cabin cleaning is done by the crew members, who perform a superficial cleaning removing mainly objects and garbage that are scattered around the cabin. This simplified procedure allows the two airlines to start boarding passengers earlier than Lufthansa.

Although the cabin cleaning on Wizz Air is performed by the cabin crew, the garbage collection from the incoming flight is performed by the company "Spitfire" which at Lisbon Airport is licensed by ANAC to provide category 2 ground handling services to third parties, which includes the provision of departure, arrival, transit or transfer passenger handling services, such as ticket control, travel documents, baggage registration and transport to the sorting systems and as well in the subcategories 6.1 and 6.3. allowing it to perform external and internal aircraft cleaning, toilets service, water service, and also the conditioning of the cabin with its own equipment and storage of the same. Due to the nature of the service is only the collection of garbage bags prepared by the crew, this service is performed very quickly and is done, in all observed turnarounds, between the twenty and twenty-fifth minute of the turnaround.

Fuel

The fuel truck can approach the aircraft as soon as it is safe, and the operator can also start connecting the truck hoses to the airport's fuel distribution system pit and to the aircraft's fuel inlets, starting fueling as soon as safety conditions are met.

The analysis of fuel duration in this paper is carried out in a simple way because it depends mainly on factors related to the amount of fuel requested by the aircraft captain and not by variables of the fuel supplier, making its analysis a little imprecise because, even on the same route the amount of fuel may be different due to operational variables that are not directly linked to ground handling services. Nevertheless, the author identifies two factors that impact the amount of fuel that is filled, they are the distance of the route that the aircraft will operate and the takeoff weight.

Boarding

Once the fuel and cabin preparation services are complete, the boarding of passengers onto the aircraft can begin. In the cases of Lufthansa and Vueling, this procedure is mostly done through the telescopic bridges in Terminal 1. However, if the aircraft is positioned in a remote stand, boarding is done following the same procedure (in reverse) as disembarking in remote positions.

At Wizz Air, passenger boarding happens mostly on foot, where passengers move from terminal 2 to the aircraft following the guidance of an operator and the paths outlined on the sidewalk. As mentioned earlier, although the company has a higher density of seats in its aircraft, Wizz Air has shorter boarding times for passengers, starting boarding between the twenty-fifth and thirtieth minute of the turnaround taking about 12 minutes as a rule. Lufthansa and Vueling have longer boarding times, around 20 minutes, and in addition, due to longer preceding activities, they start boarding later. Vueling starts boarding around the half hour of the turnaround and Lufthansa starts around the thirty-fifth minute of the rotation.

The fact that Wizz Air has shorter boarding times is directly linked to the characteristics of terminal 2 at Lisbon Airport. This terminal has a post-boarding gate area that allows the ground handling provider to perform, as soon as the aircraft lands, the usual passenger control procedures at the boarding gate. In this way, with passengers just waiting for the aircraft to be ready to be boarded, their boarding on the aircraft ends up being much faster because the usual boarding procedures have been performed beforehand. On the other hand in Terminal 1 boarding at the gate usually only starts about 5 minutes before the aircraft is allowed to board, because there is no post-boarding gate area that allows passengers to wait.

Pushback and Off-blocks

The final phase of the turnaround comprises, in a first step, the unhooking of the stairs and/or telescopic bridges and disconnection of the external power source. Next, the connection of the pushback tractor can be performed, and then the removal of the safety cones and wheel chocks when the aircraft is cleared to leave the parking position.

This last set of tasks is normally completed in 5-10 minutes, and it has been observed that at Vueling they are accomplished slightly faster due to the fact that the telescopic bridge is maneuvered by an operator outside the team allocated to the aircraft, giving availability to the thre team members to simultaneously disconnect the GPU/GPS, connect the pushback tractor, and remove the signage from the aircraft. Because this is the last set of tasks to be performed, these tasks are hopelessly dependent on the completion of the previous activities.

In total, Wizz Air completes the turnaround tasks between 35-45 minutes, Vueling between 45-55 minutes, and Lufthansa between 55-65 minutes.

Figure 3. Service schedule on Lufthansa, Vueling and Wizz Air turnarounds

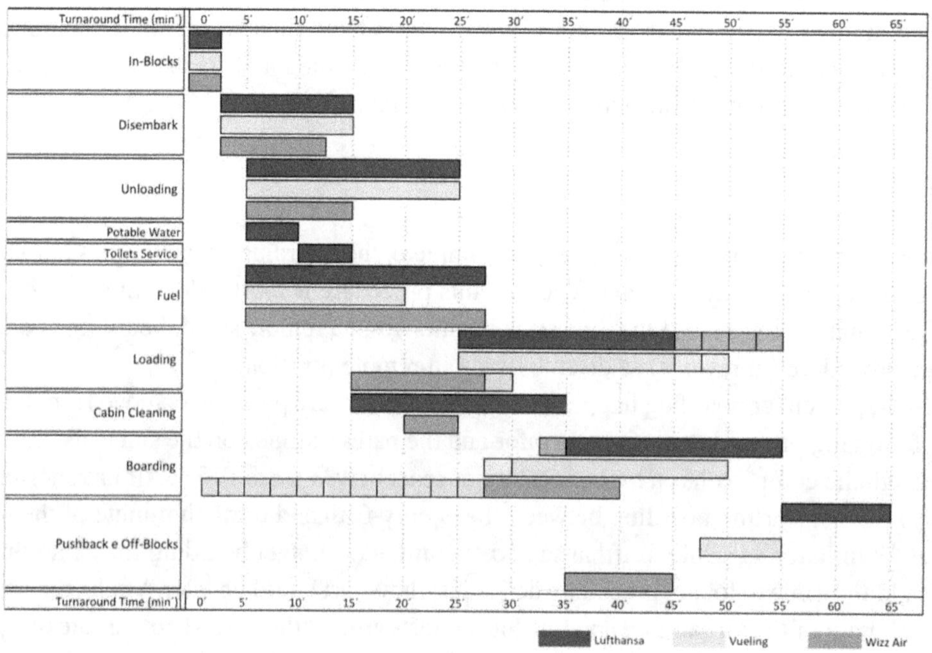

Impact of aircraft turnaround characteristics on ramp operators' teams

As Padrón and Guimarans (2019) point out, the variables of a turnaround depend on several factors, including the type of aircraft (family, size, number of seats, hold typology, etc.) and the parking position (near the terminal or in a remote position). As noted in the previous analysis of the three airlines, although they operate the same type of aircraft, the characteristics of their turnarounds are different. Because they have different characteristics, the way the tasks are performed as well as the requirements and resources needed to perform them are also different, and it is up to ground handling service providers to manage and allocate the resources needed to meet the needs and service levels of the airline that hires them.

Table 2. Main tasks of an aircraft turnaround

		Task
In-Blocks	A	Chock the Wheels
	B	Signalize the Edges
	C	Connect GPU/GPS
Desimbark	D	Stairs Approach - Door 1
	E	Stairs Approach - Door 2
Unloading	F	Open the Holds
	G	Pull Over the Conveyor Belt
	H	Unload the Bulk
	I	Unload the Holds
Loading	J	Load the Holds
	K	Load the Bulk
	L	Remove the Conveyor Belt
	M	Close the Holds
Boarding	N	Stairs Disapproach - Door 2
	O	Stairs Disapproach - Door 1
Off-Blocks	P	Disconnect the GPU/GPS
	Q	Remove Chocks
	R	Remove the Signage

That said, the author now proposes three schemes that aim to demonstrate the ground handling tasks performed by the aircraft handling ramp operator teams based on the turnarounds observed in the case study. Table 1 summarizes the tasks of a turnaround to be performed by aircraft handling ramp operators in a remote position (no telescopic bridge) and is divided into 6 main activities, the in-blocks, landing, unloading, loading, boarding and off-blocks. In the elaborated schematics, each arrow represents a ramp operator, and each ball represents a task identified according to table 1. The start of the ramp operators' tasks is given from the moment the aircraft beacon light turns off and the end of the tasks is given when the aircraft is ready to be towed by the pushback. Depending on the type of pushback allocated to the task, it may be necessary for a ramp operator from the aircraft assistance team to accompany the pushback in order to assist this task.

It should be noted that the examples below, although they can be taken as standard models, represent only a hypothetical performance of the tasks of ramp operators allocated to the aircraft. Depending on the available resources and conditions at each moment, the performance of the tasks on a turnaround from the same airline may present differences. In summary, the presented schemes try to demonstrate and offer the reader a better understanding of the relationship between the operational characteristics of aircraft turnarounds and the accomplishment of the tasks performed by the ramp operator teams.

In addition, it is important to highlight that tasks performed by the baggage terminal team operators and tasks such as pushback and, if necessary, the supply of air conditioning through the ACU - Air Conditioner Unit or of compressed air to start engines through the ASU - Air Starter Unit, are not included in this set of tasks because they are tasks that are performed by specific teams of ramp operators allocated solely to perform those tasks. Other aircraft support services, such as fuel, cabin cleaning services, potable water, and dumping, are also not considered in this model.

Lufthansa

Figure 4 shows the route of the two ramp operators allocated to the Lufthansa aircraft turnarounds.

Figure 4. Example of task distribution with 2 ramp operators allocated to a Lufthansa turnaround

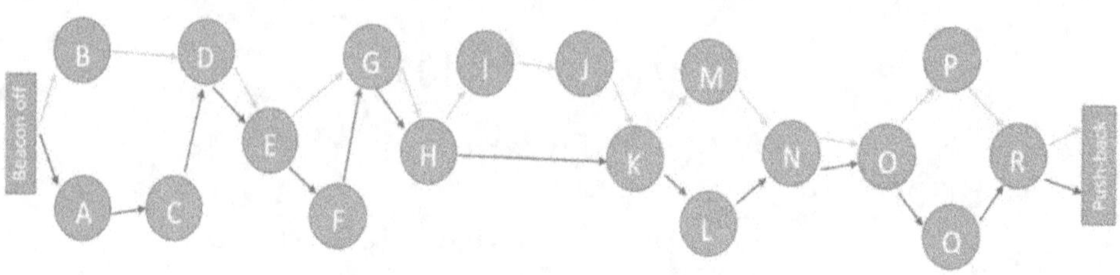

The operators start the aircraft turnaround tasks by placing chocks (A) and signaling the aircraft (B) in order to ensure the safety. Upon completion of these two tasks, one element proceeds to connect the GPU/GPS (C) that provides external power to the aircraft while the other element prepares for the approach of the passenger stairs to the aircraft door 1 (D). Once the GPU/GPS connection is complete, the operator assists the approach of the stair maneuvered by the other element by performing marshalling. The same procedure is then performed for the stair to aircraft door 2 (E). After the stairs have been placed, an operator begins to open the holds (F) while the other operator prepares for the approach and the placing of the conveyor belt to the "Bulk" hold (G), which, as with the stairs, is done with marshalling. With all the holds open and the belt against the "bulk", the operators start the unloading of the holds. Initially the two ramp operators unload the baby strollers and baggage labeled with "Delivery to the Aircraft (DA)" that are carried in the bulk (H). After the DA baggage is delivered, one of the operators begins unloading the rear and front hold (I), using a loader to unload the ULD's and pallets that hold the baggage, cargo and mail carried by the airline. The use of the loader allows the unloading of luggage to be carried out by only one operator, leaving the other operator responsible for finalizing the unloading of any luggage, cargo and mail that needs to be unloaded from the bulk. The loading (J) of the departure ULD's is started as soon as the holds are completely unloaded and checked by the LSGP supervisor present in the rotation (if not present, the board is responsible for checking). The "bulk" hold loading (K) must wait for the operator loading the ULD's to finish the task and be available to load the conveyor belt with the baggage already prepared and collected by the other element. The unstrapping of the belt (M) and closing of the holds (N) is performed as soon as the loading of the aircraft is completed. After that, the two elements perform the unhooking of the door ladder 2 (O) and then door 1 (P). The preparation of the aircraft for off-blocks begins with the crew indicating to the operators that the GPU/GPS (P) can be disconnected from the aircraft. When the pushback vehicle, maneuvered by a specialized operator, is connected to the aircraft, the chocks (Q) and signaling (R) can be removed thus completing the rotation tasks.

Taking into account the safety requirements and operational restrictions inherent to ground handling tasks, with a team of two operators allocated to the aircraft turnaround, the performance of simultaneous tasks is more restricted, since there are tasks that to be completed require the teamwork of two operators, thus postponing or putting other tasks on stand-by.

The most "critical" case is the unloading of bulk which implies the postponement of the unloading of ULD's and may cause delays in the delivery of luggage to passengers or in the case of luggage transfer there is the loss of connection to the other flight. Cases like these have costs and reprisals for the service provider due to the costs associated with the resolution of problems and failure to meet the times agreed upon in the SLA - Service Level Agreement.

Vueling

Figure 5 shows the path of the three ramp operators allocated to Vueling aircraft turnarounds.

Figure 5. Example of task distribution with 3 ramp operators allocated to a Vueling turnaround

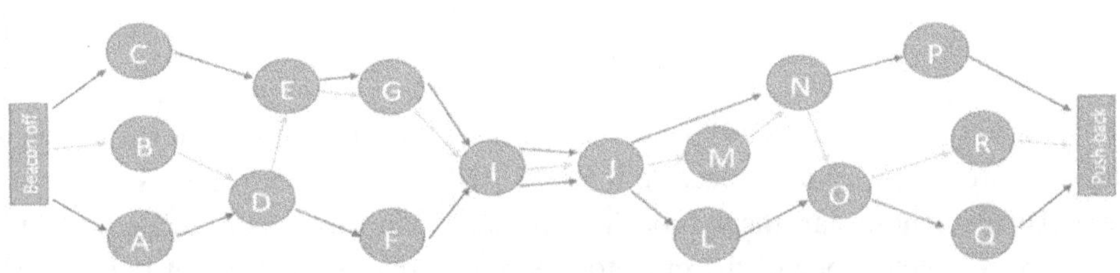

With a team of three ramp operators allocated to the aircraft turnaround it is possible to simultaneously perform the tasks of chocking (A), signaling the aircraft (B) and connecting the GPU/GPS (C). This is followed by the passenger stairs at door 1 (D) and door 2 (E). The approach of the two stairs can be performed simultaneously, however, the operator who maneuvers the stair at the door 2 must wait for the completion of the marshaling of the ladder at gate 1, in order to have someone to do the marshalling. While the approach of the stair at gate 2 is in progress, the operator who finished the stair approach in the door 1 proceeds to open the aircraft holds (F), allowing time for the other two elements to finish the stair approach and perform the approach of the conveyor belt (G) to the aircraft holds. In the case of Vueling, due to the fact that the company carries smaller amounts of baggage (due to the fares charged) and because it does not carry cargo and mail, it is common not to carry baggage in the hold "Bulk", stowing the baggage only in the front and/or rear hold of the aircraft. For this reason, the hold "Bulk" unloading and loading tasks (H and K) are not represented in the diagram. However, in order to avoid occurrences, operators must visually check all aircraft holds to ensure that they unload all baggage in the aircraft holds. The unloading (I) and loading (J) of baggage from the holds is done in bulk at Vueling, so it requires teamwork of the three operators, with two people inside the hold and one at the bottom of the conveyor belt. After loading is complete, two operators perform the disapproach of the belt (L) and the closing of the holds (M) while the third operator prepares the disapproach of the stair in door 2 (N) which will require marshalling by one of the other two operators. The same requirements apply for the disapproach of the ladder at door 1 (O). The remaining tasks, namely disconnecting the GPU/GPS (P), removing the chocks (Q) and the signaling (R) can in this case, where there are 3 operators, be performed simultaneously or distributed among the three operators.

Wizz Air

Figure 6 demonstrates the path of the four ramp operators allocated to Wizz Air aircraft turnarounds.

Figure 6. Example of task distribution with 4 ramp operators allocated to a Wizz Air turnaround

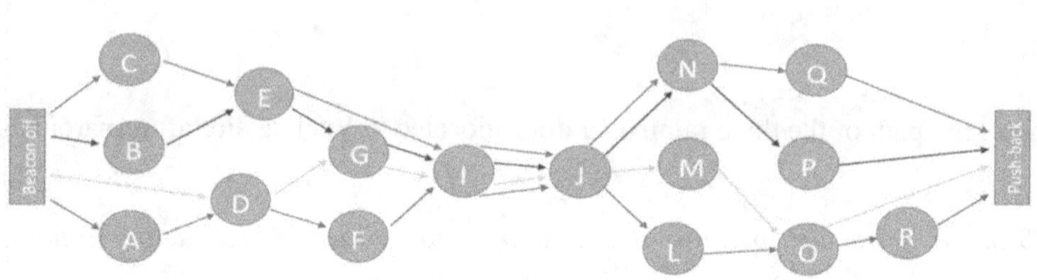

With a team of four ramp operators allocated to the aircraft turnaround, the tasks of placing chocks (A), signaling the aircraft (B) and connecting the GPU/GPS (C) are performed simultaneously, also allowing the fourth element to start the approach of the stair to the door 1 of the aircraft (D). After the In-blocks tasks are completed, one of the operators performs the marshalling of the stair approach to door 1, while the other two operators approach the stair to the door 2 (E). As soon as the operator marshalling the stair at door 1 is free, he proceeds to open the aircraft holds (F), while two other operators start approaching the conveyor belt (G) to the aircraft holds. In terms of load control, Wizz Air has the same characteristics as Vueling, carrying bulk baggage only in the front and/or rear hold of the aircraft. Although the unloading (I) and loading (J) of the holds can be performed by only three operators, on

Wizz Air all four operators are do the unloading and loading, with two operators remaining in the hold and two on the bottom of the conveyor belt. After loading is complete, the disapproach of the belt (L), closing of the holds (M), and the disapproach of the stair on door 2 (N) can be performed at the same time due to the number of operators available. The disapproach of the stair in door 1 is performed as soon as two operators complete the previous tasks, and the remaining tasks, namely, the disconnection of the GPU/GPS (P), the removal of the chocks (Q) and the signaling (R) can be performed simultaneously or distributed among the operators.

The allocation of more staff to these tasks and the fact that Wizz Air carries smaller amounts of baggage, enables the airline to have shorter unloading and loading completion times compared to Lufthansa and Vueling.

CONCLUSION

The development of the proposed theme focused initially on airlines' business models and ground handling services in general, directing, as the work was developed, its attention to more specific issues of these services, with main focus on the impact of airlines' business models on aircraft turnarounds and on the impact of their characteristics on the tasks performed by ramp operators' teams in aircraft handling. This contributes to an increased understanding of the impact of airline business models on ground handling services.

The results of the case study corroborate the data suggested in the literature review, showing a more complete and higher cost ground handling service structure at Lufthansa (FSC) and a more basic cost and ground handling service structure at the operational level at Wizz Air (LCC). At Vueling (Hybrid) it is noticeable when comparing with Lufthansa and Wizz Air that the service and operational cost level is in a "middle" zone with typical FSC and LCC characteristics.

Lufthansa shapes ground handling services in order to add value to its product, focusing above all on differentiation through personalized service where the company's presence and investment in ground handling services is evident, having its own subsidiary, LGSP which performs the services of Representation, Station Management, Passenger Handling, Plaid Service Supervision and Ticket Sales. In addition, Lufthansa has a larger number of services provided to passengers and aircraft as well as a larger number of providers, Groundforce in aircraft handling, Portway in cargo and mail handling and Iberlim in cleaning services.

Wizz Air shapes ground handling services so as to minimize the cost structure as well as the time the aircraft is on the ground, reducing and simplifying passenger handling services and aircraft handling services as much as possible. The airline operates in terminal 2, which offers fewer amenities and a more basic service to passengers due to more basic facilities when compared to terminal 1 facilities, but which allows the airline to increase its operational efficiency mainly through the boarding process where, due to the characteristics of the terminal, it is possible to open the boarding gate prematurely (sometimes before the aircraft has landed) reducing boarding times which contribute, together with other simplified processes such as manual and bulk baggage processing and no cabin cleaning, drinking water replenishment and toilet services at Lisbon Airport, that the company has a low cost structure and shorter turnaround times.

Vueling shapes ground handling services through a symbiosis between a differentiation strategy in services directly linked to passengers and a cost reduction strategy in direct aircraft handling services.

The company offers passengers, in pre-flight and post-flight, some facilities typical of FSC companies such as the operation in the main terminal which offers passengers a better airport experience, where they can enjoy restaurant and leisure areas, better access and also, in general, more comfortable procedures, such as boarding/disembarkation by telescopic bridge avoiding that they are exposed to adverse weather conditions, noise and hazards arising from the operations that take place around the aircraft. However, in aircraft handling operations, the airline opts for a cost reduction strategy with simpler operational procedures, such as manual and bulk baggage processing, also opting to perform only the essential services to prepare the aircraft between the arrival flight and the departure flight, giving up services such as potable water service, toilets and cabin cleaning at Lisbon Airport.

The tasks related to the unloading/loading of holds, cabin cleaning and the disembarkation/embarkation of passengers are the tasks which present the greatest differences between the three business models, leading to the conclusion that the strategy of creating value and increasing the competitive advantage of airlines by means of ground handling services is based above all on these services and their adaptation to the airline's business model.

Study Limitations

The confidentiality of the Service Level Agreements and of most of the documents related to airline operations represented a barrier to the collection and analysis of reliable and accurate data describing the intended and agreed service levels with the ground handling service providers for the handling.

In view of the difficulty in obtaining documents that would provide valid data for the achievement of the study's objectives, the author opted, benefiting from access to the reserved areas of Lisbon Airport, to carry out data collection based on direct observation, being aware of the limitations and disadvantages that this method has, such as, for example, the time required for observation in order for data collection to be solid, as well as the restriction in the quantity and services that can be observed at the same time.

Thus, and being that the ground handling services expand from the parking position to the passenger and baggage terminals, the author decided to exclude the in-depth analysis of ground handling services performed in the passenger terminal and baggage terminal, focusing his main analysis on the parking positions and the services directly linked to the aircraft.

Future Research Directions

There are several issues that can be further addressed in future research, both in analyzing the impact of business models on ground handling services, as well as the relationship between the characteristics of ground handling services and the work of ramp operator teams.

A study with a larger sample of airlines will contribute to a better understanding of the impact of airlines' business models on ground handling services. Likewise, an approach that extends its analysis to airline services at the passenger terminal will benefit the understanding of how airlines shape ground handling services in order to create and enhance their competitive advantage.

In studying the relationship between the characteristics of ground handling services and the work of ramp operator teams, including variables to the examples, such as an average completion time for each task across different airlines, is one measure that will provide a better understanding of the impact of the characteristics of a turnaround on the teamwork of ramp operators. An approach based on a macro perspective, including the teams of baggage terminal and passenger terminal operators, is also something to consider in future research.

REFERENCES

Adler, N., & Golany, B. (2001). *Evaluation of deregulated airline networks using data envelopment analysis combined with principal component analysis with an application to Western Europe*. www.elsevier.com/locate/dsw

Aires, L. (2015). Paradigma qualitativo e práticas de investigação educacional [Qualitative paradigm and educational research practices.]. Research Gate.

Airport Research Center. (2009). *Study on the Impact of Directive 96/67/EC on Ground Handling Services 1996-2007 Final Report Content*. Airport Consultants. www.airport-consultants.com

Akbar, Y. H., & Kisilowski, M. (2020). To bargain or not to bargain: Airlines, legitimacy and nonmarket strategy in a COVID-19 world. *Journal of Air Transport Management, 88*, 101867. doi:10.1016/j.jairtraman.2020.101867 PMID:32834693

Alves, C. F., & Barbot, C. (2007). Do low cost carriers have different corporate governance models? *Journal of Air Transport Management, 13*(2), 116–120. doi:10.1016/j.jairtraman.2006.11.001

Amaeshi, F., Lamba, S., & Choudhary, N. (2013). Impact Of Hrm Practices On Organizational Commitment Of Employees. *International Journal of Advancements in Research & Technology, 2*.

Andersson, K., Carr, F., Feron, E., & Hall, W. D. (2000). *Analysis and Modeling of Ground Operations at Hub Airports*. MIT.

Ashford, N. (2013). *Airport operations*. McGraw-Hill.

Assaf, A. G., & Josiassen, A. (2012). European vs. U.S. airlines: Performance comparison in a dynamic market. *Tourism Management, 33*(2), 317–326. doi:10.1016/j.tourman.2011.03.012

Atems, B., & Yimga, J. (2021). Quantifying the impact of the COVID-19 pandemic on US airline stock prices. *Journal of Air Transport Management, 97*, 102141. doi:10.1016/j.jairtraman.2021.102141 PMID:34566280

Baggia, A., & Rodič, B. (2010). *Dynamic Airport Ground Crew Scheduling*. Research Gate. https://www.researchgate.net/publication/256789547

Balk, A. D. (2007). *Safety of ground handling*. NLR. www.nlr.nl

Barbot, C., Roberto Frias, R., Costa, Á., & Sochirca, E. (2008). *Airlines performance in the new market context: a comparative productivity and efficiency analysis*. MIT.

Barrett, S. D. (2004). How do the demands for airport services differ between full-service carriers and low-cost carriers? *Journal of Air Transport Management, 10*(1), 33–39. doi:10.1016/j.jairtraman.2003.10.006

Bauer, L. B., Bloch, D., & Merkert, R. (2020). Ultra Long-Haul: An emerging business model accelerated by COVID-19. *Journal of Air Transport Management*, *89*, 101901. doi:10.1016/j.jairtraman.2020.101901 PMID:32839647

Bogdan, R., & Biklen, S. (1994). Investigação Qualitativa e Educação [Qualitative Research and Education.]. Scientific Research Publishing.

Carmona, R. (2008). *Literature review of the passenger airline business models*. Semantic Scholar.

Chang, Y., & Chen, L. (2007). *Solve The Vehicle Routing Problem With Time Windows Via A Genetic Algorithm*. AIMS. www.AIMSciences.org

Chesbrough, H. (2007). Business model innovation: It's not just about technology anymore. *Strategy and Leadership*, *35*(6), 12–17. doi:10.1108/10878570710833714

Chiou, Y. C., & Chen, Y. H. (2010). Factors influencing the intentions of passengers regarding full service and low cost carriers: A note. *Journal of Air Transport Management*, *16*(4), 226–228. doi:10.1016/j.jairtraman.2009.11.005

Clausen, T. (2010). Airport Ground Staff Scheduling. Technical University of Denmark. In DTU Management (Issue 2). APA.

Cowling, P. I., Colledge, N. J., Dahal, K. P., & Remde, S. M. (2006). *The trade off between diversity and quality for multiobjective workforce scheduling*. In Evolutionary computationin combinatorial optimization. Proceedings of the 6th EuropeanConference (EvoCOP 2006), Budapest, Hungary.

Daft, J., & Albers, S. (2014). *An Empirical Analysis of Airline Business Model Convergence Working Paper Series*. Department Of Business Policy And Logistics.

Deshaies, B. (1997). *Metodologia da Investigação em Ciências Humanas* [Research Methodology in Human Sciences.]. Scribd.

Diaconu, L. (2012). The Evolution of the European Low-cost Airlines'Business Models. Ryanair Case Study. *Procedia: Social and Behavioral Sciences*, *62*, 342–346. doi:10.1016/j.sbspro.2012.09.054

Díez, I. L. (2021). *Análisis económico-financiero de Vueling Airlines S.A.* [Economic-financial analysis of Vueling Airlines S.A.]

Dobruszkes, F. (2006). An analysis of European low-cost airlines and their networks. *Journal of Transport Geography*, *14*(4), 249–264. doi:10.1016/j.jtrangeo.2005.08.005

Ek, Å., & Akselsson, R. (2007). Aviation on the Ground: Safety Culture in a Ground Handling Company. *The International Journal of Aviation Psychology*, *17*(1), 59–76. doi:10.1080/10508410709336937

Ernst, A. T., Jiang, H., Krishnamoorthy, M., & Sier, D. (2004). Staff scheduling and rostering: A review of applications, methods and models. *European Journal of Operational Research*, *153*(1), 3–27. doi:10.1016/S0377-2217(03)00095-X

Europen Comission. (1996). *DIRECTIV 96/67/CE*. EC. https://eur-lex.europa.eu/legal-content/PT/ALL/?uri=CELEX:31996L0067

Fethi, M. D., Jackson, P. M., & Weyman-Jones, T. G. (2001). European Airlines: a Stochastic DEA study of efficiency with market. *NC-ND 4.0 REPOSITORY RECORD*. https://hdl.handle.net/2134/420

Flight Safety Foundation. (n.d.). *Ground Accident Prevention (GAP)*. Ground Accident Prevention (GAPhttps://flightsafety.org/toolkits-resources/past-safety-initiatives/ground-accident-prevention-gap/

Gillen, D., & Lall, A. (1997). Developing Measures Of Airport Productivity And Performance: An Application Of Data Envelope Analysis. *Science Direct*.

Good, D. H., Roller, L.-H., & Sickles, R. C. (1995). Airline efficiency differences between Europe and the US: Implications for the pace of EC integration and domestic regulation. In European Journal of Operational Research, 80.

Hunter, L. (2006). Low Cost Airlines. Business Model and Employment Relations. *European Management Journal*, *24*(5), 315–321. doi:10.1016/j.emj.2006.08.001

IAG. (2022). *International Airlines Group*. Nuestras Marcas - Vueling.

IATA. (2018). *Airport Service Level Agreement (SLA)*, 3. IATA.https://www.iata.org/policy/infrastructure/Documents/airport-service-level-agreement.pdf

İnan, T. T. (2018). Yer Hizmetleri Yönetiminin Ramp Hizmetleri Sürecinde Performans Ölçüm Sistemi (PSM) ve Kurumsal Karne (BSC) Stratejilerinin İncelenmesi [Investigation of Performance Measurement System (PSM) and Corporate Scorecard (BSC) Strategies in the Ramp Services Process of Ground Handling Management.]. *Journal of Aviation*. doi:10.30518/jav.401511

Inglada, V., Rey, B., Rodríguez-Alvarez, A., & Coto-Millan, P. (2006). Liberalisation and efficiency in international air transport. *Transportation Research Part A, Policy and Practice*, *40*(2), 95–105. doi:10.1016/j.tra.2005.04.006

Katou, A. A. (2008). Measuring the impact of HRM on organisational performance. *Journal of Industrial Engineering and Management*, *1*(2), 119–142. doi:10.3926/jiem.2008.v1n2.p119-142

Kierzkowski, A., & Kisiel, T. (2017). *The Human Factor in the Passenger Boarding Process at the Airport*. Science Direct.

Kökény, L., Kenesei, Z., & Neszveda, G. (2022). Impact of COVID-19 on different business models of European airlines. *Current Issues in Tourism*, *25*(3), 458–474. doi:10.1080/13683500.2021.1960284

Kolukısa, A. (2011). *Evaluating Aircraft Turnaround Process in the Framework of Airport Design and Airline Behaviour*.

Land, K. C., Lovell, K., & Thore, S. (1993). *Chance-constrained data envelopment analysis New efficiency and productivity indexes*. Research Gate. https://www.researchgate.net/publication/313058771

Lawton, T., Rajwani, T., & O´Kane, C. (2011). Strategic Reorientation and Business Turnaround: The Case of Global Legacy Airlines. *Journal of Strategy and Managment*, 215–237. doi:10.1108/17554251111152252

Lee, B. L., & Worthington, A. C. (2010). *The Relative Efficiency of International, Domestic, and Budget Airlines: Nonparametric Evidence*. Research Gate.

Lordan, O. (2014). Study of the full-service and low-cost carriers network configuration. *Journal of Industrial Engineering and Management*, *7*(5), 1112–1123. doi:10.3926/jiem.1191

Lufthansa Group. (2022). *Annual Report 2021*. LG.

Mahmood, A.-L., Ammar, A.-B., & Roxani, A. (2016). Improving_Workforce_Planning_of_Aircraft. *Proceedings of the 3rd International Aviation Managemnt Conference, IAMC*, (pp. 34–43). EAU.

Mazareanu, E. (2021, October 26). *Low-cost carrier market share in Europe 2001-2020*. Statista. https://www.statista.com/statistics/1117218/low-cost-carrier-market-share-europe/

Mrázová, M., & Kazda, A. (2021). The direction of airlines business models - up or down? (Post Covid-19 crisis - case study). *INCAS Bulletin*, *13*(1), 263–274. doi:10.13111/2066-8201.2021.13.1.27

Munisamy, S., Tahir Jan, M., Asraf Abdullah, M., & Mohd Satar, N. (2013). *A Critical Review of Literature on Performance of Airlines Antecedents of Loyalty in the Airline Industry of Malaysia:* Academia.

O'Connell, J. F., & Williams, G. (2005). Passengers' perceptions of low cost airlines and full service carriers: A case study involving Ryanair, Aer Lingus, Air Asia and Malaysia Airlines. *Journal of Air Transport Management*, *11*(4), 259–272. doi:10.1016/j.jairtraman.2005.01.007

P., D. B., J., H., & M, D. S. (1975). Dynamique de la recherche en sciences sociales [Dynamics of social science research]. *Revue Française de Sociologie*, *16*(3), 421–421. https://www.persee.fr/doc/rfsoc_0035-2969_1975_num_16_3_5811

Padrón, S., & Guimarans, D. (2019). An Improved Method for Scheduling Aircraft Ground Handling Operations from a Global Perspective. *Asia-Pacific Journal of Operational Research*, *36*(4), 1950020. doi:10.1142/S0217595919500209

Padrón, S., Guimarans, D., Ramos, J. J., & Fitouri-Trabelsi, S. (2015). *A bi-objective approach for scheduling ground-handling vehicles in airports*. Science Direct.

Popova, E. (2015). *Staff Scheduling in Ground Handling Company*. MathMods. https://www.mathmods.eu/resources/downloads/master-theses/finish/37-master-s-theses/251-elena-popova-thesis

Porter, M. E. (1980). *Competitive strategy: techniques for analyzing industries and competitors*. Harvard Business School.

Quivy, R., & Van Campenhoudt, L. (1998). Manual de Investigação em Ciências Sociais (Gradiva (ed.); 2nd ed.).

Rathnaweera, T. (2010). *Do hrm practices impact employee satisfaction, commitment or retention?* Research Gate. https://www.researchgate.net/publication/345764147

Reynolds-Feighan, A. (2010). Characterisation of airline networks: A North American and European comparison. *Journal of Air Transport Management*, *16*(3), 109–120. doi:10.1016/j.jairtraman.2009.07.009

Roegiers, X., & Ketele, J.-M. (1993). *Metodologia da Recolha de Dados*. IPIA.

Rose, N. L., Barnett, A., Borenstein, S., Dionne, G., Fisher, F., Hausman, J., Kahn, A., Oster, C., Papke, L., Pickrell, D., Poterba, J., Reiss, P., & Rosen, S. (1990). Profitability and Product Quality: Economic Determinants of Airline Safety Performance. *Journal of Political Economy*, *98*(5), 944–964. https://about.jstor.org/terms. doi:10.1086/261714

Rozenberg, R., Szabo, S., & Šebeščáková, I. (2014). Comparison of FSC and LCC and their market share in aviation. *International Review of Aerospace Engineering*, *7*(5), 149–154. doi:10.15866/irease.v7i5.4439

Saggar, S., Tomasella, M., Cattaneo, G., & Matta, A. (2021). *Enhanced Operational Management Of Airport Ground Support Equipment For Better Aircraft Turnaround Performance*. IEEE.

Savaneviciene, A., & Stankeviciute, Z. (2011). *Human Resource Management Practices Linkage With Organizational Commitment And Job Satisfaction*. Research Gate.

Savenkovs, K., & Sztore, M. (2021). *Features Of Development Of The Market Of Airlines -Lowcosters At The Present Stage Institūts*. RePec.

Scheraga, C. A. (2004). Operational efficiency versus financial mobility in the global airline industry: A data envelopment and Tobit analysis. *Transportation Research Part A, Policy and Practice*, *38*(5), 383–404. doi:10.1016/j.tra.2003.12.003

Schmidberger, S., Bals, L., Hartmann, E., & Jahns, C. (2009). Ground handling services at European hub airports: Development of a performance measurement system for benchmarking. *International Journal of Production Economics*, *117*(1), 104–116. doi:10.1016/j.ijpe.2008.10.006

Schultz, M., Evler, J., Asadi, E., Preis, H., Fricke, H., & Wu, C. L. (2020). Future aircraft turnaround operations considering post-pandemic requirements. *Journal of Air Transport Management*, *89*, 101886. doi:10.1016/j.jairtraman.2020.101886 PMID:33013010

Schultz, M., & Fricke, H. (2011). *Managing Passenger Handling at Airport Terminals Managing Passenger Handling at Airport Terminals Individual-based Approach for Modeling the Stochastic Passenger Behavior*. Research Gate. https://www.researchgate.net/publication/262567357

Semenick Alam, I. M., & Sickles, R. C. (1998). The Relationship between Stock Market Returns and Technical Efficiency Innovations: Evidence from the US Airline Industry. *Journal of Productivity Analysis*, *9*(1), 35–51. doi:10.1023/A:1018368313411

Sezgen, E., Mason, K. J., & Mayer, R. (2019). Voice of airline passenger: A text mining approach to understand customer satisfaction. *Journal of Air Transport Management*, *77*, 65–74. doi:10.1016/j.jairtraman.2019.04.001

Shank, J. K., & Govindarajan, V. (1993). *Strategic Cost Management: The New Tool for Competitive Advantage*. The Free Press.

Suidan, J., Cicetti, M., & Fleming, P. (n.d.). *IATA Ground Operations & Audits*. IATA. www.iata.org/ground-operations

Tabares, D. A., & Mora-Camino, F. (2017). Aircraft ground handling: Analysis for automation. *17th AIAA Aviation Technology, Integration, and Operations Conference, 2017*. 10.2514/6.2017-3425

Ter Mors, A., Mao, X., Roos, N., Witteveen, C., & Salden Almende, A. B. (2007). *Multi-Agent Support System for Airport De-icing Scheduling*.

Tomová, A., & Ramajová, L. (2014). Frequent Flyer Programs and Low-cost Airlines: Ongoing Hybridization? *Procedia: Social and Behavioral Sciences*, *110*, 787–795. doi:10.1016/j.sbspro.2013.12.923

Türeli, N., Durmaz, V., Bahçecik, Y. S., & Akay, S. S. (2019). An Analysis of Importance of Innovatice Behaviors of Ground Handling Human Resources in Ensuring Customer Satisfaction. *Procedia Computer Science*, *158*, 1077–1087. doi:10.1016/j.procs.2019.09.149

Vidović, A., Štimac, I., & Vince, D. (2013). Development Of Business Models Of Low-Cost Airlines. *IJTTE. International Journal for Traffic and Transport Engineering*, *3*(1), 69–81. doi:10.7708/ijtte.2013.3(1).07

Vlaar, P., De Vries, P., & Willenborg, M. (2005). Why incumbents struggle to extract value from new strategic options: Case of the European Airline. *European Management Journal*, *23*(2), 154–169. doi:10.1016/j.emj.2005.02.009

WizzAir. (n.d.). *About Wizz*. WizzAir. https://wizzair.com/en-gb/information-and-services/about-us/about-wizz

Yager, T. J. (1983). *Factors Influencingaircraft Ground Handling Performance. Research Gate.*

Chapter 10
Impacts of COVID-19 on Flights Operated at Lisbon and Porto Airports (Portugal):
Comparative Analysis of TAP Air Portugal, Ryanair, and easyJet

Jorge Abrantes
https://orcid.org/0000-0003-4692-907X
Estoril Higher Institute for Tourism and Hotel, Portugal & Hotel Studies, Universidade Aberta, Portugal

Rui Castro e Quadros
Instituto Superior de Educação e Ciências, Portugal

ABSTRACT

The pandemic had a very negative and penalizing impact on service activity sectors, including tourism and transport, to which Portugal was not immune. TAP Air Portugal, like Ryanair and easyJet, felt these adverse effects in their operations, forcing them to readjust their operating network. The present investigation aims to assess the effects of the pandemic on these airlines at Lisbon and Porto airports between 2019 (pre-pandemic) and 2022. In methodological terms, the analysis will be exploratory, descriptive, and mostly qualitative, based on the collection of information on the departure flights operated from these airports by the selected airlines in selecting dates. The results obtained show that TAP was the airline company most penalized given its international operation with many countries and their borders closed. In Porto, the results of this company were even more severe, reducing more than 50% of the airports served.

DOI: 10.4018/978-1-6684-6835-7.ch010

INTRODUCTION

The COVID-19 pandemic had a strong impact on the services sector, in particular on businesses with strong mobility such as tourism and transport, in particular air transport. The number of air passengers carried dropped 60.2% in 2020 to 1,808 million (4,543 million in 2019), with global revenue down 54.4%. Losses reached record values in 2020 (USD -137.7 billion) after net profits in 2019 of USD 26.4 billion. Regarding tourism, the behavior was identical, with a 72.3% drop in international tourist arrivals (405 million in 2020 versus 1,466 in 2019), with revenues dropping by 63.2% in the same period (IATA, 2022; UNWTO, 2022).

The current situation in 2022, with a strong recovery of tourism and post-pandemic air activity, has however been hampered by factors related to airport infrastructure (in particular Lisbon airport, along with other constraints at European and American airports), delays with missed connections for transit passengers and luggage left behind, lack of human resources at airports, in air navigation and in airlines.

Likewise, the increase in fuel prices, forcing an increase in air travel prices, together with cancellations for operational reasons, whether due to airport limitations or lack of crews, will have a negative impact on the fragile profitability of airlines (Russell, 2022).

The present investigation aims to assess the effects the pandemic had on the air operations and respective networks of TAP Air Portugal (TAP) as a full service carrier and of Ryanair and easyJet as low cost carriers at the Portuguese airports of Lisbon and Porto, at the end of July and beginning of August from Friday to Monday. The option for these days and for that period has to do with their greater demand on those dates traditionally dedicated to the tourism market and to avoid repetitive patterns over the remaining days of the week. The analysis takes into account the years 2019 (pre-pandemic) and 2021 and 2022, the first in a pandemic season and the second in a period of strong recovery in tourism and air transport.

Thus, the analysis will focus on traditional business models based, in this case, on the national airline TAP and on its hub at Lisbon airport, as well as the low cost models of the low cost airlines Ryanair and easyJet, both with operational bases at Lisbon and Porto airports. Conceptually, attention will be addressed to the evolution of these main business models for greater market hybridization, as these models adopted each other's features and characteristics, making their differentiation more difficult.

A characterization of the main statistical data regarding the evolution of traffic at Lisbon and Porto airports will be also carried out, allowing to measure the impacts of COVID-19 in Portugal.

In methodological terms, the analysis will be exploratory, descriptive and mostly qualitative, based on the collection of information on the departing flights operated from these airports on the indicated dates and mentioned days. The information was collected based on the flights advertised in ANA - Aeroportos de Portugal website.

The results show greater drops at TAP Air Portugal rather than at Ryanair or easyJet given its hub & spoke business model in Lisbon, based on an international network with many of the destinations and borders closed. On the other hand, the loss of a large part of that transfer traffic limits the number of operations to the remaining destinations. In the case of Porto, the reduction in TAP's operation is significant, reaching half of the flights operated in 2019.

LITERATURE REVIEW

Airline Business Models

Despite the greater development of the concept and definitions of business models starting in the early 1990s, mostly associated with the development of new technology-based businesses, there is still a lack of consensus on the subject (Seddon & Lewis, 2003). This situation is also a reality in the aviation sector, where there is no standardization in relation to the classification of airlines business models (Almeida, Costa, & Abrantes, 2020, p. 218).

Since the definitions of business models are many and varied, there are some that gained greater acceptance given their objectivity towards the creation of value and business, with a high focus on strategic or economic-financial aspects. Tucker (2001) defined business model as "*(…) a description of how your company creates value for customers that in turn generated revenue and profits for your company*". Likewise, Osterwalder and Pigneur (2010) considers that "*business model describes the rationale of how an organization creates, delivers, and captures value*". The authors defend a model based on nine blocks (the Business Model Canvas): customer segments, value proposition, channels, customer relationship, revenue streams, key resources, key activities, key partnership e cost structure, which, even today, is widely accepted and used in the market.

In the same sense, in commercial aviation, the concept tends to focus on the more economical and value creation aspects. Bakir, Akan, and Atalik (2020) defined airline business models as a "*basic framework that directs how airlines produce value and how they do business*".

Looking at the different business models in commercial aviation and despite the systematization carried out by the International Civil Aviation Organization (ICAO) regarding the terminology that characterizes the different types of airlines (ICAO, 2018), full service carriers (FSCs), low cost carriers (LCCs), regional carriers, charter airlines, hybrid models and specialists tend to be the most recognized business models by the market, although depending on the characteristics of the products and services they offer. Other authors broaden the spectrum to include cargo and aviation business, which also play an important role in the aviation business (Almeida, Costa, & Abrantes, 2020; Bitzan & Peoples, 2016; Camilleri, 2018; ICAO, 2018; Magdalina & Bouzaima, 2021; Whyte & Lohmann, 2016).

For Magdalina and Bouzaima (2021) airlines have a diversity of business models and do not operate in a watertight or monolithic way, since each company has distinctive characteristics that help to differentiate it from the others. But, as mentioned by the authors, at the end there is a strong concentration of companies in traditional (FSC) or low cost (LCC) models.

In the same way that the technological aspects were essential for the development of business models in general, the liberalization of air transport was also relevant for the emergence of new businesses in aviation, as happened with the emergence and consolidation of low cost carriers while charters airlines were losing importance in the market. In the European case, after the process initialized in the USA in 1978, Europe started a more complex process between 1987 and 1997, and from April 1997 airlines could have free access to markets, including cabotage rights, with direct effects on the increase in routes and competitiveness between airlines, and decrease in air tariffs which resulted in a rise in the number of passengers transported and development of aviation (Abrantes, 2016; Almeida, Costa, & Abrantes, 2020; Button, 1998; Francis, Fidato, & Humphreys, 2003).

In the same way fusions, acquisitions and different types of alliances between airlines were also responsible of the entry of new business models (Barrett, 2004; Button, 2001; Carmona-Benítez & Lodewijks, 2008; Dobruszkes, 2009; Doganis, 2006; Neufville & Odoni, 2003).

The emergence of low cost carriers brought new competitiveness to aviation, including new functionalities and dematerialization of the product value chain, with many of the services paid and simplicity in the booking process, which forced the traditional airlines to rethink and readjust their business strategies, incorporating many of the innovations introduced by the low cost carriers. Currently, there is an opposite movement with many of the low cost carriers adopting many of the features offered to their customers by traditional airlines, in terms of promoting business passengers, using Global Distribution Systems (GDS) or even creating conditions for the transfer and routing of its passengers (Abrantes, 2016; Mrázová & Kazda, 2021). This new reality led to a so-called "hybridization" of the market with many of the concepts and many of the attributes and functionalities no longer being an exclusive of one or another business model, as defended by Magdalina and Bouzaima (2021) as airlines tend to adopt attributes of different airline business models (Whyte & Lohmann, 2016) or combine different strategies of traditional airlines and low cost airlines in a single business model (Jean & Lohmann, 2016; Lawton & Solomko, 2005; Tomová & Kirschnerová, 2015). As mentioned by Stoenescu and Gheorghe (2017, p. 586) *"In the future, the "hybrid" model might apply to all airlines and, due to this highly competitive sector, will lead to other new business models."*.

In 2020, low cost airlines were responsible for 35.2% of installed capacity worldwide, increasing their presence in a terrible year for aviation and world tourism (CAPA, 2020). The greater commercial aggressiveness during the COVID-19 pandemic, the greater financial support, better withstanding times of crisis, the more than proportional retreat of traditional airlines and lower cost levels were the reasons given for this growth of more than four percentage points compared to 2019, where they already represented 31% of the capacity offered (CAPA, 2020). In Europe, low cost carriers were responsible for 44.5 percent of the total seats in the region (Statista, 2022), although this value slipped to 40.5% in 2021 (CAPA, 2022).

Brief Characterization of Airlines Involved

TAP Air Portugal

In operation since 1945, TAP Air Portugal (TAP) is the main Portuguese airline. Member of the Star Alliance since 2005 has its hub in Lisbon, a privileged and strategic location between Europe, Africa and Americas, mainly South America with more than 80 weekly flights (76 in 2022) to 11 cities in Brazil (São Paulo, Rio de Janeiro, Belém, Belo Horizonte, Brasilia, Fortaleza, Natal, Maceió, Porto Alegre, Recife and Salvador da Bahia).

The shareholder process in the company has been long and winding. The company was privatized in 1953, and nationalized in 1975 after the Revolution of April, 25th in 1974. Between November 2015 and August 2016, TAP became a private company again, with a 61% stake in the share capital of TAP SGPS by Atlantic Gateway SGPS, a consortium between Humberto Pedrosa – a Portuguese private group with a strong position in road and rail transports in Portugal – and David Neeleman, owner of Azul (Brazil) and JetBlue (USA) airlines.

However, the change in Portuguese government in 2016 led it to choose to (re)buy a part of the capital from the private sector in order to regain "strategic control" of the company. Of this control, Atlantic Gateway SGPS took a 45% stake, the Portuguese State 50% and TAP workers the remaining 5% (TAP, 2018).

As a result of the pandemic and the already fragile economic and financial situation of TAP, the company received support in the context of COVID-19 and state aid for its Restructuring, situations duly approved by the European Commission (TAP, 2022) and which led to new changes in shareholders with the concentration of capital in the Portuguese State. The Restructuring Plan presented included a significant transformation of TAP's operation in order to ensure its economic viability in 2023 and its sustainability in 2025. This restructuring includes measures to improve operational efficiency, a fleet resizing and a reduction in labor costs in order to ensure the survival and sustainability of the TAP Group.

Thus, as a result of the capital injected by the Portuguese State to save the company, the Government decided in October 2020 to increase its stake to 72.5% by exiting the shareholder structure of Atlantic Gateway SGPS (David Neelemen was "invited" to leave the company), which sold the remaining 22.5% to HPGB SGPS (TAP, 2022) of Humberto Pedrosa. The proposed capital increase for November 2022 will remove the remaining shareholders from the capital structure (HPGB, SGPS and employees), leaving the Portuguese State with the entire capital of TAP SGPS, as it had already done with the airline (TAP, SA) throughout 2021 (Veríssimo, 2022).

In 2019, TAP reached a new record of passengers transported, with an increase of 1.3 million passengers compared to the same period of the previous year (+8.2%) to 17.1 million passengers, with a strong contribution of the North American market as a result of the company's expansion to that region. The Brazilian market reinforced its importance, growing again compared to the previous year (3%). The company served 95 destinations in 38 countries in 2019, 4 more than in 2018, operating more than 130 thousand flights, with a capacity increase of more than 11%. Overall, the company opened 11 new routes this year, with an annual Load Factor of 80.1% (-0.9 p.p. compared to 2018), due to the increase in installed capacity.

The airline has been accumulating losses year after year but the years of 2020 and 2021 were terrible for the company. In 2019, net income stood at negative € 95.6 million, values strongly impacted negatively by the pandemic, with losses of € -1,230.3 and € -1,599.1 million in 2020 and 2021, respectively, creating strong pressures and doubts about the future economic and financial viability of the company. Portuguese Prime Minister is actually accelerating the privatization process of the company, which he expects to be concluded during 2023, even if the Portuguese State may not recover all the capital invested with the sale (Campos, 2022b).

Ryanair

The opening of intra-community airspace in Europe, which began in 1983, led to the establishment of Ryanair in 1985, with the launch of its first route in July of that year between Waterford (Ireland) and London-Gatwick (United Kingdom), in an Embraer Bandeirante plane with capacity to carry 15 passengers (Creaton, 2007). The following year, the liberalization of air service agreements between the United Kingdom and Ireland led to the opening of flights between Dublin (Ireland) and London-Luton (United Kingdom), with a 46-seat aircraft (BAE Hawker Siddeley HS 748) and with a fare of 99 pounds sterling, less than half that charged by the main airlines that operated this route.

As a result of this liberalization and the entry into service of 3 BAC One-Eleven leased aircrafts, the number of passengers on the main route between London and Dublin between 1986 and 1987 increased by 64.9% and fares fell by 54%, largely due to at the price charged by Ryanair (Barrett, 1997).

Despite the success in terms of passengers carried, Ryanair accumulated negative results over the years. To reverse this situation, the company had a new CEO – Michael O'Leary – and decided to change its business model, copying the successful case of Southwest Airlines, a profitable American airline company over the years and with a model business based on a low cost strategy (Creaton, 2007).

In 2001, Ryanair opened its first operational base outside the British Isles, in Charleroi (Belgium), after the bases already created in Dublin (company headquarters), London-Luton (later transferred to London-Stansted in 1991) and Glasgow-Prestwick. Since that year, Ryanair has had sustained growth over the years, both in the number of passengers and in the number of operational bases created. In 2003, Ryanair has completed the acquisition of Buzz from KLM (Creaton, 2003). Currently, the Ryanair Group includes Ryanair and Ryanair UK (created in 2019, due to Brexit, allowing the airline to maintain its operation and operational bases in the United Kingdom), Buzz (rebrand in 2019 of from Ryanair Sun), Lauda and Malta Air.

The effects of the pandemic led the company to return to negative results, after successive years of profits until 2019. The company has been exploring new post-pandemic opportunities with the opening of 15 new operational bases throughout 2022, increasing in March 2023 with the reopening of the base at Belfast International airport after its closure in October 2021 (Goodbody, 2021; Ryanair Group, 2022a).

Ryanair's main indicators in the last three years (accounts ended in March of each year, so 2022 is also presented) show the impacts resulting from the pandemic. Even so, the year 2021 showed an increase in activity compared to the previous year, with an increase in the number of bases and fleet (which is expected to reach 500 aircraft throughout 2022). The results for the first six months of the year (March to September 2022) show a strong growth in activity (11% more passengers over pre-Covid reaching 95.1 million passengers) and a record net income of € 1,370.8 million (Ryanair Group, 2022b)

When confronted with the question of what factors are Ryanair's success factors, Ryanair's Chief Marketing Officer (CMO) was quite objective: "*The lowest cost and a simple business model.*" (ATN, 2015). However, rising fuel and new environmental charges in Europe have already led to the CEO from Ryanair to admit in an interview with the Financial Times that "*(...) fares will rise for the next five years because flying has become "too cheap" to make profits as industry costs spiral.*" (Giorgiadis, 2022), in what some analysts have already called the end of low cost or at least "*the days of €9.99 fares are probably coming to an end*" (Weber & Alabi, 2022).

Over the years, Ryanair has created several operational bases in Portugal, the first of which in Porto (2009), followed by Faro (2010), Lisbon (2014), Ponta Delgada (2015) and, more recently, Funchal (2022). In 2019, the 3.9 million tourists who traveled with Ryanair spent €2,109 million in Portugal, a PwC study shows. By 2027, the British airline plans to increase passenger traffic to Portugal by 36%, from 11 to 15 million passengers per year, and increase annual spending "*to three or four billion euros*" (Tnews, 2022).

easyJet

easyJet, founded by Stelios Haji-Ioannou, began operating in the United Kingdom in 1995. The company was one of the first to use the "low cost, low fare" concept, based on existing models in the United States of America (Southwest) as in Europe, with the example of the company Ryanair that flew from

Ireland to the United Kingdom. The initial fleet was two Boeing 737-200s, with flights from London Luton to Glasgow in Scotland.

In March 1998 easyJet acquired a 40% stake in a Swiss charter airline called TEA Switzerland that later became a wholly owned subsidiary called easyJet Switzerland, and in 2002 easyJet acquired Go Fly, a low-cost airline created by British Airways (easyJet, n.d.). That year, easyJet ordered 120 Airbus aircraft, with additional 120 aircraft as an option (CNN, 2002).

In 2007, the company acquired GB Airways, which in 2010 allowed it to become the UK's largest airline with over 500 routes operated.

Due to the Brexit and in order not to lose the possibility of operating within the European Union, a new airline was created in Austria in 2017 – easyJet Europe (Hardiman, 2021).

With the pandemic, the company canceled its operation in March 2020, having gradually resumed it in June of the same year, while securing a £600 million loan from the Treasury and Bank of England's emergency coronavirus fund (Topham, 2020). The results were quite negative in 2020 (accounts are reported in September of each year), exceeding £1 billion, and in 2022, easyJet is anticipating losses of over £200 million (Bodell, 2022).

In its most operational aspects, easyJet has adjusted its network, although there is some balance in its operation even in the years most affected by the pandemic.

In September 2017, easyJet launched "*Worldwide by easyJet*", announced as "*the first global airline connections service by a European low fares airline*", with West Jet and Norwegian as launch partners. Its main objective is to capture and promote traffic to destinations from London Gatwick to North and South America and the Far East (easyJet, 2017).

easyJet started flying to Portugal in 1998 and launched its first base in Lisbon (2012), followed by Porto (2015) and Faro (2021). In the last 24 years, the company has transported around 70 million passengers to and from Portugal. With the transfer of slots from TAP to EasyJet in 2022, the company becomes number two at Lisbon airport, behind TAP (Pestana, 2022), allowing it to launch 13 new routes and reinforcing capacity in eight destinations, starting to operate to ten countries (Suspiro, 2022).

Globally, the information collected in Annual Reports of the airlines permits to compare the main operational and economic indicators. The information of Ryanair is related to end of March as easyJet closes its accounts in September (TAP in end of December):

Table 1. Operational and economic indicators

	TAP Air Portugal			Ryanair Group				easyJet		
Indicators	2019	2020	2021	2019	2020	2021	2022	2019	2020	2021
Total Revenue (Mill. €/£.)	3,298.8	1,060.2	1,388.5	7,697	8,495	1,636	4,801	6,385	3,009	1,458
Net Result (Mill. €/£.)	-95,6	-1,230.3	-1,599.1	885	648,7	-1,015.1	-240,8	349	-1,079	-858
ASM's (Mill. kms)	52 527	18 376	23 663	115 524	49 271	192 165	184 325	116 056	62 380	33 348
RPM's (Mill. kms)	42 065	11 876	14 917	109 976	34 982	182 556	176 952	107 741	58 914	23 594
Load Factor (%)	80,1	64,6	63	96	95	71	82	91,5	87,2	72,5
Number of passengers (Mill.)	17,1	4,7	5,8	142	149	27,5	97	96,1	48,1	20,4
Number of operational bases	n.a.	n.a.	n.a.	86	79	86	90	30	27	28
Airports/Destinations Served	95	68	93	219	242	225	223	159	154	153
Fleet	105	96	94	431	440	422	471	331	342	308
Number of routes (aprox.)	n.a.	n.a.	n.a.	n.a	1,051	981	927	1000	927	1100

Notes: Economic Information of easyJet in pounds (£).
Ryanair fleet excludes Airbus aircrafts.
n.a. - not available/not applicable

Source: Own elaboration based on TAP Annual Reports (TAP, 2022), Ryanair Annual Reports (Ryanair Group, 2019-2022a) and easyJet Annual Reports (EasyJet, 2022)

Impacts of COVID-19 on the movement of Lisbon and Porto airports

As previously mentioned, the COVID-19 pandemic has had harmful and adverse effects on the commercial aviation sectors. Statistical information on passenger traffic at Lisbon and Porto airports clearly shows the loss of passengers in 2020, with drops of more than 70% in Lisbon and 66% in Porto. The recovery in 2021 (provisional data), in the order of 31% at both airports, shows however that there were a long road to traffic recovery to 2019 levels. In Lisbon, passenger data were 61% below levels of 2019, and in Porto this value stood at 55.4% (Pordata, 2022).

Table 2. Passengers at Lisbon and Porto airports (2017-2021)

	Passengers			
Years	**Lisbon Airport**	**Var. (%)**	**Porto Airport**	**Var. (%)**
2017	26 676 552	18,8%	10 790 271	15,1%
2018	29 045 733	8,9%	11 942 333	10,7%
2019	31 184 594	7,4%	13 112 453	9,8%
2020	9 267 968	-70,3%	4 436 370	-66,2%
2021 (P)	12 160 468	31,2%	5 846 732	31,8%

Source: Own elaboration based on Pordata (2022)

The analysis through the "Monthly Air Traffic Monitoring in National Airport Infrastructures" published by the National Civil Aviation Authority (ANAC, 2022), also allows to verify the effects and evolution of traffic throughout 2022, contributing to understand the recovery of air transport at these airports.

Table 3. Passengers at Lisbon and Porto airports (2002)

	Passengers					
Month	**Lisbon Airport**	**Var. 22/21 (%)**	**Var. 22/19 (%)**	**Porto Airport**	**Var. 22/21 (%)**	**Var. 22/19 (%)**
January	1 213 631	189,7%	-38,7%	476 119	138,7%	-42,9%
February	1 412 039	1046,7%	-25,4%	619 745	930,3%	-22,9%
March	1 893 132	863,4%	-20,5%	855 009	785,8%	-11,9%
April	2 367 116	562,8%	-11,8%	1 137 323	574,8%	-0,8%
May	2 531 429	350,1%	-8,3%	1 202 973	302,1%	1,2%
June	2 717 165	216,9%	-6,7%	1 231 699	177,8%	1,6%
July	2 935 953	136,0%	-5,5%	1 312 229	98,5%	1,2%
August	2 948 501	73,9%	-5,8%	1 367 020	56,1%	1,3%

Source: Own elaboration based on ANAC (2022)

The data presented above, both in comparison with the previous year and with the year 2019, shows a strong recovery of traffic at Porto airport with values above 2019 since the month of May. In the case of Lisbon, the recovery has been slower but sustained over the months, with only approximately 5% below the pre-pandemic year. The first months of the year show an impressive growth compared to the previous year, as 2021 saw a worsening of the pandemic situation in Portugal, forcing a period of strong pandemic confinement and high mobility restrictions, with the closure of borders and cancellation of air operations.

METHODOLOGY

The infinity of methodological research approaches allows the researcher to observe and study his object through various lenses and different angles of vision, enriching his study and the reliability of the results obtained, and the investigation strategies in the qualitative, quantitative, or multi-method approaches they help the exploration of reality and the understanding of social phenomena (Massukado, 2008).

In methodological terms, the analysis will be exploratory, descriptive and mostly qualitative, based on the collection of information on the departing flights operated from these airports on the indicated dates. Although the incidence is mainly qualitative, the application of descriptive statistics will also allow a quantitative framework, in line with Patton (2002) in reaffirming that the approaches are not mutually exclusive.

The information was collected based on the flights advertised on ANA - Aeroportos de Portugal website on those specific days, that is, between Friday and Monday at the end of July and beginning of August, the traditionally busiest period at European airports.

The analysis focused on the years 2019 (pre-pandemic) and years 2021 and 2022 where the effects of the pandemic were more felt (2021) and in a clear recovery of tourism and travel (2022). No information was collected for the year 2020, which could be a limitation in the continuity of the series, but the existing information on that date was quite small and not very expressive, given the limitations and restrictions imposed by the confinements and closing of borders in many international markets, with many of the flights operated as cargo for the transport of masks, ventilators, and other medical materials and equipment.

The analysis includes the airline TAP Air Portugal with a traditional operation based on its hub in the city of Lisbon and on the low cost carriers Ryanair and easyJet, with operational bases in the airports under study, with the aim of evaluating and comparing the potential differences between business models in this period.

Comparative analysis is the process to determine and quantify relationships of observing and comparing items to one another and distinguishing their similarities and differences (Bukhari, 2011; Indeed, 2021). Comparative analysis makes it possible to compare several items trying to identify common points and differences that help to give a better explanation of the phenomena under study.

There are several reasons for the use of comparative analysis (Pickvance, 2001) such as: a) To explore a theoretically postulated relationship; b) To examine whether a relationship reported in a study in one society also holds in another; c) To examine whether a condition which is given or fixed for one society is influential or not; d) To examine a small number of empirical cases holistically to grasp the causal processes leading to observed similarities and differences.

The comparative analysis will focus on the number of airports served and the number of flights operated to these destinations, in order to be able to understand which ones were definitively or temporarily closed and which airlines' new flights are betting at both airports.

ANALYSIS AND RESULTS

Comparative Evaluation Between Airlines at Lisbon and Porto Airports

The information collected over the three years under analysis, as mentioned above, made it possible to verify the evolution of the operation of each of the airlines on the selected operating days.

The results obtained, firstly by airport and by airline, will make it easier to highlight its main aspects, followed by a more detailed analysis by airline in each one airport.

TAP's operation, because supported by its main hub in Lisbon, has a greater presence in this airport than the main low cost carriers operating in Portugal. In 2019, TAP operated to 96 airports (Fuerteventura (Spain) was excluded of the analysis, because operated on a day – Tuesday – that did not match with the sample), while Ryanair and easyJet operated to 29 and 20 airports, respectively. TAP's operation includes, in addition to flights to Europe (58 airports served), intercontinental flights to Africa (19 airports) and the Americas (19 airports), while Ryanair and easyJet operate, that year, only European routes.

Table 4. Airports served by airline and airport per year

	Lisbon					
Airports Served	2019	2021	2022	Var. 21/19	Var. 22/21	Var. 22/19
TAP Air Portugal	96	80	90	-16,7%	12,5%	-6,3%
Ryanair	29	29	33	0,0%	13,8%	13,8%
easyJet	20	16	19	-20,0%	18,8%	-5,0%
Total	145	125	142	-13,8%	13,6%	-2,1%
	Porto					
	2019	2021	2022	Var. 21/19	Var. 22/21	Var. 22/19
TAP Air Portugal	21	11	11	-47,6%	0,0%	-47,6%
Ryanair	56	55	64	-1,8%	16,4%	14,3%
easyJet	17	18	24	5,9%	33,3%	41,2%
Total	94	84	99	-10,6%	17,9%	5,3%

Notes: TAP (2019) - excludes Fuerteventura both in Lisbon and Porto
Ryanair (2019) - excludes Gran Canaria (Porto)

Source: Own elaboration based on ANA – Aeroportos de Portugal website information

At Porto airport, the domain goes to Ryanair with 56 airports operated in 2019, while TAP only operated 21 and easyJet (17), excluding operations this year by TAP to Fuerteventura and Ryanair as they are carried out on a different day to those of the sample. easyJet's operation was exclusively European, while Ryanair operated also to Marrakech (Morocco). TAP included intercontinental flights departing from Porto to Newark (USA) and Brazil (Rio de Janeiro and São Paulo).

The pandemic years were more adverse for TAP, given its greater network and diversity of markets, many of them with borders closed for extended periods of time. In the summer of 2021, TAP stopped operating at 16 airports in Lisbon operation, recovering part of this loss in 2022 (operating, this summer, to 90 airports). easyJet had slight adjustments to its network in Lisbon, although more visible in 2021, ceasing to operate to 4 of the previously served airports. In 2022, the operation reached levels almost equal to those of 2021. In turn, Ryanair maintained its operation in 2021 (19 airports as in 2019), increasing it to 33 airports served (an increase of 4 destinations), being the only company to increase its presence in Lisbon airport when compared to 2019.

In Porto, TAP was the airline that suffered the most with the loss of almost half of the airports served in 2021 (-47.6% representing 10 fewer destinations that summer), a value that it maintained in 2022. This drastic reduction in the operation in Porto led to a strong criticism from various sectors of Portuguese civil society, with the Mayor of Porto stating that "*TAP is of no use to Porto*" (Porto, 2021) or even from the President of the Regional Tourism Entity of Porto and Northern Portugal, referring that "*TAP seems to have given up on this part of the country*" (Diário de Notícias, 2021). Ryanair practically maintained its operation in 2021 (55 airports, that is, only one destination stopped operating this year), reinforcing it in 9 more airports in 2022, for a total of 64 airports operated this summer. In turn, easyJet was the only one to increase its presence over the years with the addition of 7 new airports in 2022 compared to 2019.

By evaluating the number of flights carried out by airline and by airport on the selected days (Friday to Monday), it is possible to identify patterns of behavior in line with the airports served. TAP lost 251 flights in Lisbon in the period under review, not only because of the reduction in the airports operated but also because of the reduction in operations at many of the other airports. The recovery in 2022 is, even so, 10% below 2019. Even maintaining flights in 2021, Ryanair adjusted its operation in this airport, with a slight decrease in operated flights (-3.7%). The growth that started in 2022 allowed it to take more flights (+12% compared to 2019, that is, 13 more flights in the selected period, in line with the increase in the number of airports, which grew by 13.8%). easyJet reduced its operation in Lisbon in 2021 (-17.7% of flights), recovering in 2022 to 2019 levels (with a minimum negative variation of 2.1%, that is, 2 flights less).

Table 5. Flights operated by airline and airport per year

	Lisbon					
Flights operated	2019	2021	2022	Var. 21/19	Var. 22/21	Var. 22/19
TAP Air Portugal	751	500	676	-33,4%	35,2%	-10,0%
Ryanair	108	104	121	-3,7%	16,3%	12,0%
easyJet	96	79	94	-17,7%	19,0%	-2,1%
Total	955	683	891	-28,5%	30,5%	-6,7%
	Porto					
	2019	2021	2022	Var. 21/19	Var. 22/21	Var. 22/19
TAP Air Portugal	160	57	75	-64,4%	31,6%	-53,1%
Ryanair	152	132	162	-13,2%	22,7%	6,6%
easyJet	68	77	98	13,2%	27,3%	44,1%
Total	380	266	335	-30,0%	25,9%	-11,8%

Source: Own elaboration based on ANA – Aeroportos de Portugal website information

Despite a greater presence at Porto airport, Ryanair operated fewer flights than TAP in 2019 (152 Ryanair flights versus 160 TAP flights in the selected period), even if Ryanair operated 56 destinations against 21 for TAP. The greater density of TAP's operation to European destinations and its air bridge with Lisbon (13 daily flights) explain the results obtained. Even having only reduced presence at one airport in 2021, flights operated by Ryanair show higher drops (20 flights) due to adjustments in their operation at other operated airports. easyJet maintained growth in the number of flights in line with the increase in its presence at this airport.

TAP, as previously mentioned, has significantly reduced its presence at this airport. In 2022, its operation was 53.1% lower than in 2019, operating only 75 flights departing from Porto, of which almost half to Lisbon (36 flights in the four selected days).

A more detailed analysis by airport and by airline makes it possible to identify the main changes observed in the airlines' operating network below.

Evaluation of Operational Performance at Lisbon Airport

TAP Air Portugal

Its hub at Lisbon airport shows that most of the flights are operated on this airport. The scope of its network makes it possible to identify different flights operated with the highest concentration in Europe (58 airports operated in 2019). The connection to many of the former Portuguese-speaking countries and the focus on medium-haul flights in the Maghreb and West Africa region has allowed the company to maintain a diversified and stable network and operation throughout the pandemic in African continent (19 airports operated in 2019, moving to 17 in the following years). The company's commitment to South America, especially Brazil, having been the main European operator for several years for this market, will permit to operate to 12 airports in this region in 2022 (11 in Brazil and 1 in Venezuela), after the reinforcement of the operation in Maceio (Brazil). On the other hand, the commitment and development of the network for North and Central America means that TAP currently operates at 11 airports on this region, reinforcing its presence since 2019 (8 in 2019 and 9 airports in 2020).

Table 6. Number of airports served to/from Lisbon Airport (TAP Air Portugal)

	Lisbon		
TAP Air Portugal	**2019**	**2021**	**2022**
Europe	58	47	50
Africa	19	17	17
North & Central America	8	9	11
South America	11	7	12
Total	**96**	**80**	**90**

Source: Own elaboration based on ANA – Aeroportos de Portugal website information

Detailed information with the different airports operated in the selected years and days of operation is detailed in Figure 1 (Europe), Figure 2 (Africa), Figure 3 (North and Central America) and Figure 4 (South America).

Looking at the flights operated, the concentration of operations in the medium-haul network (Europe) to allow a greater capacity of feeding/defeeding in the different waves throughout the day for the mostly long-haul traffic, explains the greater number of frequencies daily. For the 58 airports operated in 2019, the company performed, on average, 151.5 flights in Europe in the four days of analysis, while for the other continents these figures were much lower (16.75 in Africa, 12 in South America and 7.5 in North America).

Table 7. Number of daily flights per continent (TAP Air Portugal)

	TAP Air Portugal - Lisbon Airport											
	2019				2021				2022			
Flights operated	Friday	Saturday	Sunday	Monday	Friday	Saturday	Sunday	Monday	Friday	Saturday	Sunday	Monday
Europe	152	149	153	152	95	105	109	80	131	136	142	120
Africa	16	17	19	15	16	15	18	9	15	16	18	13
North & Central America	7	7	8	8	10	6	9	7	12	10	12	12
South America	11	14	11	12	6	4	7	4	8	13	7	11
Total	**186**	**187**	**191**	**187**	**127**	**130**	**143**	**100**	**166**	**175**	**179**	**156**

Source: Own elaboration based on ANA – Aeroportos de Portugal website information

The effects of the pandemic are evident on the European continent, with losses of 35.8% of flights in 2021 (97.25 flights on average). However, it was in South America, due to pandemic restrictions, that the drop was more pronounced, with -56.3% of flights (5.25 flights operated on average to 7 airports, all in Brazil). Regarding the African continent, the harmful effects of the COVID-19 pandemic were minimized with a drop of -13.4% compared to 2019. Only the American continent showed growth in its operation, with an increase of 6.7% compared to 2019, reinforcing this value in 2022, with a growth of 43.8% compared to 2021 and 53.3% versus 2019 (11.5 flights operated on average in the 4 days under analysis in 2022 when in 2019 this figure was 7.5 flights).

In addition to the new destinations to Montreal (Canada, in July 2020), Cancun (Mexico, in March 2021) and Punta Cana (Dominican Republic, in December 2021), TAP reinforced the operation in all airports in the USA (with exception of Chicago and San Francisco), as shown in Figure 3.

In South America, where the drop was most evident (-56.3%), the operation was interrupted in Venezuela (with flights only to be resumed in June 2022) and also in Belem (operation return in October 2021), Maceió (opened in October 2020 and resumed, after interruption due to the pandemic, in October 2021), Natal (November 2021) and Porto Alegre (resumed in March 2022). With the exception of Belo Horizonte and Fortaleza, all the remaining operation to Brazil are below its 2019 threshold (Figure 4). In 2016, TAP operated 86 weekly flights to Brazil, including the Porto operation (Silva, 2022).

TAP's operation for Africa, as shown in Figure 2, remained almost constant during the pandemic, with 19 airports operated in 2019 and 17 in the following years. Even so, the effects of the pandemic dictated some inconstancy in its network. Abidjan (Ivory Coast) after its recovery in 2017 (after 17 years of absence) saw its operation canceled in 2022, as did Lomé (Togo) and Fez (Morocco). Banjul (Gambia) with the route opening in October 2019 saw the route resumed after the airport reopened, as did Accra (Ghana), although in the selected week there were no flights to this destination in 2021 and 2022. Some operations of a more seasonal and charter nature (as is the case of Oudja in Morocco or even Djerba in Tunisia) reflect the evolution of the tourist market itself.

The operation in Europe, as shown in Figure 1, was the one that, due to its proximity to Lisbon, showed the greatest reductions in its operation, going from the 58 airports served (59 with the operation in Fuerteventura in 2019 held on Tuesdays), to the 47 airports in 2021 (a reduction of 12 destinations) and a slight recovery for the 50 airports operated in 2022.

In a more detailed analysis of the various airports and the period under analysis, it is concluded that:

- As of 2019, TAP ceased operating, in its own operation, at airports of Basel (Switzerland), Bordeaux and Nantes (France), Budapest (Hungary), Cologne and Stuttgart (Germany), Helsinki (Finland) and London City (United Kingdom). Airports operated on seasonal or charter to Heraklion (Greece) and Olbia (Italy) should also be included.
- In 2022, TAP stopped operating in Moscow (Russia), as a result of the sanctions imposed by the European Union after the Russian invasion of Ukraine;
- New operations to Zabreg (Croatia, March 2021), Ibiza (Spain, May 2021) and Fuerteventura (Spain, June 2021), although the latter already operated in 2019 in a seasonal charter operation;
- Resumption of interrupted and non-operated operations in 2021 for Alicante (Spain), London Gatwick (United Kingdom), Oslo (Norway) and Tel Aviv (Israel);
- 11 of the destinations are already at the 2019 level, with 14 destinations above the 2019 values (led by Nice (France) with 4 additional flights – 16 flights – compared to 2019, followed by Amsterdam (Netherlands) and Madrid and Malaga (Spain) with 3 additional flights to the 2019 operation.

Ryanair

Ryanair created its operational base in Lisbon in April 2014. In 2019 the company flew to 29 European airports, a figure it maintained in 2021, having increased its presence to 33 destinations in 2022 (Figure 5).

Although there are no changes between 2019 and 2021, the company has been, over time, looking for new destinations and continuously adapting its network. Thus, the company ceased operations after 2019 to Clermont-Ferrand and Lourdes (France), Frankfurt Main, Hamburg, and Karlsruhe (Germany) and Wroclaw (Poland), in addition to the operation between Lisbon and Porto (Portugal). In 2021, new destinations appear on its network, as in the cases of Budapest (Hungary), Cologne and Memminghem (Germany, the latter of which was no longer served in 2022), Malaga (Spain) and Vienna (Austria), an operation that was reinforced in 2022 with new flights to Barcelona (Spain), Fez and Marrakech (Morocco) and Funchal.

The company's continuous commercial and marketing action with continued announcements of new destinations does not always translate into operations actually launched on the market, reinforced with modifications and threats to market withdrawal. In August 2021 Ryanair announced that it would reinforce its Lisbon base with 3 additional aircrafts and open 22 new routes to Italy (Bari, Palermo, Alghero), Morocco (Agadir, Fez, Marrakech, Oujda), France (Brest, Perpignan, Tours), Spain (Alicante, Barcelona, Fuerteventura, Madrid, Tenerife), Poland (Wroclaw), United Kingdom (Birmingham, Bournemouth), Denmark (Billund), Germany (Cologne, Karlsruhe-Baden-Baden) and Malta (Publituris, 2021), and in March 2022 it informed the market that it would withdraw these same planes and cancel 19 routes due to the fact that TAP continues to block slots at Lisbon airport (Publituris, 2022).

In fact, slots were once again a topic when commemorating the 20th anniversary of Ryanair's arrival in Portugal, in October 2022, where the company's CEO stated that "*The continuous misuse of slots at Portela, both TAP and easyJet that have slots in Portela are not using them, they are blocking those slots and preventing Ryanair from growing, creating more jobs and bringing more passengers to Lisbon*", which would allow Ryanair to place more aircraft in the country, create more employment and, over the next four or five years, increase annual spending by Ryanair passengers in Portugal from two billion to three or four billion euros (Tnews, 2022).

easyJet

easyJet created its operational base in Lisbon in April 2012. In 2019 the company flew to 20 European airports, but the effects of the pandemic dictated a reduction of the operation to 16 destinations in 2021, having recovered its presence to 19 destinations in 2022 (Figure 6).

The greater consistency in the operation over the years shows that the company ceased to operate, in 2019, definitively, flights to Lille (France) and Zurich (Switzerland), with operations to Berlin (Germany) and Edinburgh (United Kingdom) were resumed in 2022, with Porto Santo being the only bet until the summer of 2022, reinforced with Paris Beauvais (France) from November 2022.

However, the new slots provided by TAP, within the scope of public aid to this airline, will allow it to become the second operator in Lisbon (led by TAP) starting new flights from the IATA winter to 13 new destinations and reinforcing 8 of those already operated. The new airports to operate will be Barcelona, Valencia, Bilbao, Fuerteventura, Gran Canaria and Tenerife (Spain), Marseille, Toulouse and Limoges (France), Birmingham (United Kingdom), Milan Bergamo (Italy), Zurich (Switzerland) and Marrakech (Morocco). The operation will be reinforced in Porto Santo, London (United Kingdom), Geneve (Switzerland), Milan Malpensa (Italy) and Lyon, Nice, Nantes and Bordeaux (France) (Suspiro, 2022).

Evaluation of Operational Performance at Porto Airport

TAP Air Portugal

At Porto airport, TAP's operation (Figure 7) has suffered a significant drop in recent years. If, in 2019, the company operated to 21 airports (22 including the charter operation to Fuerteventura on Tuesdays), it dropped to 11 airports in 2021 and 2020. In addition to European destinations, TAP maintained its international operations to Newark (USA) and Rio de Janeiro and São Paulo (Brazil).

After the pandemic, the company abandoned traditional operations to Agadir (Morocco), Amsterdam (Netherlands), Barcelona (Spain), Brussels (Belgium), London City (United Kingdom), Milan Malpensa (Italy) and Porto Santo, in addition to Lamezia (Italy) and Oudja (Morocco) operated in charter operations. Of the 160 flights operated in 2019, the company saw its operation reduced by 64.4% in 2021 (57 flights operated), with a recovery in 2020 (+31.6%, that is, 75 flights operated). Compared to 2019, the operation in 2022 represents a drop of 53.1% of the flights operated.

Ryanair

With the creation of its operational base in Porto in September 2009, Ryanair has been increasing its presence at this airport over the years (Figure 8).

The airline's operation did not undergo major changes between 2019 and 2021 (only one less airport served, in a total of 55 destinations, with the exception of Gran Canaria airport in 2019 operated on a different day), being reinforced in 2022 to the 64 operated destinations. All operations are carried out in Europe, with the exception of three airports in Morocco (Marrakesh, Agadir, and Fez).

The greater diversification of markets, many of them operated with two or three flights per week, allows it to have a number of destinations far superior to its competition, as previously identified.

In its 2022 operation, the new destinations operated were Bari and Verona (Italy), Bergerac (France), Billund (Denmark), Agadir and Fez (Morocco) and Gran Canaria (resumed in September 2021). Likewise, the pandemic led to the closure of several operations, including flights to/from Lisbon, Napoli (Italy), Stuttgart (Germany) and Villacoublay (France). In 2021 Ryanair announced other proposals for new flights, such as Brivé and Toulouse (France, both in April 2021), Budapest (Hungary, in June 2020), Malaga (Spain, as of October 2019) and Vienna (Austria, as of October 2019, although previously operated by Lauda Air).

easyJet

easyJet created its operational base in Porto in March 2015, being in any case its biggest challenge at Lisbon airport. Even so, as in the Portuguese capital, easyJet has also been increasing its presence in Porto. The company operated to 17 airports in 2019 but in 2021 the operation was reinforced with one more destination (18), with greater visibility in 2022 with the focus on 24 airports served (Figure 9).

The new routes in summer 2022 were Berlin Brandenburg and Cologne (Germany), Madrid (Spain), Milan Malpensa (Italy), Porto Santo, Prague (Czech Republic) and Rennes (France). In 2019, the company discontinued flights to Malaga (Spain) and Stuttgart (Germany), having in the meantime started operations to Palma de Mallorca (Spain, in July 2021) and Zurich (Switzerland, from March 2018) and Montpellier (France), as of December 2019, however already discontinued).

In summary, TAP's operation, based on a traditional business model, in "*hub and spoke*", requires a balanced management of medium-haul destinations with long-haul capacity. The closure of many international markets limited operations to these destinations and affected operations on its European network and on the air bridge between Lisbon and Porto. In fact, it is in Porto that the greatest effects of the pandemic are felt, with the company divesting from this market, concentrating its operation in Lisbon. The strong recovery of the tourist market in Porto and North of Portugal in 2022, above 2019, has been visible in the tourism sector data (Turismo de Portugal, 2022), with increases in guests (+1.8%) and overnight stays (5.6%) between January and August this year. Likewise, despite the still negative values at the beginning of the year compared to 2019, between May and August 2022, Porto airport recorded a growth of 1.3% above the values of 2019, representing more than 5.1 million passengers in that period (ANAC, 2022).

On the other hand, the operations of low cost airlines maintained a remarkable degree of stability, with slight decreases in 2021 and a strong recovery in 2022. In the case of easyJet, the slight reduction in Lisbon (-2.1% of flights in 2022 when compared to 2019), contrasts with its growth in the operation in Porto (+44.1% of flights in 2022 compared to 2019), with 6 more new airports operated. In average terms, there is a balance between operations in 2022 in Lisbon, with 23.5 flights in the capital and 24.5 flights on average in Porto. The new routes to operate from Winter Iata, as a result of the allocation of slots transferred by TAP, will allow it to consolidate its position at this airport.

Ryanair's focus on served destinations, even if only on flights two to three times a week, means that the company has a very wide route plan when compared to easyJet. Overall, the operation remained stable even during the pandemic, with flights maintained in Lisbon in 2021 and strong growth in operations at both airports in 2022 (9 new airports operated in Porto and 4 in Lisbon compared to 2019 values). The successive information about new aircraft reinforcements and new routes not always implemented makes it difficult to assess their operational performance with many of the announcements of new flights on routes previously discontinued or already in operation.

The information collected previously allows us to assess the impacts of the pandemic on the operating network of the different airlines and their business models. The necessary flexibility imposed by changing market and public health conditions had different effects on the operation and management of destinations.

CONCLUSION

The present investigation aimed, through its application to Lisbon and Porto airports, to assess the impacts of the pandemic on the operating network of the traditional airline TAP Air Portugal and the two main low cost carriers operating in Portugal.

The differences in the business models of these airlines help to understand in each case how these companies operate and how they promote, serve and make their services available to customers.

The traditional operation based more on transfer traffic – *hub and spoke* – tends to contrast with low cost carriers, which are much more focused on a European network and on a passenger who prefers direct, point-to-point flights. The greatest difficulties in differentiating business models has led to the hybridization of the market, through a combination of different existing models.

The network, which translates into the product made available by the airlines to their customers, as assessed at the Lisbon and Porto airports, made it possible to identify a strong operational flexibility in the different airlines, seeking new routes and destinations and adjusting their network according to the characteristics of the market, restrictions imposed for reasons of public health and mobility due to confinement and border closures.

TAP, due to the size of its network and its presence at the main airport in Portugal, was the one that most felt the effects of the pandemic. The reduction in its operation was influenced by restrictions at international airports both in Africa and in the Americas, which also affected the installed capacity on European routes, given the loss of much of its transfer traffic. In the case of Porto airport, the drastic reduction of its operation left open wounds in the relationship with the North of Portugal, whose attitude was severely criticized and whose confidence with the market was greatly shaken.

In turn, low cost airlines had only slight adjustments to their network with very noticeable increases in 2022, in a bet on new destinations and reinforcement of the existing operation. The new routes operated by easyJet in Lisbon and the reinforcement of its operation, as a result of the assignment of slots by TAP, together with the increased presence of Ryanair in new routes, bring greater pressure on the national airline.

One of the great future challenges concerns the airport infrastructure in Lisbon, whose discussion on the study of the location of the new Lisbon airport began 50 years ago (GNAL, 1972). This situation has brought enormous operational and economic constraints to the city of Lisbon, with the airport being a brake on the development of tourism, systematic refusal of slots for new flights and huge losses due to the successive postponements of the new airport. According to the Confederation of Tourism of Portugal (CTP) the postponement of the airport to 2027 will cost the country €6.8 billion in gross value-added, representing a loss of 5.3 million tourists, and the option for Alcochete in 2034 will represent a loss of additional wealth of around €16.8 billion (Campos, 2022a).

The evolution shows, nowadays, a high competitiveness for passengers, where the most intangible aspects are increasingly important in the purchase decision and choice of an airline. The successive increases in the price of airport fees, where both business models are sensitive, together with strong operational restrictions at Lisbon airport, put greater pressure on airlines, on their costs and on their operation and where the current moment of recovery may prove insufficient to generate additional growth and enhanced business opportunities for the evaluated airlines.

REFERENCES

Abrantes, J. (2016). *O contributo das companhias aéreas de baixo custo para o desenvolvimento dos hostels nas cidades de Lisboa e Porto [The contribution of low cost carriers for the development of hostels in the cities of Lisbon and Oporto].* [PhD thesis, Universidade de Lisboa, Instituto de Geografia e Ordenamento do Território, Lisboa, Portugal].

Almeida, C., Costa, V., & Abrantes, J. (2020). Airline Business Models and Tourism Sector. In L. Carvalho, L. Calisto, & N. Gustavo (Eds.), *Strategic Business Models to Support Demand, Supply, and Destination Management in the Tourism and Hospitality Industry* (pp. 216–239). IGI Global., doi:10.4018/978-1-5225-9936-4.ch012

ANAC. (2022). *Monitorização Mensal do Tráfego Aéreo nas Infraestruturas Aeroportuárias Nacionais. [Monthly Air Traffic Monitoring in National Airport Infrastructures].* Autoridade Nacional de Aviação Civil. https://www.anac.pt/vPT/Generico/PublicacoesINAC/newslettersdetrafego/Paginas/Newsletters-deTrafego.aspx

Bakir, M., Akan, S., & Atalik, O. (2020). An evaluation for Long-Haul Low-Cost Carriers sing User-Generated Content: The impact of perceived service quality on Value for Money. In C. Ramos, C. R. Almeida, & P. O. Fernandes (Eds.), *Handbook of Research on Social Media Applications for the Tourism and Hospitality Sector* (pp. 231–251). IGI Global., doi:10.4018/978-1-7998-1947-9.ch014

Barrett, S. (1997). The implications of the Ireland-UK airline deregulation for an EU internal market. *Journal of Air Transport Management*, *3*(2), 67–73. doi:10.1016/S0969-6997(97)00003-3

Barrett, S. (2004). How do the demands for airport services differ between full-service carriers and low-cost carriers? *Journal of Air Transport Management*, *10*(1), 33–39. doi:10.1016/j.jairtraman.2003.10.006

Bitzan, J., & Peoples, J. (2016). A comparative analysis of cost change for low-cost, full-service, and other carriers in the US airline industry. *Research in Transportation Economics*, *56*, 25–41. doi:10.1016/j.retrec.2016.07.003

Bodell, L. (2022). *Despite a good 4th Quarter easyJet expects a Loss for 2022*. Simple Flying. https://simpleflying.com/easyjet-2022-loss-despite-good-q4/

BukhariS. (2011). What is Comparative Study? SSRN. doi:10.2139/ssrn.1962328

Button, K. (1998). Opening U.S. Skies to Global Airline Competition – Executive Summary. Center for Trade Policy Studies, CATO Institute.

Button, K. (2001). Deregulation and liberalisation of European air transport markets. *Innovation (Abingdon), 14*(3), 255–275. doi:10.1080/13511610120102619

Camilleri, M. (2018). *Travel Marketing, Tourism Economics and the Airline Product.* Springler International., doi:10.1007/978-3-319-49849-2

Campos, A. (2022a, July 14th). Adiar o aeroporto até 2027 vai custar ao país 6,8 mil milhões, diz CTP. Esperar por Alcochete resultará numa perda de 21,4 mil milhões. [Postponing the airport until 2027 will cost the country 6.8 billion, says CTP. Waiting for Alcochete will result in a loss of 21.4 billion]. *Expresso.* https://expresso.pt/economia/2022-09-29-Costa-espera-privatizacao-da-TAP-nos-proximos-12-meses-e-admite-que-Estado-possa-perder-dinheiro-d4e66e83

Campos, A. (2022b, September 20th). Costa espera privatização da TAP nos próximos 12 meses e admite que Estado possa perder dinheiro. [Costa expects privatization of TAP in the next 12 months and admits that the State could lose money]. *Expresso.* https://expresso.pt/economia/2022-09-29-Costa-espera-privatizacao-da-TAP-nos-proximos-12-meses-e-admite-que-Estado-possa-perder-dinheiro-d4e66e83

CAPA. (2020). *Aviation 2020 to 2030. A COVID reality check for airlines.* Centre for Aviation. https://centreforaviation.com/analysis/reports/aviation-2020-to-2030-a-covid-reality-check-for-airlines-545006

CAPA. (2022). *Europe's low cost carriers continue to drive region's capacity recovery.* Centre for Aviation. https://centreforaviation.com/analysis/reports/europes-low-cost-carriers-continue-to-drive-region-capacity-recovery-604992

Carmona-Benítez, R., & Lodewijks, G. (2008). *Literature review of the passenger airline business models.* TRAIL Research School.

CNN. (2002). EasyJet selects Airbus over Boeing. *CNN.com.* http://edition.cnn.com/2002/BUSINESS/10/14/airbus.easyjet/

Creaton, S. (2003, April 12th). Ryanair saves €3.8m on Buzz deal. *The Irish Times.* https://www.irishtimes.com/business/ryanair-saves-3-8m-on-buzz-deal-1.355488

Creaton, S. (2007). *Ryanair. The full story of the controversial low-cost airline.* Aurum Press.

Diário de Notícias. (2021, December 4th). "TAP parece ter desistido desta parte do país", acusa Turismo do Norte. ["TAP seems to have given up on this part of the country", accuses Turismo do Norte]. *Diário de Notícias.* https://www.dn.pt/dinheiro/tap-parece-ter-desistido-desta-parte-do-pais-acusa-turismo-do-norte-14379732.html

Dobruszkes, F. (2009). New Europe, new low-cost air services. *Journal of Transport Geography, 17*(6), 423–432. doi:10.1016/j.jtrangeo.2009.05.005

Doganis, R. (2006). *The airline business.* Routledge.

EasyJet. (2022). *Reports and Presentations.* EasyJet. https://corporate.easyjet.com/investors/reports-and-presentations/2022

easyJet (2017). *easyJet announces 'Worldwide by easyJet'*. easyJet. https://mediacentre.easyjet.com/story/4000/easyjet-announces-worldwide-by-easyjet

easyJet (n.d.). *Our Journey*. easyJet. https://corporate.easyjet.com/about/our-journey

Francis, G., Fidato, A., & Humphreys, I. (2003). Airport–airline interaction: The impact of low-cost carriers on two European airports. *Journal of Air Transport Management*, *9*(4), 206–273. doi:10.1016/S0969-6997(03)00004-8

Georgiadis, P. (2022, August 8). *Ryanair chief warns fares will rise for 5 years because flying is 'too cheap'*. Financial Times. 8 August https://www.ft.com/content/32108696-9ef5-49ad-8a07-fa30375702bf

GNAL. (1972). *Estudo da Localização do Novo Aeroporto de Lisboa* [Study for the Location of the New Lisbon Airport]. Ministério das Comunicações.

Goodbody, W. (2021, August 9th). *Ryanair to cease flying to and from Northern Ireland*. Rté.ie. https://www.rte.ie/news/business/2021/0824/1242608-ryanair-to-cease-flying-to-and-from-northern-ireland/

Hardiman, J. (2021, August 9). *Why Does easyJet Have Swiss And Austrian Subsidiaries?* Simple Flying. 9 August https://simpleflying.com/easyjet-swiss-austrian-subsidiaries/

IATA. (2022). *Airline Industry Economic Performance – June 2022*. International Air Transport Association. https://www.iata.org/en/iata-repository/publications/economic-reports/airline-industry-economic-performance---june-2022---data-tables/

ICAO. (2018). *Manual on the regulation of International Air Transport* (3rd ed.). International Civil Aviation Organization.

Indeed (2021). *What Is Comparative Analysis and How Is It Used?* Indeed. https://www.indeed.com/career-advice/career-development/comparative-analysis

Jean, D., & Lohmann, G. (2016). Revisiting the airline business model spectrum: The influence of post global financial crisis and airline mergers in the US (2011–2013). *Research in Transportation Business & Management*, *21*, 76–83. doi:10.1016/j.rtbm.2016.06.002

Lawton, T., & Solomko, S. (2005). When being the lowest cost is not enough: Building a successful lowfare airline business model in Asia. *Journal of Air Transport Management, 11*(6), 355–362. doi:.jairtraman.2005.03.001 doi:10.1016/j

Magdalina, A., & Bouzaima, M. (2021). An empirical investigation of European airline business models: Classification and hybridization. *Journal of Air Transport Management*, *93*, 102059. 10.1016/j.jairtraman.2021.102059

Massukado, M. S. (2008). Análise comparativa de estratégias qualitativas de investigação: Possibilidades para a pesquisa em turismo. [Comparative analysis of qualitative research strategies: possibilities for tourism research]. *Turismo e Sociedade*, *1*(1), 9–27.

Mrázová, M., & Kazda, A. (2021). The direction of airlines business models – up or down? (Post Covid-19 crisis – case study). *Incas Bulletin*, *13*(1), 263–274. doi:10.13111/2066-8201.2021.13.1.27

Neufville, R., & Odoni, A. (2003). *Airport Systems Planning, Design and Management*. McGraw Hill.

Osterwalder, A., & Pigneur, Y. (2010). *Business Models Generation*. John Wiley and Sons.

Patton, M. Q. (2002). *Qualitative research and evaluation methods* (3rd ed.). Sage Publications, Inc.

Pestana, J. F. (2022, June 16th). *EasyJet ganha 18 novos slots em Lisboa e torna-se a companhia aérea n.º 2 no aeroporto de Lisboa*. [EasyJet wins 18 new slots in Lisbon and becomes the #2 airline at Lisbon airport]. Diário de Notícias. https://www.dnoticias.pt/2022/6/16/315790-easyjet-ganha-18-novos-slots-em-lisboa-e-torna-se-a-companhia-aerea-no-2-no-aeroporto-de-lisboa/

Pickvance, C. G. (2001). Four varieties of comparative analysis. *Journal of Housing and the Built Environment*, *16*(1), 7–28.

Pordata (2022). *Tráfego de passageiros nos principais aeroportos: Lisboa, Porto e Faro*. [Passenger traffic at the main airports: Lisbon, Porto and Faro]. Pordata. https://www.pordata.pt/portugal/trafego+de+passageiros+nos+principais+aeroportos+lisboa++porto+e+faro-3248

Porto (2021). *TAP "não tem nenhuma utilidade para o Porto", critica Rui Moreira*. [TAP "has no use for Porto", criticizes Rui Moreira]. Porto.pt. https://www.porto.pt/pt/noticia/tap-nao-tem-nenhuma-utilidade-para-o-porto-critica-rui-moreira

Publituris (2021, August 5th). *Ryanair abre 22 novas rotas e coloca mais três aviões em Lisboa no inverno*. [Ryanair opens 22 new routes and places three more planes in Lisbon in winter]. Publituris. https://www.publituris.pt/2021/08/05/ryanair-abre-22-novas-rotas-e-coloca-mais-tres-avioes-em-lisboa-no-inverno

Publituris (2022, March 8th). *Ryanair retira três aviões de Lisboa e cancela 19 rotas devido a falta de slots*. [Ryanair withdraws three planes from Lisbon and cancels 19 routes due to lack of slots]. Publituris. https://www.publituris.pt/2022/03/08/ryanair-retira-tres-avioes-de-lisboa-e-cancela-19-rotas-devido-a-falta-de-slots

Russell, E. (2022). *Whats behind the airlines operational mess this summer*. Skift.com. https://skift.com/2022/06/27/whats-behind-the-airlines-operational-mess-this-summer/

Ryanair Group. (2019). *Annual Report 2019*. Ryanair. https://www.annualreports.com/HostedData/AnnualReportArchive/r/NASDAQ_RYAAY_2019.pdf

Ryanair Group. (2020). *Annual Report 2020*. Ryanair. https://www.annualreports.com/HostedData/AnnualReportArchive/r/NASDAQ_RYAAY_2020.pdf

Ryanair Group. (2021). *Annual Report 2021*. Ryanair. https://investor.ryanair.com/wp-content/uploads/2021/08/FINAL_Ryanair-Holdings-plc-Annual-Report-FY21.pdf

Ryanair Group. (2022a). *Annual Report 2022*. Ryanair. https://investor.ryanair.com/wp-content/uploads/2022/07/Ryanair-2022-Annual-Report.pdf

Ryanair Group. (2022b). Ryanair reports Half-Year profits of €1.37BN. Ryanair. https://investor.ryanair.com/wp-content/uploads/2022/11/H1-FY23-Results.pdf

Seddon, P., & Lewis, G. (2003). Strategy and Business Models: What's the difference? *In proceedings of the 7th Pacific Asia Conference on Information System*, (pp. 236-248). https://core.ac.uk/download/pdf/301340932.pdf doi10.1.1.200.7881&rep=rep1&ty pe=pdf

Silva, B. (2022, July 8th). A cidadania brasileira da TAP. [TAP's Brazilian citizenship]. Isto é Dinheiro. https://www.istoedinheiro.com.br/a-cidadania-brasileira-da-tap/

Statista. (2022). *Market share of low cost carriers (LCCs) in selected regions from 2008 to 2020*. Statista. https://www.statista.com/statistics/1097337/low-cost-carrier-market-regional-share/

Stoanescu, C., & Gheorghe, C. (2017). "Hybrid" airlines – Generating value between low-cost and traditional. *Proceedings of the International Conference on Business Excellence*, (pp. 577-587). doi:10.1515/picbe-2017-0062

Suspiro, A. (2022, July 27th). *Para onde vai a Easyjet voar com os slots entregues pela TAP: 13 novas rotas na Europa e Marrocos. [Where is Easyjet going to fly with the slots delivered by TAP: 13 new routes in Europe and Morocco].* Observador. https://observador.pt/2022/07/27/para-onde-vai-a-easyjet-voar-com-os-slots-entregues-pela-tap/

TAP. (2022). *Relatórios Anuais*. [Annual Reports] TAP. https://www.tapairportugal.com/pt/sobre-nos/relatorios-anuais

Tnews (2022, October 25th). Ryanair prevê aumentar tráfego de passageiros para Portugal em 36% até 2027. [Ryanair expects to increase passenger traffic to Portugal by 36% by 2027]. Tnews. https://tnews.pt/ryanair-preve-aumentar-trafego-de-passageiros-para-portugal-em-36-ate-2027/

Tomová, A., & Kirschnerová, I. (2015). Network strategies of European low cost and hybrid(ized) airlines: Is Ryanair unique? *Ekonomiczne problemy usług, 890*, 120-137.

Topham, G. (April, 6th). EasyJet secures £600m coronavirus loan from UK Treasury and Bank. *The Guardian*. https://www.theguardian.com/business/2020/apr/06/easyjet-secures-600m-coronavirus-loan-from-uk-treasury-and-bank

Tucker, R. (2001). Strategy Innovation takes Imagination. *The Journal of Business Strategy*, *22*(3), 23–27. https://dx.doi.org/10.1108/eb040168

Turismo de Portugal. (2022). Turismo em números – Agosto 2022. [Tourism in numbers – August 2022]. Turismo de Portugal. https://travelbi.turismodeportugal.pt/turismo-em-portugal/turismo-em-numeros-agosto-2022/

UNWTO. (2022). Word Tourism Barometer – May 2022: Vol. 20. *Issue 3*. World Tourism Organization.

Veríssimo, A. (2022, October 11th). *Capital increase will remove Pedrosa and workers from TAP SGPS*. [Capital increase will remove Pedrosa and workers from TAP SGPS]. Eco. https://eco.sapo.pt/2022/10/11/aumento-de-capital-tira-humberto-pedrosa-da-tap-sgps/

Weber, J., & Alabi, L. O. (2022, August 9th). Ryanair warns days of cheap fares are coming to an end. *Financial Times*. https://www.ft.com/content/1da8eff7-3da8-45db-98de-866b1f68817c

Whyte, R., & Lohmann, G. (2016). Airline business models. In L. Budd & S. Ison (Eds.), *Air Transport Management: An international perspective* (pp. 107–121). Routledge.

KEY TERMS AND DEFINITIONS

Cabotage: Possibility of operation on domestic routes by an airline from a third country (normally known as 8th or 9th freedom of the air).

Full Service Carrier: Airline whose business model tends to operate on hub and spoke, with a strong focus on transfer traffic, serving different classes of services onboard and offering various inflight facilities, mainly in long-haul flights

Hub & Spoke: Operation centralized at an airport (hub) where flights from markets converge (spokes), allowing their transfer between destinations and enlarging their flights network.

Hybrid Carrier: Airline whose business model tends to mix the most used models incorporating features from both full service carriers and low cost carriers.

Low Cost Carrier: Airline whose business model tends to operate on point-to-point flights, with a strong focus on cost reduction and where ancillary revenues represent a strong component of its business.

APPENDIX

Figure 1. Number of European airports served to/from Lisbon Airport (TAP Air Portugal)
Source: Own elaboration based on ANA – Aeroportos de Portugal website information

	TAP Air Portugal - Europe											
	2019				2021				2022			
Airports served	Friday	Saturday	Sunday	Monday	Friday	Saturday	Sunday	Monday	Friday	Saturday	Sunday	Monday
Alicante	1	1	1	1					1		1	1
Amsterdam	3	3	3	3	3	3	2	2	4	4	4	3
Barcelona	6	6	6	6	4	3	4	3	6	6	6	5
Basel	2	2	2	2								
Berlin Tegel	2	1	2	2								
Berlin Bradenburg					3	3	2	1	3	2	2	2
Bilbao	2	2	2	2	2	2	2	1	2	2	2	2
Bologna	1	1	1	1	1	1		1	1	1	1	1
Bordeaux	1	1		1								
Brussels	4	4	4	4	3	3	2	2	4	4	4	4
Budapest	1	1	2	1								
Cologne Bonn	2	2	2	2								
Copenhagen	2	2	2	2	2	2	2	1	2	3	2	2
Dublin	2	2	2	2	2	1	2	1	3	1	3	2
Dusseldorf	2	2	2	1	1	2	2	1	2	2	2	2
Faro	4	4	4	4	4	4	4	4	3	3	3	3
Florence	2	2	2	1	1	1	1		1	2	2	1
Frankfurt	4	3	3	4	2	3	2	1	3	3	3	3
Fuerteventura					1	1	1			1	1	
Funchal (Madeira)	8	8	8	8	5	5	5	5	7	7	7	7
Geneva	3	3	3	3	3	3	3	2	3	3	3	3
Gran Canaria	1	1	2	1	1	1	1	1	1	1	1	
Hamburg	2	2	2	2	2	2	1	1	2	2	2	2
Helsinki	1	1	2	1								
Heraklion				1								
Ibiza						1	1		1	1		
London City	2	1	1	2								
London Gatwick	1	2	1	2					1	2	1	2
London Heathrow	6	6	7	6	3	5	4	4	6	6	7	6
Luxembourg	1	1	1	1	1	1	2	1	1	1	1	1
Lyon	3	2	3	3	1	3	3	2	2	3	3	1
Madrid	7	6	6	6	5	5	5	5	7	7	7	7
Malaga	2	2	2	2	2	2	3	1	3	3	4	1
Manchester	2	2	2	2	1	1	1	1	3	2	3	2
Marseille	2	2	2	2	1	3	3	2	2	3	3	2
Milan Malpensa	4	4	4	4	3	3	3	2	4	4	4	4
Moscow	1	1	1	1				1				
Munich	3	2	3	3	2	2	2	1	3	3	3	3
Nantes	1	1	1	1								
Naples	1	1	1	1	1		1	1	1	2	1	2
Nice	3	3	3	3	1	3	3	2	3	4	5	4
Olbia			1									
Oslo	1	1		1					1	1	2	1
Paris Orly	7	8	8	7	5	5	5	4	8	8	8	8
Ponta Delgada (Azores)	4	4	4	4	3	3	3	1	3	3	3	3
Porto	13	13	13	13	7	7	8	7	9	10	10	8
Porto Santo (Madeira)	2	1	2	2	1	1	2	1	1	1	1	1
Prague	1	3	1	2	1	1	2	1	1	1	1	1
Rome Fiumicino	5	5	5	5	3	3	3	3	4	4	4	3
Seville	3	3	3	3	1	1	2	1	2	3	3	3
Stockholm	2	2	1	2	1	1	1	1	1	1	1	1
Stuttgard	2	2	2	2								
Tel Aviv	1	1	1	1					1		1	1
Tenerife Sur	1	1	1	1	1	1	1	1	1	1	1	
Terceira (Azores)	2	2	2	2	1	1	1	1	2	2	2	2
Toulouse	2	3	3	3	1	3	3	2	2	3	3	1
Valencia	2	2	2	2	1	2	2	1	2	3	3	2
Venice	2	2	2	2	1	2	2	1	2	1	2	2
Vienna	2	2	2	2	1	1	1	1	1	1	1	1
Warsaw	2	1	2	1	2	1	2	2	1	1	1	1
Zagreb					1		1		1	1	1	
Zurich	3	3	3	3	3	3	3	2	3	3	3	3
Total	**152**	**149**	**153**	**152**	**95**	**105**	**109**	**80**	**131**	**136**	**142**	**120**

Figure 2. Number of African airports served to/from Lisbon Airport (TAP Air Portugal)
Source: Own elaboration based on ANA – Aeroportos de Portugal website information

	TAP Air Portugal - Africa											
	2019				2021				2022			
Airports served	Friday	Saturday	Sunday	Monday	Friday	Saturday	Sunday	Monday	Friday	Saturday	Sunday	Monday
Abidjan	1	1		1	1		1					
Accra			1									
Agadir					1	1	1	1	1	1	1	1
Banjul										1		
Bissau		1				1				1		
Boavista		1	1							1		
Casablanca	1	1	1	1	2	1	2	2	2	2	2	2
Conakry	1		1		1	1			1		1	
Dakar	1	1	2	1	2	1	2	1	1	1	1	1
Djerba							1				1	
Fez	1	1	1	1								
Lome	1			1								
Luanda	2	1	2	1		1	1		1	1	1	1
Maputo	1		1	1	1		1		1		1	
Marrakech	2	2	2	2	2	2	2	2	2	2	2	2
Monastir				1	1						1	
Oujda		1	1			1	1					
Praia	2	2	2	2	2	2	2	2	3	2	2	2
Sal	1	2	1	1		1	1		1	1	2	1
Sao Tome		1	1		1	1	1			1	1	1
Sao Vicente	1	1	1	1	1	1	1	1	1	1	1	1
Tanger	1	1	1	1	1	1	1		1	1	1	1
Total	**16**	**17**	**19**	**15**	**16**	**15**	**18**	**9**	**15**	**16**	**18**	**13**

Figure 3. Number of North and Central America airports served to/from Lisbon Airport (TAP Air Portugal)
Source: Own elaboration based on ANA – Aeroportos de Portugal website information

	TAP Air Portugal - North and Central America											
	2019				2021				2022			
Airports served	Friday	Saturday	Sunday	Monday	Friday	Saturday	Sunday	Monday	Friday	Saturday	Sunday	Monday
Boston	1	1	1	1	1	1	1	1	2	2	1	1
Cancun					1	1			1	1		1
Chicago	1	1	1	1	1		1		1		1	1
Miami	1	1	1	1	1		1	1	1	1	1	1
Montreal					1		1	1	1		1	1
New York JFK	1	1	1	1					1	1	1	1
Newark	1	1	1	1	2	2	2	2	2	2	2	2
Punta Cana										1		1
San Francisco		1	1	1	1		1		1		1	
Toronto	1	1	1	1	1	1	1	1	1	1	2	2
Washington D.C.	1		1	1	1	1	1	1	1	1	2	1
Total	**7**	**7**	**8**	**8**	**10**	**6**	**9**	**7**	**12**	**10**	**12**	**12**

Figure 4. Number of South America airports served to/from Lisbon Airport (TAP Air Portugal)
Source: Own elaboration based on ANA – Aeroportos de Portugal website information

	TAP Air Portugal - South America											
	2019				2021				2022			
Airports served	Friday	Saturday	Sunday	Monday	Friday	Saturday	Sunday	Monday	Friday	Saturday	Sunday	Monday
Belem	1		1	1						1		
Belo Horizonte	1	1	1	1		1	1		1	1	1	1
Brasilia	1	1	1	1			1		1		1	1
Caracas		1								1		
Fortaleza	1	1	1	1	1		1	1	1	1	1	1
Maceió										1		
Natal		1	1						1			
Porto Alegre		1		1						1		
Recife	1	2	1	1	1	1		1	1	1	1	1
Rio de Janeiro	2	2	1	2	1	1	1		1	2	1	2
São Salvador da Bahia	1	1	1	1	1		1		1	1		1
São Paulo	3	3	3	3	2	1	2	2	1	3	2	4
Total	**11**	**14**	**11**	**12**	**6**	**4**	**7**	**4**	**8**	**13**	**7**	**11**

Figure 5. Number of airports served to/from Lisbon Airport (Ryanair)
Source: Own elaboration based on ANA – Aeroportos de Portugal website information

	Ryanair											
	2019				2021				2022			
Airports served	**Friday**	**Saturday**	**Sunday**	**Monday**	**Friday**	**Saturday**	**Sunday**	**Monday**	**Friday**	**Saturday**	**Sunday**	**Monday**
Barcelona									1		1	1
Berlin Schonefeld	1	1	1	1								
Berlin Brandenburg					1		1	1	1	1	1	
Bologna		1		1		1		1	1	1	1	2
Bordeaux	1		1	1	1	1	1	1	1		1	
Brussels	2	2	1	2	2	2	1	2	1	1	1	1
Brussels Charleroi	1	1	1	1	1	1	1	1	2	2	1	1
Budapest						1				1		
Clermont-Ferrand		1										
Cologne Bonn							1				1	
Dublin	2	2	2	2	2	2	2	2	2	2	2	3
Edinburg		1				1				1		
Eindhoven	1		1	1	1		1	1	1			1
Fez											1	
Funchal (Madeira)									2	2	2	2
Frankfurt	1	1	1	1								
Hamburg		1										
Karsruhe		1										
Krakow			1				1				1	
London Stansted	4	3	4	4	4	2	4	4	3	4	2	4
Lourdes		1										
Luxembourg		1	1	1		1	1	1	1	1		1
Malaga							1		1		1	
Manchester	1	1	1	1	1	1		1		1	1	
Marrakech									1			1
Marseille	1	1	1	1	1	1	1	1	1	1	1	1
Memmingen							1					
Milan Bergamo	1	1	1	1	1	1	1	1	2	2	2	2
Naples	1			1	1			1			1	
Paris Beauvais	1	1	1	1	1	1	1	1	1	1	1	1
Pisa		1				1				1	1	
Ponta Delgada (Azores)	3	2	2	2	3	2	2	2	3	2	2	2
Porto	2		2	1								
Rome Ciampino	2	2	2	2	2	1	2	2	2	2	1	2
Seville					1			1	1			1
Terceira (Azores)	1		1	1	1		1	1			1	1
Toulouse	1	1	1	1	1	1	1	1	1	1	1	1
Valencia						1				1		
Venice									1	1		1
Vienna					1	1	1	1	1	1	1	1
Warsaw Modlin				1	1			1			1	
Wroclaw		1										
Total	**27**	**28**	**26**	**28**	**27**	**23**	**26**	**28**	**31**	**30**	**30**	**30**

Figure 6. Number of airports served to/from Lisbon Airport (easyJet)
Source: Own elaboration based on ANA – Aeroportos de Portugal website information

	easyJet											
	2019				2021				2022			
Airports served	**Friday**	**Saturday**	**Sunday**	**Monday**	**Friday**	**Saturday**	**Sunday**	**Monday**	**Friday**	**Saturday**	**Sunday**	**Monday**
Amsterdam	1		1	1	1		1	1	1		1	1
Basel	1	1	1	1	1	1	1	1	1	1	1	1
Berlin Schonefeld	1			1								
Berlin Brandenburg									1			1
Bordeaux	2	1	1	2	1	1	1	1	2	1	1	1
Bristol	1		1		1		1		1		1	
Edinburg			1								1	
Funchal (Madeira)	3	2	3	3	4	4	5	4	3	3	4	2
Geneva	3	3	3	3	3	3	3	3	3	3	3	3
Lille		1										
London Gatwick	2	2	2	2	1	1	1		2	3	3	2
London Luton	3	2	2	3	2	1	2	1	3	2	3	3
Luxembourg	1		1	1			1				1	
Lyon	1	1	1	1	1	1	1	1	1	1	1	1
Madrid	2	1	2	2	1		1	1	1		1	1
Manchester	1		1	1	1		1	1	1		1	1
Milan Malpensa	1	1	1	1	1	1	1	1	2	2	2	2
Nantes	1	1	1	1	1	1	1	1	1	1		1
Nice		1				1				1		
Paris Charles de Gaulle	2	2	2	2	2	2	2	2	2	2	2	2
Porto Santo (Madeira)											1	
Zurich		1	1									
Total	**26**	**20**	**25**	**25**	**21**	**17**	**23**	**18**	**25**	**20**	**27**	**22**

Figure 7. Number of airports served to/from Porto Airport (TAP Air Portugal)
Source: Own elaboration based on ANA – Aeroportos de Portugal website information

	TAP Air Portugal											
	2019				2021				2022			
Airports served	**Friday**	**Saturday**	**Sunday**	**Monday**	**Friday**	**Saturday**	**Sunday**	**Monday**	**Friday**	**Saturday**	**Sunday**	**Monday**
Agadir			1									
Amsterdam	1	1	1	1								
Barcelona	2	2	2	2								
Brussels	2	2	2	2								
Funchal (Madeira)	3	2	2	3	2	2	2	2	2	2	2	2
Geneva	2	2	2	2	1		1		1			1
Lamezia	1											
Lisbon	13	13	13	13	6	6	6	7	9	10	9	8
London City	1		1	1								
London Gatwick	2	2	2	2	1	1	1	1	2	2	2	2
Luxembourg	1	1	1	1	1		1	1	1		1	1
Madrid	3	2	2	3								
Milan Malpensa	2	2	2	2								
Newark	1	1	1	1	1		1				1	
Oujda		2	1									
Paris Orly	3	3	3	3	2	2	2	2	3	3	3	3
Ponta Delgada (Azores)	1	1	1	1		1				1		
Porto Santo (Madeira)				1								
Rio de Janeiro	1		1					1	1			
Sao Paulo		1				1				1		
Zurich	2	2	2	2		1		1		1		1
Total	**41**	**39**	**40**	**40**	**14**	**14**	**14**	**15**	**19**	**20**	**18**	**18**

Figure 8. Number of airports served to/from Porto Airport (Ryanair)
Source: Own elaboration based on ANA – Aeroportos de Portugal website information

	Ryanair											
	2019				2021				2022			
Airports served	Friday	Saturday	Sunday	Monday	Friday	Saturday	Sunday	Monday	Friday	Saturday	Sunday	Monday
Agadir										1		
Alicante			1				1				1	
Barcelona	3	2	3	3	1	1	1	1	3	2	3	3
Bari											1	
Bergerac											1	
Berlin Schonefeld	1		1	1								
Berlin Brandenburg					1		1		1		1	1
Billund											1	
Birmingham			1				1				1	
Bologna		1				1				1		
Bordeaux		1				1			1		1	1
Bremen		1			1			1	1			1
Brest		1				1						
Brive					1			1	1			1
Brussels	1	1	2	1	1	1	1	1	1	1	1	1
Brussels Charleroi	1	1	1	1	1	1	1	1	1	1	1	1
Budapest						1				1		
Cagliari	1				1			1	1			1
Carcassonne			1				1				1	
Clermont-Ferrand			1								1	
Cologne Bonn	1	1	1	1	1		1	1	1	1	1	1
Copenhagen	1			1	1			1	1			1
Dole		1				1				1		
Dortmund			1				1				1	
Dublin	1	1	1	1	1	1	1	1	1		1	1
Dusseldorf Weeze			1				1				1	
Edinburg		1				1				1		
Eindhoven	1		1	1	1		1	1	1	1	1	1
Faro	2	1	2	2	2	1	2	2	2	2	2	2
Fez											1	
Frankfurt Hahn	1	1	1	1	1	1	1	1			1	
Funchal (Madeira)									1	1	1	1
Gran Canaria									1			
Hamburg		1				1				1		
Karsruhe	1			1	1			1	1			1
Krakow	1			1	1			1	1			1
La Rochelle		1				1				1		
Lille		1				1				1		
Lisbon	2		2	1								
Liverpool	1			1	1			1		1		
London Stansted	3	3	3	3	2	3	2	2	3	3	3	3
Luxembourg	1	1	1	1	1	1	1	1	1	1	1	1
Madrid	3	2	2	2	1	1	1	1	2	2	2	2
Malaga						1				1		
Malta	1			1	1			1	1			1
Manchester		1			1		1	1			1	1
Marrakech	1			1	1			1	1			1
Marseille	1	1	1	1	1	1	1	1	2	1	1	1
Memmingen		1				1				1		
Milan Malpensa	1		1	1		1	1	1	1		1	
Milan Bergamo	2	1	2	1	1	1	1	1	1	1	1	1
Naples			1									
Nurenberg		1									1	
Palma de Mallorca		1			1			1	1	1		1
Paris Beauvais	3	2	2	3	2	2	2	2	3	3	2	3
Paris Vatry			1				1				1	
Ponta Delgada (Azores)	1		1	1	1	1	1	1	1	1	1	1
Rome Ciampino	1			1	1				1	1		1
Seville	1		1	1	1		1		1	1		1
Strasbourg	1			1								
Tenerife Sur			1				1				1	
Terceira (Azores)		1			1			1		1		
Toulouse						1		1		1		
Tours		1				1		1		1		
Valencia	1		1	1	1		1			1	1	
Venice		1			1					1		
Verona												1
Vienna					1		1	1	1		1	1
Villacoublay	1			1								
Warsaw Modlin		1				1				1		
Total	**41**	**35**	**39**	**37**	**36**	**31**	**31**	**34**	**40**	**41**	**42**	**39**

Figure 9. Number of airports served to/from Porto Airport (easyJet)
Source: Own elaboration based on ANA – Aeroportos de Portugal website information;

	easyJet											
	2019				2021				2022			
Airports served	**Friday**	**Saturday**	**Sunday**	**Monday**	**Friday**	**Saturday**	**Sunday**	**Monday**	**Friday**	**Saturday**	**Sunday**	**Monday**
Basel	2	2	2	2	2	2	2	2	2	2	2	1
Berlin Brandenburg									1		1	1
Bordeaux			1		1		1	1	1		1	1
Bristol		1				1				1		
Cologne Bonn									1			1
Madeira (Funchal)	1		1		2	2	2	2	2	3	2	2
Geneva	4	4	4	4	5	4	5	4	5	4	5	5
Ibiza		1		1		1				1		1
London Gatwick	1	2	2	1		1	1		1	2	2	1
London Luton		1	1			1	1		1	1	1	
Luxembourg			1		1		1		1	1	1	1
Lyon	2	1	2	2	2	1	2	2	2		2	2
Madrid									1		1	1
Malaga		1										
Manchester	1		1		1		1		1		1	
Milan Malpensa									1		1	
Montpellier						1						
Nantes	1	1	1	1	1	1	1	1		1	1	1
Nice		1				1				1	1	1
Palma de Mallorca							1				1	
Paris Charles de Gaulle	2	2	2	3	2	2	2	3	1	2	3	3
Porto Santo											1	
Prague										1		
Rennes										1		
Stuttgard	1											
Toulouse	1	1	1	1	1	1	1	1	1	1	1	1
Zurich					1		1	1	1		1	1
Total	**16**	**18**	**19**	**15**	**19**	**19**	**22**	**17**	**23**	**22**	**29**	**24**

Chapter 11
An Analysis of Factors Influencing Green Purchase Intention in Airlines:
An Interpretive Structural Modeling Approach

Mahmut Bakır
https://orcid.org/0000-0002-3898-4987
Samsun University, Turkey

Nadjim Mkedder
https://orcid.org/0000-0002-4523-2038
Anadolu University, Turkey

ABSTRACT

This chapter intends to investigate the factors influencing the green purchase intention of passengers in the context of the airline industry and to develop a hierarchical interdependencies model among these factors. To this end, consensus-based interviews with academic experts were conducted. Drawing on a total of 8 factors defined, interpretive structural modeling (ISM) created a hierarchical model. The Matrice d'Impacts Croisés Multiplication Appliquée á un Classement (MICMAC) analysis was additionally applied to validate the ISM model and determine the driving and dependence power of each factor. In terms of influencing green purchasing intention, the results demonstrated that perceived consumer effectiveness, subjective norm, environmental knowledge, environmental awareness, and environmental concern were at the root of the hierarchical model, and green image was at the top. In order to manage passengers' green purchase intentions, policymakers and marketers could benefit from the ISM model designed to comprehend the interdependencies between factors.

DOI: 10.4018/978-1-6684-6835-7.ch011

INTRODUCTION

Since deregulation in the U.S. in 1978, the airline industry has seen substantial growth. For example, from 2010 to the outbreak of the pandemic in 2020, the number of passengers carried by airlines doubled to 4.56 million (World Bank, 2021). Indeed, the airline business is a significant economic industry in terms of the number of passengers carried, its contribution to the global gross domestic product (GDP), and employment. Nevertheless, despite its economic benefits, this industry is detrimental to natural resources and the environment. Beyond minor effects, like noise pollution, the aviation industry is accountable for significant long-term effects, such as carbon emissions (Baumeister, Zeng, et al., 2022).

The global aviation industry produces 2.1% of all carbon dioxide (CO2) emissions resulting from human activity. Furthermore, aviation activities are responsible for 12% of CO2 emissions of all modes of transport (ATAG, 2020). Consequently, the aviation industry is currently in a position where both economic contributions and losses must be accounted for. In response to these concerns, the aviation industry plans to achieve zero carbon emissions by 2050 through the Fly Net Zero project (IATA, 2021). Furthermore, the concern for sustainability in the civil aviation industry has been raised in the academic arena. For the Journal of Air Transport Management (JATM), Merkert (2022) analyzed the status of the air transport management research area between 2015 and 2021. Although sustainability is a long-established theme of JATM, Merkert (2022) asserts that sustainability has room for growth. Rocha et al. (2022) highlighted that sustainability is disregarded as a quality concern in airport terminals. Therefore, in the future, airport management will place a greater emphasis on the energy efficiency of terminals, green spaces, and recycling (Rocha et al., 2022).

The level of consumer awareness and understanding about the degradation of the environment and its effect on social life is constantly increasing. This condition has spawned a procedure in which consumers examine their purchase decisions in light of environmental and human health considerations (Varshneya et al., 2017). Consequently, consumers seek eco-friendly products due to rising environmental consciousness and associated concerns (Zhuang et al., 2021). Therefore, it is crucial to comprehend the purchasing patterns of environmentally concerned consumers in the airline industry, a major source of CO2 emissions.

Consumers' green purchase intention is the best manifestation of green purchase behavior. A better understanding of this phenomenon is of great importance for businesses to formulate marketing strategies. To support environmentally friendly products, marketers must think about consumer preferences and decision-making procedures (Zhuang et al., 2021). Various studies have been conducted on the factors affecting consumers' green purchase intention (Mishal et al., 2017; Ruangkanjanases et al., 2020; Sreen et al., 2018). Given the importance of sustainability in the airline industry, as discussed above, a deeper understanding of green purchase intention is critical.

Many studies focusing on green purchase intention on airlines have resorted to techniques such as regression analysis and structural equation modeling (SEM) (Baumeister, Nyrhinen, et al., 2022; Mohd Suki, 2016; Wong et al., 2020). However, such methods fail to discover causal relationships between factors and provide a multi-level solution by decomposing the problem into smaller substructures. To bridge this gap, this research aims to identify the factors affecting green purchasing intention in the airline industry and to shed light on their causal relationships through the use of Interpretive Structural Modeling (ISM) and Matrix Impact Cross-Reference Multiplication Applied to a Classification (MICMAC) approach. ISM establishes a hierarchical model for the effective application of factors while exploring the interrelationships of factors in a research problem (Khan et al., 2020). To the authors' knowledge, no studies have explored causal relationships between factors influencing green purchase intention using ISM in the airline context.

The above-mentioned issues have not yet been comprehensively addressed in the airline industry. Therefore, this research seeks answers to the following research questions:

RQ1. What are the factors that affect green purchase intention for airlines business, and what kind of relationship is there between these factors?

RQ2. Can these relationships be hierarchically ordered in a structural model?

This research aims to contribute to the existing literature in various aspects. From a managerial perspective, this research helps practitioners understand the critical factors that stimulate green purchase intention and develop marketing strategies accordingly. On the other hand, the integrated application of ISM and MICMAC enables the relationships between factors affecting green purchase intention to be revealed on a systematic level in the context of the airline industry. Therefore, this study is the first to examine the relationship between factors affecting green purchase intention in the airline business.

The remainder of the chapter is addressed as follows: Section 2 focuses on the literature review and outlines pertinent factors affecting green purchase intention. Section 3 demonstrated the application of ISM and MICMAC in developing the research model. Section 4 provides the findings of integrated ISM and MICMAC methodology. Next, the research findings are discussed in Section 5, and various implications are presented, followed by the last section that concludes the present chapter with limitations.

LITERATURE REVIEW

Green Purchase Intention

In the literature, terms such as "green purchase," "green acquisition," and "environmentally responsible purchasing" are used to describe the green purchasing behavior of consumers (Zaremohzzabieh et al., 2021). As outlined by Hwang and Lyu (2020), green purchasing is the acquisition of services and products that have a slight environmental impact. It is most often represented by green purchase desire, which is the intention of customers to purchase and pay for green goods and services. Recently an enormous number of consumers are shifting their purchasing patterns to favor green products and services (Zaremohzzabieh et al., 2021). According to Chen and Chang (2012), green products satisfy the needs and desires of customers without negatively impacting the environment. Baumeister et al. (2022) stated that a green product minimizes its environmental effect over its entire life cycle. In this study, the authors identify the green purchase intention as the consumers' intention to go green in their consumption decisions to protect the environment. Furthermore, based on a thorough literature review, the authors acknowledge green perceived value, perceived consumer effectiveness, consumer attitudes, subjective norm, green image, environmental knowledge, environmental awareness, and environmental concern as factors that motivate airline consumers to engage in green purchasing behavior.

Factors Affecting Green Purchase Behavior

Green Perceived Value

Perceived value is defined as the consumers' overall evaluation of what they give and what they receive from using a product or service based on their environmental preferences, aspirations of sustainability, and green requirements (Chen & Chang, 2012; Zeithaml, 1988). Previous research has demonstrated that perceived value is vital to marketing performance since it enables businesses to foster customers' purchase intentions (Zhuang et al., 2021).In the context of eco-friendliness, environmentally conscious customers will participate in green products and services for their own environmental benefits (H.-C. Wu et al., 2018). As a high standards state variable in the consumer purchasing process, perceived value can be regarded as an indicator for consumer evaluation and a major predictor of purchase intention in the airline industry (Zhuang et al., 2021). Thus, the more consumers perceive the value of green airline products and services, the greater the consumer intention to purchase green.

Perceived Consumer Effectiveness

The extent to which consumers feel their actions help resolve issues like environmental issues are defined as their perceived consumer effectiveness (Zhuang et al., 2021). Perceived consumer effectiveness is the consumer's belief that it may participate in solutions and decrease the negative influence on the environment (Naz et al., 2020). Prior research has demonstrated that perceived consumer effectiveness is directly related to consumers' attitudes toward green products and is a significant indicator of purchase intention (Sharma & Foropon, 2019; Zhuang et al., 2021). In the present chapter, the authors consider perceived consumer effectiveness as the consumers believe that their green purchase behaviors will contribute significantly to solving the environmental problems in the airline industry.

Consumer Attitude

Consumer attitude refers to the consumer's overall evaluation and judgment toward products or services (Ajzen, 1991). Consumer attitudes can be regarded as a significant element in the development of effective strategic choices (Hwang & Lyu, 2020). Prior studies elucidated the significant positive effect of consumer attitude in predicting brand attachment and green purchase behavior (Albayrak et al., 2013; Tan, 2011; Zhuang et al., 2021). In this research, consumer attitudes are regarded in a way that may help the consumer to enhance airline consumers' evaluations, judgments, and intentions to use and purchase green products and services.

Subjective Norms

Subjective norm refers to the social pressure experienced by people in determining whether or not to engage in a certain action (Ajzen, 1991). Individuals are often impacted by the people around them while making decisions. It represents how people are influenced in society and how their reference group will regard them if they engage in certain actions (Ajzen, 1991). Previous studies have demonstrated that individuals conform to subjective norms because of fear of social pressure from their referents or because their referents give direction on suitable or advantageous social behavior, and it was found that subjective norms have a significant effect on consumers' green purchase intention (Albayrak et al., 2013; Sharma & Foropon, 2019; Zhuang et al., 2021). In the context of airlines' green purchases, the subjective norms can be regarded as the consumer's engagement to purchase green airline products in order to satisfy the reference group and feel like members of their environmentally interested reference group.

Green Image

Green image is the customers' perception of environmental issues, such as global warming and habitat loss, and their desire to purchase environmentally friendly products and services (Hwang & Lyu, 2020). Several studies addressed the crucial role of the companies' green image in influencing consumer green purchase intention (Chen, 2008; Hagmann et al., 2015; H.-C. Wu et al., 2018). In this research, the green image refers to consumer awareness and perception of environmental issues and their tendency to purchase green airline products and services.

Environmental Knowledge

Environmental knowledge is regarded as the capacity of individuals to recognize environmental and sustainability-related concerns (Naz et al., 2020). Environmental knowledge can further be characterized as the quantity of information a person remembers that impacts their evaluation on and interpretation of the available selection of products and services (Tan, 2011). The existing literature concluded that environmental knowledge significantly influences environmentally friendly purchase intention (Albayrak et al., 2013; Naz et al., 2020; Zhuang et al., 2021). This research defines environmental knowledge as an individual's impression of their knowledge about extensive environmental problems.

Environmental Awareness

Environmental awareness is defined as "consumers' understanding of the effects of human actions on the environment" (Nayeri & Moradi, 2019). Environmental awareness consists of cognitive, knowledge-based, and emotive, perceptual components (Kollmuss & Agyeman, 2002). Previous studies emphasize the crucial role of consumers' environmental awareness in influencing their green purchase decision (Lee et al., 2012; Nayeri & Moradi, 2019). In this research, environmental awareness can be regarded as the consumer green purchase choice that stems from cognitive awareness of environmental issues.

Environmental Concern

Environmental concern is the level of concern for environmental issues and reflects efforts to solve these issues (Albayrak et al., 2013; Kollmuss & Agyeman, 2002). Environmental concern is a strong behavior toward protecting the environment (Nayeri & Moradi, 2019). Previous studies indicated a strong association between environmental concern and green purchase intention (Lee et al., 2012; Nayeri & Moradi, 2019). In the present research, the authors identify environmental concern as the consumer's strong attitude towards airlines' green products and services, influencing consumer's decision to select green products and services.

Table 1. Constructs list along with their definitions and references

No	Constructs	Definition	Reference(s)
F1	Green perceived value	Refers to consumers' overall evaluation of what they give and what they receive from using a product or service based on their environmental preferences, aspirations of sustainability, and green requirements.	(Biswas, 2017; H.-C. Wu et al., 2018; Zhuang et al., 2021)
F2	Perceived consumer effectiveness	Refers to the extent to which customers believe their own activities help to solve issues and to their self-perception of their contribution to their own effort.	(Sharma & Foropon, 2019; Tan, 2011; Zhuang et al., 2021)
F3	Consumer attitudes	Refers to the attitudes regarding a certain product or service that are seen as a significant element in the development of effective strategic choices.	(Hwang & Lyu, 2020; Woo & Kim, 2019; Zaremohzzabieh et al., 2021)
F4	Subjective norm	Refers to the societal pressure people sense in deciding whether or not to engage in a certain behavior, such as going green.	(Albayrak et al., 2013; Zaremohzzabieh et al., 2021; Zhuang et al., 2021)
F5	Green image	Refers to customers' awareness of environmental issues, such as global warming and habitat loss, and their desire to purchase environmentally friendly products and services.	(Chen, 2008; Hagmann et al., 2015; Hwang & Lyu, 2020; Mayer et al., 2012)
F6	Environmental knowledge	Refers to the understanding of the environment, the important linkages that lead to environmental problems, and the individual's environmental duty that contributes to a sustainable future.	(Armstrong et al., 2021; Kollmuss & Agyeman, 2002; Zhuang et al., 2021)
F7	Environmental awareness	Refers to "knowing the impact of human behavior on the environment."	(Dangi et al., 2020; Kollmuss & Agyeman, 2002; S.-I. Wu & Chen, 2014)
F8	Environmental concern	Refers to the level of concern for environmental issues and reflects efforts to solve these issues.	(Albayrak et al., 2013; Dangi et al., 2020; Nguyen et al., 2019; Zhuang et al., 2021)

RESEARCH METHODS

In order to seek answers to the research questions, a hybrid research methodology was designed for this research. First, an extensive literature search identified factors that influence green purchase intention in airlines. Using the ISM method, the authors modeled the identified factors in a structural model according to their causal networks. Finally, MICMAC analysis categorized these factors according to their driving and dependence power. This section elaborates on the steps of the ISM technique and its successor analysis, the MICMAC analysis.

ISM Methodology

Developed by Warfield (1974), ISM is a mathematical method used to simplify complex problems into smaller fundamental factors using the knowledge of field experts (Gupta & Dhingra, 2022). This method provides insight into the causal relationships between the components by dividing the problem into different factors and establishing a multi-layered structural model (Mondal & Chakrabarti, 2021; Qarnain et al., 2021).

ISM is an interpretive approach since it relies on expert opinions to explain the relationships between the various components (Gan et al., 2018). Furthermore, it is structural as it directly or indirectly links variables for a transparent and well-articulated model (Gupta & Dhingra, 2022; Raut et al., 2018). This methodology is also widely employed in consumer behavior research (Khan et al., 2020; Mondal & Chakrabarti, 2021; S. Singh & Gupta, 2021; Xiao, 2018). To the authors' knowledge, no previous research has utilized this methodology to examine the factors influencing green purchase intention in the airline business. Therefore, it is useful to examine the interplay among the factors affecting green purchase intention and to divide this problem into various layers. In order to determine the driving force (DRP) and dependence power (DNP) of factors, the ISM technique also utilizes the MICMAC diagram (Raut et al., 2018).

The following are the stages of the ISM approach to investigate the factors influencing airlines' green purchasing intention (Gan et al., 2018; S. Singh & Gupta, 2021). Figure 1 also depicts the stepwise flowchart of the ISM approach.

Step 1. Identify the factors that influence green purchasing intention in airlines through a comprehensive literature review.
Step 2. Create interrelationships between the identified factors.
Step 3. Develop a structural self-interaction matrix (SSIM) to establish the pairwise relationships between the factors.
Step 4. Create an initial reachability matrix from the SSIM matrix.
Step 5. Create the final reachability matrix by applying the transitivity rule to the initial reachability matrix. Transitivity in mathematics implies that if factor A affects factor B and factor B affects factor C, then factor A necessarily affects factor C (Khan et al., 2020).
Step 6. Perform level partitioning of the final reachability matrix.
Step 7. Establish the conical form of the final reachability matrix.
Step 8. Develop a diagram and interpret the structural model.
Step 9. Establish the driving power and dependence power of factors employing MICMAC analysis.

MICMAC Analysis

MICMAC technique uses the final reachability matrix data from ISM as input. MICMAC analysis creates a graph that plots the factors on the x-axis and y-axis according to their driving and dependence power (Khan et al., 2020). On the graph, the dependence power of each factor is plotted on the x-axis, while the driving powers are plotted on the y-axis. The driving force of a factor is the total number of other factors that it influences, whereas dependence power represents the number of other factors that it is influenced by. The sum of the rows and columns in the final reachability matrix represents the driving power and dependency power of the factors, respectively (Gan et al., 2018). According to DRP and DNP, each factor can be divided into four categories, namely autonomous, dependent, linkage, and driving factors (S. Singh & Gupta, 2021):

- Autonomous factors (I): The factors represent factors with weak driving and dependence power. This quadrant contains factors that are largely isolated from the system.
- Dependent factors (II): Although the driving force of these factors is weak, the dependence force is strong. These variables serve as dependent variables inside the system.
- Linkage factors (III): Linkage factors have a high degree of driving and dependence power. These factors serve as a linkage between any two other factors.

- Driving factors (IV): These factors serve as drivers in the system and are characterized by a strong driving force and a weak dependence force.

Figure 1. Flowchart of the research methodology based on Kannan et al. (2009)

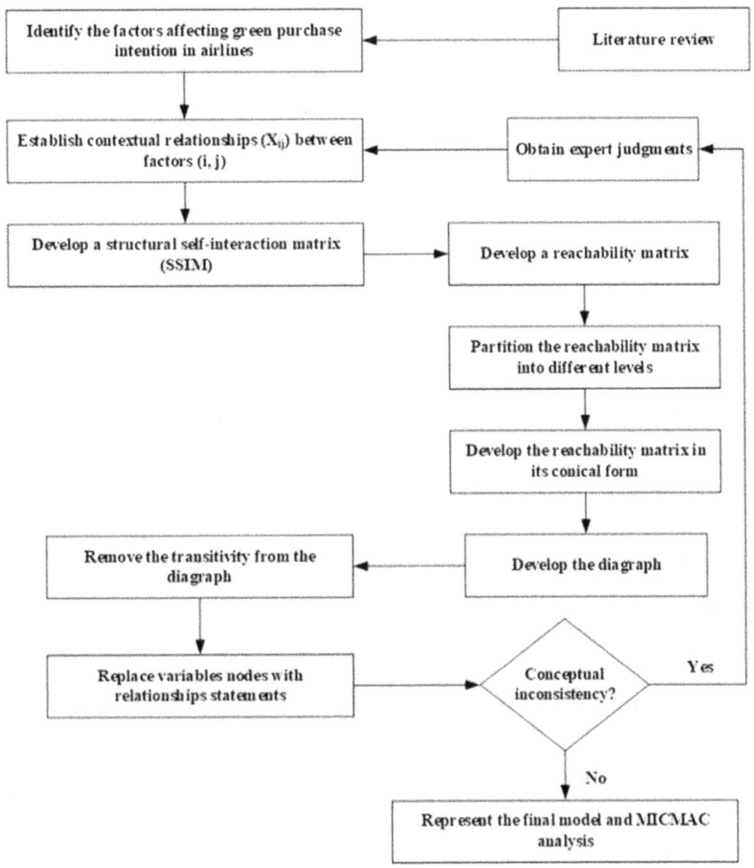

RESULTS OF ISM AND MICMAC ANALYSIS

Data Collection Phase

In the ISM methodology, the opinions of professionals, practitioners, or academics are employed when transforming complex problems into smaller structures (Ahmad & Qahmash, 2021). In this present research, four academics were questioned to establish the interrelationships between the factors that affect green purchase intention in airlines. The minimum number of experts for the ISM has yet to be discussed in the literature (Ma et al., 2019), but previous research has witnessed numbers of experts between 4-10 (Liu et al., 2016). Therefore, four experts are deemed sufficient for this ISM application despite the small sample size. Four PhD academics with at least five years of experience in aviation management or consumer behavior provided responses. Thus, this research ensures that experts have the necessary experience and knowledge of the field and topic to establish relationships between factors (Mondal & Chakrabarti, 2021).

To obtain the opinions of the experts, a structured questionnaire based on a pairwise comparison of identified factors was developed. However, the ISM method relies on the use of a single response form rather than aggregated solutions (Mondal & Chakrabarti, 2021). Therefore, the authors employed the consensus decision-making technique to reach an agreement among all members of the expert team to collect data. The experts answered the question "Do you think factor *i* directly influence factor *j*?" in an online meeting for 28 pairwise comparisons of eight criteria.

Structural Self-interaction Matrix

For pairwise comparisons, the ISM methodology employs a 'leads to' perspective between two factors (e.g., *i* and *j*). Four symbols are used to represent the relationships between factors *i* and *j* (Mondal & Chakrabarti, 2021):

- *V* – factor *i* directly influences factor *j*.
- *A* – factor *i* is directly influenced by factor *j*.
- *X* – factor *i* and factor *j* influence each other.
- *O* – factor *i* and factor *j* has no relationship.

Table 2 displays the SSIM matrix for the factors influencing green purchase intention in airlines. This matrix describes the direction of relationships between factors *i* and *j*.

Table 2. Structural self-interaction matrix (SSIM)

No	Factors	F8	F7	F6	F5	F4	F3	F2	F1
F1	Green perceived value	A	A	A	V	A	A	A	
F	Perceived consumer effectiveness	O	X	A	V	A	V		
F3	Consumer attitudes	A	A	A	V	A			
F4	Subjective norm	A	A	A	V				
F5	Green image	A	A	A					
F6	Environmental knowledge	X	V						
F7	Environmental awareness	X							
F8	Environmental concern								

Initial Reachability Matrix

In this step, the initial reachability matrix seen in Table 3 was developed from the SSIM matrix. In doing so, the following rules apply (Qarnain et al., 2021):

- If the cell (*i*, *j*) in the SSIM is *V*, then the cell (*i*, *j*) in the initial reachability matrix is 1, and the cell (*j*, *i*) is 0.
- If the cell (*i*, *j*) in the SSIM is *A*, then the cell (*i*, *j*) in the initial reachability matrix is 0, and the cell (*j*, *i*) is 1.
- If the cell (*i*, *j*) in the SSIM is *X*, then the cells (*i*, *j*) and (*j*, *i*) in the initial reachability matrix become 1.
- If the cell (*i*, *j*) in the SSIM is *O*, then the cells (*i*, *j*) and (*j*, *i*) in the initial reachability matrix become 0.

The transitivity rule was then applied to the initial reachability matrix, and the changes were incorporated. Thus, the final reachability matrix was obtained. In the final reachability matrix, transitivity cells were represented as 1*, while driving and dependence powers of the factors were also included (Khan et al., 2020). The final reachability matrix is shown in Table 4.

Table 3. Initial reachability matrix

No	Factors	F1	F2	F3	F4	F5	F6	F7	F8
F1	Green perceived value	1	0	0	0	1	0	0	0
F2	Pereived consumer effectiveness	1	1	1	0	1	0	1	0
F3	Conumer attitudes	1	0	1	0	1	0	0	0
F4	Subective norm	1	1	1	1	1	0	0	0
F5	Gren image	0	0	0	0	1	0	0	0
F6	Envronmental knowledge	1	1	1	1	1	1	1	1
F7	Envronmental awareness	1	1	1	1	1	0	1	1
F8	Envronmental concern	1	0	1	1	1	1	1	1

Table 4. Final reachability matrix

No	Factors	F1	F2	F3	F4	F5	F6	F7	F8	Driving Power
F1	Green perceived value	1	0	0	0	1	0	0	0	2
F2	Perceived consumer effectiveness	1	1	1	1*	1	1*	1	1*	8
F3	Consumer attitudes	1	0	1	0	1	0	0	0	3
F4	Subjective norm	1	1	1	1	1	1*	1*	1*	8
F5	Green image	0	0	0	0	1	0	0	0	1
F6	Environmental knowledge	1	1	1	1	1	1	1	1	8
F7	Environmental awareness	1	1	1	1	1	1*	1	1	8
F8	Environmental concern	1	1*	1	1	1	1	1	1	8
	Dpendence Power	7	5	6	5	8	5	5	5	

Level Partitions

After constructing the final reachability matrix, level partitioning was performed to determine the hierarchy of factors. By doing so, the reachability set and antecedent set were obtained for each critical factor from the final reachability matrix. The reachability set of each factor consists of itself and other factors it can help achieve. The antecedent set consists of the factor itself and other factors that help to construct it (Gan et al., 2018). The intersection of these two sets is obtained from the intersection of the reachability and antecedent set for all other factors (Khan et al., 2020). If the reachability and intersection set of a factor are the same, this factor is at the top of the hierarchy. In other words, this factor is driven by other factors and has no influence over other factors (Raut et al., 2018).

Subsequently, the top-level factors are separated from the others, and this procedure is repeated for the remaining factors downward to identify which factors belong to each level of the hierarchy. These levels result in the creation of the digraph and the final ISM model (M. D. Singh & Kant, 2008). Table 5 shows the reachability set, antecedent set, intersection set, and level of each factor. With four iterations, the level partitioning of eight factors was achieved.

Table 5. Level partition of the reachability matrix (iterations 1-4)

No	Factors	Reachability set	Antecedent set	Intersection set	Level
F1	Green perceived value	1	1,2,3,4,6,7,8	1	2
F2	Perceived consumer effectiveness	2,4,6,7,8	2,4,6,7,8	2,4,6,7,8	4
F3	Consumer attitudes	3	2,3,4,6,7,8	3	3
F4	Subjective norm	2,4,6,7,8	2,4,6,7,8	2,4,6,7,8	4
F5	Green image	5	1,2,3,4,5,6,7,8	5	1
F6	Environmental knowledge	2,4,6,7,8	2,4,6,7,8	2,4,6,7,8	4
F7	Environmental awareness	2,4,6,7,8	2,4,6,7,8	2,4,6,7,8	4
F8	Environmental concern	2,4,6,7,8	2,4,6,7,8	2,4,6,7,8	4

Final ISM-based Model

Finally, a structural model was developed for the factors affecting green purchasing intention for airline business based on level partitions in Table 5. To show the relationship between factors *i* and *j*, an arrow appears from *i* to *j* in each pairwise comparison, and the resulting graph is called the first digraph (Raut et al., 2018). As depicted in Figure 1, the transitivity in the digraph was removed and converted to the final ISM model. The final structural model shown in Figure 2 shows the dependencies and interdependencies of critical factors at four levels. Finally, the validity of the ISM model was checked and validated by using expert opinions and empirical studies in the literature.

At the bottom level of the structural model, perceived consumer effectiveness, subjective norm, environmental knowledge, environmental awareness, and environmental concern are the most influential driving factor influencing green purchase intention in the airline business. On the other hand, green image, which is placed at the top of the model, is the dependent factor that influences the factors in the entire network the least but is influenced by them all.

Figure 2. ISM-based model of factors affecting green purchase intention

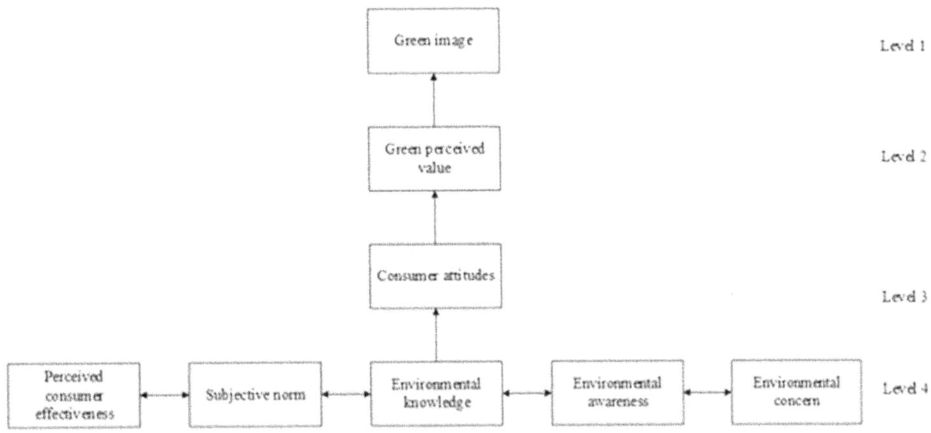

MICMAC Analysis

This study also employed MICMAC analysis to determine the driving and dependence power of factors influencing green purchase intention in an airline setting. In doing so, the MICMAC analysis input was obtained from the final reachability matrix (Raut et al., 2018; Sarkar et al., 2021). The diagram showing the driving and dependence power of factors is shown in Figure 3.

Based on Figure 3, there is no autonomous and driver factors in this research. On the other hand, there are three dependent factors, namely consumer attitude, green perceived value, and green image. Furthermore, surprisingly other factors are linkage factors.

Figure 3. The driving and dependence power diagram of factors

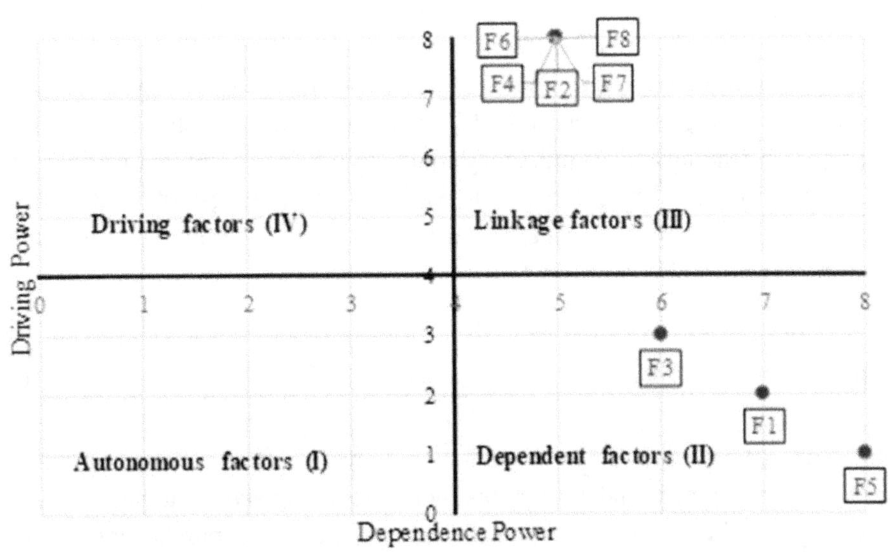

DISCUSSION AND IMPLICATIONS

The current research aims to identify the factors that affect green purchasing intention in the airline industry and shed light on their causal relationships through the use of the hybrid ISM and MICMAC methodology.

The findings reveal that perceived consumer effectiveness, subjective norms, environmental knowledge, environmental awareness, and environmental concern are crucial factors that impact consumer green purchase intention in the airline industry. Our findings are consistent with previous studies (Albayrak et al., 2013; Sharma & Foropon, 2019; S.-I. Wu & Chen, 2014; Zhuang et al., 2021) that considered the five factors as important elements in predicting the green purchase intention. Therefore, airline companies' managers are advised to give great attention to these factors because consumers who care more about the environment, have appropriate knowledge, are aware of environmental issues, and hold environmentally concerned perceptions are more likely to switch to green purchases. Thus, airline managers should provide essential information about environmental issues to educate their customers to encourage them to go green.

Our findings further demonstrated that consumer attitude, perceived value, and green image are independent factors also influenced by the bottom factors (e.g., perceived consumer effectiveness, subjective norms, environmental knowledge, environmental awareness, and environmental concern) and influence consumers' green purchase intention in the airline industry. These findings are consistence with several previous studies (Chen, 2008; Hagmann et al., 2015; Mayer et al., 2012; Zhuang et al., 2021) that illustrated the significant effect of consumer attitude, green perceived value, and green image in influencing consumer green purchase intention. In this regard, long-term strategic analysis should pay close attention to green perceived value, customer attitude, and green image if airline managers wish to increase consumers' green purchasing intentions. First, airlines must increase the perceived value of green products in order to increase consumers' intention of purchasing green products. They should develop marketing tactics that enhance customers' perceptions of green value. Second, regarding consumer attitudes and green image, businesses can use promotion and other tools such as advertisement to attract consumers' attention to green products and enhance the corporate image from the perspective of consumers. Therefore, providing consumers with an exceptional experience, and establishing a good corporate image of the usefulness of green product consumption, will encourage consumers to go green, which in turn increases their green purchase intention for the airline industry.

Finally, based on the findings, the authors may conclude that the combination of all proposed factors offers a good opportunity for airline companies to establish a good long-term strategy that may protect the environment as well as increase consumers' awareness about environmental issues, which in turn leads to a boost in their profitability.

CONCLUSION

Environmental issues have witnessed great attention, and several airline corporations have taken several steps to preserve the environment. Indeed, actions such as reducing waste campaigns and energy-saving methods are raising individuals' awareness of the need to alter their purchasing habits in the airline industry. In this respect, the present research was performed to identify the factors that affect green purchasing intention in the airline industry and shed light on their causal relationships through the ISM and MICMAC approach. Significant results demonstrated the importance of perceived consumer effectiveness, subjective norms, environmental knowledge, environmental awareness, environmental concern, consumer attitude, perceived value, and green image in shaping consumer green purchase intention in the airline industry. The combination of these factors may motivate consumers to purchase green products in the airline industry. However, airline companies should implement successful strategies that educate consumers and create a sense of awareness about the importance of the environment, as well as promote green products in their advertising campaigns.

Finally, there are a few limitations to this study. The ISM model should not be generalized because it is based solely on the opinions of four experts. Future studies can analyze structural models more thoroughly by employing more experts. In addition, future research can validate the present ISM-based model empirically with data obtained from passengers using techniques such as SEM in the future.

REFERENCES

Ahmad, N., & Qahmash, A. (2021). Smartism: Implementation and assessment of interpretive structural modeling. *Sustainability, 13*(16), 1–27. doi:10.3390u13168801

Ajzen, I. (1991). The Theory of Planned Behavior. *Organizational Behavior and Human Decision Processes, 50*(2), 179–211. doi:10.1016/0749-5978(91)90020-T

Albayrak, T., Aksoy, Ş., & Caber, M. (2013). The effect of environmental concern and scepticism on green purchase behaviour. *Marketing Intelligence & Planning, 31*(1), 27–39. doi:10.1108/02634501311292902

Armstrong, B., Reynolds, C., Bridge, G., Oakden, L., Wang, C., Panzone, L., Schmidt Rivera, X., Kause, A., Ffoulkes, C., Krawczyk, C., Miller, G., & Serjeant, S. (2021). How Does Citizen Science Compare to Online Survey Panels? A Comparison of Food Knowledge and Perceptions Between the Zooniverse, Prolific and Qualtrics UK Panels. *Frontiers in Sustainable Food Systems, 4*, 1–11. doi:10.3389/fsufs.2020.575021

ATAG. (2020). *Facts & Figures*. https://www.atag.org/facts-figures.html

Baumeister, S., Nyrhinen, J., Kemppainen, T., & Wilska, T.-A. (2022). Does airlines' eco-friendliness matter? Customer satisfaction towards an environmentally responsible airline. *Transport Policy, 128*, 89–97. doi:10.1016/j.tranpol.2022.09.016

Baumeister, S., Zeng, C., & Hoffendahl, A. (2022). The effect of an eco-label on the booking decisions of air passengers. *Transport Policy, 124*, 175–182. doi:10.1016/j.tranpol.2020.07.009

Biswas, A. (2017). A consumption value-gap analysis for sustainable consumption. *Environmental Science and Pollution Research International, 24*(8), 7714–7725. doi:10.100711356-016-8355-9 PMID:28124271

Chen, Y. S. (2008). The driver of green innovation and green image - Green core competence. *Journal of Business Ethics, 81*(3), 531–543. doi:10.100710551-007-9522-1

Chen, Y. S., & Chang, C. H. (2012). Enhance green purchase intentions: The roles of green perceived value, green perceived risk, and green trust. *Management Decision, 50*(3), 502–520. doi:10.1108/00251741211216250 da Rocha, P. M., Costa, H. G., & da Silva, G. B. (2022). Gaps, Trends and Challenges in Assessing Quality of Service at Airport Terminals: A Systematic Review and Bibliometric Analysis. *Sustainability*, *14*(7), 1–17. doi:10.3390u14073796 PMID:36090804

Dangi, N., Gupta, S. K., & Narula, S. A. (2020). Consumer buying behaviour and purchase intention of organic food: A conceptual framework. *Management of Environmental Quality*, *31*(6), 1515–1530. doi:10.1108/MEQ-01-2020-0014

Gan, X., Chang, R., Zuo, J., Wen, T., & Zillante, G. (2018). Barriers to the transition towards off-site construction in China: An Interpretive structural modeling approach. *Journal of Cleaner Production*, *197*, 8–18. doi:10.1016/j.jclepro.2018.06.184

Gupta, S., & Dhingra, S. (2022). Modeling the key factors influencing the adoption of mobile financial services: An interpretive structural modeling approach. *Journal of Financial Services Marketing*, *27*(2), 96–110. doi:10.105741264-021-00101-4

Hagmann, C., Semeijn, J., & Vellenga, D. B. (2015). Exploring the green image of airlines: Passenger perceptions and airline choice. *Journal of Air Transport Management*, *43*, 37–45. doi:10.1016/j.jairtraman.2015.01.003

Hwang, J., & Lyu, S. O. (2020). Relationships among green image, consumer attitudes, desire, and customer citizenship behavior in the airline industry. *International Journal of Sustainable Transportation*, *14*(6), 437–447. doi:10.1080/15568318.2019.1573280

IATA. (2021). *Net-Zero Carbon Emissions by 2050*. IATA. https://www.iata.org/en/pressroom/2021-releases/2021-10-04-03/

Kannan, G., Pokharel, S., & Kumar, P. S. (2009). A hybrid approach using ISM and fuzzy TOPSIS for the selection of reverse logistics provider. *Resources, Conservation and Recycling*, *54*(1), 28–36. doi:10.1016/j.resconrec.2009.06.004

Khan, W., Akhtar, A., Ansari, S. A., & Dhamija, A. (2020). Enablers of halal food purchase among Muslim consumers in an emerging economy: An interpretive structural modeling approach. *British Food Journal*, *122*(7), 2273–2287. doi:10.1108/BFJ-08-2018-0528

Kollmuss, A., & Agyeman, J. (2002). Mind the Gap: Why do people act environmentally and what are the barriers to pro-environmental behavior? *Environmental Education Research*, *8*(3), 239–260. doi:10.1080/13504620220145401

Lee, C. H., Ling, H. Y., Yeow, J. A., & Hasan, M. A. (2012). Assessing Determinants of Green Purchase Intention. *International Proceedings of Economics Development and Research*, *44*, 25–29.

Liu, H., Skibniewski, M. J., & Wang, M. (2016). Identification and hierarchical structure of critical success factors for innovation in construction projects: Chinese perspective. *Journal of Civil Engineering and Management*, *22*(3), 401–416. doi:10.3846/13923730.2014.975739

Ma, G., Jia, J., Ding, J., Shang, S., & Jiang, S. (2019). Interpretive structural model based factor analysis of BIM adoption in Chinese construction organizations. *Sustainability*, *11*(7), 1–16. doi:10.3390u11071982

Mayer, R., Ryley, T., & Gillingwater, D. (2012). Passenger perceptions of the green image associated with airlines. *Journal of Transport Geography*, *22*, 179–186. doi:10.1016/j.jtrangeo.2012.01.007

Merkert, R. (2022). Quo vadis air transport management research? *Journal of Air Transport Management*, *100*, 102205. doi:10.1016/j.jairtraman.2022.102205

Mishal, A., Dubey, R., Gupta, O. K., & Luo, Z. (2017). Dynamics of environmental consciousness and green purchase behaviour: An empirical study. *International Journal of Climate Change Strategies and Management*, *9*(5), 682–706. doi:10.1108/IJCCSM-11-2016-0168

Mohd Suki, N. (2016). Green product purchase intention: Impact of green brands, attitude, and knowledge. *British Food Journal*, *118*(12), 2893–2910. doi:10.1108/BFJ-06-2016-0295

Mondal, J., & Chakrabarti, S. (2021). The abandonment behaviour of the branded app consumer: A study using interpretive structural modelling approach. *Journal of Retailing and Consumer Services*, *63*, 102695. doi:10.1016/j.jretconser.2021.102695

Nayeri, M. D., & Moradi, M. (2019). Green Purchase Determinants Based on Interpretive Structural Modeling: Case of Iran'S Green Marketing. *ASEAN Marketing Journal*, *11*(1), 1–21. doi:10.21002/amj.v11i1.11479

Naz, F., Oláh, J., Vasile, D., & Magda, R. (2020). Green purchase behavior of university students in Hungary: An empirical study. *Sustainability*, *12*(23), 1–21. doi:10.3390u122310077

Nguyen, T. T. M., Phan, T. H., Nguyen, H. L., Dang, T. K. T., & Nguyen, N. D. (2019). Antecedents of purchase intention toward organic food in an Asian emerging market: A study of urban Vietnamese consumers. *Sustainability*, *11*(17), 1–16. doi:10.3390u11174773

Qarnain, S. S., Muthuvel, S., & Sankaranarayanan, B. (2021). Analysis of Energy Conservation Factors in Buildings Using Interpretive Structural Modeling Methodology: An Indian Perspective. *Journal of The Institution of Engineers (India): Series A*, *102*(1), 43–61. doi:10.1007/s40030-020-00483-z

Raut, R., Priyadarshinee, P., Jha, M., Gardas, B. B., & Kamble, S. (2018). Modeling the implementation barriers of cloud computing adoption: An interpretive structural modeling. *Benchmarking*, *25*(8), 2760–2782. doi:10.1108/BIJ-12-2016-0189

Ruangkanjanases, A., You, J. J., Chien, S. W., Ma, Y., Chen, S. C., & Chao, L. C. (2020). Elucidating the Effect of Antecedents on Consumers' Green Purchase Intention: An Extension of the Theory of Planned Behavior. *Frontiers in Psychology*, *11*, 1–13. doi:10.3389/fpsyg.2020.01433 PMID:32793023

Sarkar, A., Qian, L., Peau, A. K., & Shahriar, S. (2021). Modeling drivers for successful adoption of green business: An interpretive structural modeling approach. *Environmental Science and Pollution Research International*, *28*(1), 1077–1096. doi:10.100711356-020-10490-z PMID:32829426

Sharma, A., & Foropon, C. (2019). Green product attributes and green purchase behavior: A theory of planned behavior perspective with implications for circular economy. *Management Decision*, *57*(4), 1018–1042. doi:10.1108/MD-10-2018-1092

Singh, M. D., & Kant, R. (2008). Knowledge management barriers: An interpretive structural modeling approach. *International Journal of Management Science and Engineering Management*, *3*(2), 141–150. doi:10.1080/17509653.2008.10671042

Singh, S., & Gupta, A. (2021). An ISM modeling for factors affecting the purchase of green products. *Journal of Modelling in Management*, *16*(1), 103–123. doi:10.1108/JM2-05-2019-0099

Sreen, N., Purbey, S., & Sadarangani, P. (2018). Impact of culture, behavior and gender on green purchase intention. *Journal of Retailing and Consumer Services*, *41*, 177–189. doi:10.1016/j.jretconser.2017.12.002

Tan, B. C. (2011). The role of perceived consumer effectiveness on Value-Attitude-Behaviour model in green buying behaviour context. *Australian Journal of Basic and Applied Sciences*, *5*(12), 1766–1771.

Varshneya, G., Pandey, S. K., & Das, G. (2017). Impact of Social Influence and Green Consumption Values on Purchase Intention of Organic Clothing: A Study on Collectivist Developing Economy. *Global Business Review*, *18*(2), 478–492. doi:10.1177/0972150916668620

Warfield, J. N. (1974). Developing Subsystem Matrices in Structural Modeling. *IEEE Transactions on Systems, Man, and Cybernetics*, *4*(1), 74–80. doi:10.1109/TSMC.1974.5408523

Wong, L. J., Sia, J. K. M., & Ling, T. H. Y. (2020). Airline passengers' perceived sacrifice and green practices adoption behaviours. *Asian Journal of Business Research*, *10*(3), 85–110. doi:10.14707/ajbr.200092

Woo, E., & Kim, Y. G. (2019). Consumer attitudes and buying behavior for green food products: From the aspect of green perceived value (GPV). *British Food Journal*, *121*(2), 320–332. doi:10.1108/BFJ-01-2018-0027

World Bank. (2021). *Air transport, passengers carried*. World Bank. https://data.worldbank.org/indicator/IS.AIR.PSGR

Wu, H.-C., Cheng, C.-C., & Ai, C.-H. (2018). An empirical analysis of green switching intentions in the airline industry. *Journal of Environmental Planning and Management*, *61*(8), 1438–1468. doi:10.1080/09640568.2017.1352495

Wu, S.-I., & Chen, Y.-J. (2014). The Impact of Green Marketing and Perceived Innovation on Purchase Intention for Green Products. *International Journal of Marketing Studies*, *6*(5), 81–100. doi:10.5539/ijms.v6n5p81

Xiao, L. (2018). Analyzing consumer online group buying motivations: An interpretive structural modeling approach. *Telematics and Informatics*, *35*(4), 629–642. doi:10.1016/j.tele.2018.01.010

Zaremohzzabieh, Z., Ismail, N., Ahrari, S., & Abu Samah, A. (2021). The effects of consumer attitude on green purchase intention: A meta-analytic path analysis. *Journal of Business Research*, *132*, 732–743. doi:10.1016/j.jbusres.2020.10.053

Zeithaml, V. A. (1988). Consumer Perceptions of Price, Quality, and Value: A Means-End Model and Synthesis of Evidence. *Journal of Marketing*, *52*(3), 2–22. doi:10.1177/002224298805200302

Zhuang, W., Luo, X., & Riaz, M. U. (2021). On the Factors Influencing Green Purchase Intention: A Meta-Analysis Approach. *Frontiers in Psychology*, *12*, 1–15. doi:10.3389/fpsyg.2021.644020 PMID:33897545

Chapter 12
Sustainability in the Aviation Industry in the Post-COVID-19 Era

Shantanu Trivedi
University of Petroleum and Energy Studies, India

Saurav Negi
Modern College of Business and Science, Oman

Saurabh Tiwari
University of Petroleum and Energy Studies, India

ABSTRACT

The purpose of this study is to understand and identify sustainability measures applicable in the aviation industry. Further, the study also explores the enablers of sustainability in the aviation sector. The study uses a systematic literature review, published in Scopus-indexed journals. The study addresses the complexity of sustainability in the sector and identifies key indicators based on comprehensive and valid data from different stakeholders. The study focuses separately on the three pillars of sustainability—social economic, and environmental sustainability—and identifies sustainability indicators for each pillar from the views of the triple bottom line. The study finds that there is a lack of sustainability knowledge and awareness in the aviation sector despite its resistant growth and expansion. The results show that the aviation industry highlights the importance of sustainability indicators that value equitable development in the pursuit of business goals and environmental and economic efficiency.

INTRODUCTION

The coronavirus disease (COVID-19) pandemic has had a devastating impact on the aviation industry globally as nearly all air travel came to a halt in late March and April 2020. Although the COVID-19 pandemic affected the airline industry more than any other aviation subsector, the industry was not doing

DOI: 10.4018/978-1-6684-6835-7.ch012

especially well before the outbreak. The epidemic inflicted financial havoc throughout the aviation value chain, particularly among carriers (Bouwer et al., 2022). Except for freight forwarders and cargo airlines, who benefitted from an increase in demand for air cargo, every subsector recorded enormous losses in 2020. Even formerly dependable value generators, such as airports and factories, were not immune to the economic impact of the epidemic. The Russian-Ukraine crisis and the relevant sanctions will further increase complexity and operating costs. Global mobility is the norm in today's increasingly interconnected society (Bouwer, Saxon, and Tufft, 2022). However, huge global shocks will likely increase in frequency and when the global aviation sector is rebuilt, sustainability and adaptability must be two of the most important things to think about.

The usage phase dominates the whole life cycle impact in the aviation industry; consequently, component weight dictates environmental impact and costs. As a result, the effect attributable to material production and component manufacture has a tiny relative contribution to the entire life cycle impact of a material or component. This fact hinders the use of recycled components for closed-loop aircraft applications, particularly high-performance ones (Markatos and Pantelakis, 2022).

Since the beginning, aviation has had to deal with the same problems: provincial boundaries and binders, linked geographical obstacles, split airspace, expensive capital needs, and making sure planes are safe to fly. After surviving multiple waves of recurring COVID-19 pandemic, airlines are preparing for any such instances. The Russian-Ukraine crisis and the relevant sanctions will further increase complexity and operating costs. Airlines also face an attack from another direction in rising jet fuel prices. Global economic uncertainties, uncertain oil demand, and fluctuating oil production rates were driving oil prices in 2021. The Russian invasion has inflated this value even further. As per Frost & Sullivan's analysis fuel costs are the highest contributor to the operating expenditure of an airline (~30% of operating expenditure), and they have increased by almost 89% in the last three and a half years as per Energy Information Administration (Abraham, 2022). Together, these problems have made it very hard and risky to be in the aviation business (Singh et al., 2022).

Professionals in the aviation industry have always been interested in sustainability, but it seems to be becoming an even bigger concern for decision-makers as they try to restructure their business operations during the current renaissance of the industry (Connected Aviation Industry, 2021). A clear understanding of stakeholder roles is crucial though sometimes neglected part of aviation sustainability efforts (Singh et al., 2022). In addition, Song, Solaem, and Han's (2020) study stated that further research is necessary on the aviation industry's sustainability from the comparative advantage viewpoint and the policy direction perspective. The study done by Suk and Kim (2021) recommended that observing sustainability indicators and best practices from the perspective of post-pandemic government policy designs would be an interesting research topic. Considering these research directions in the field of aviation sustainability, the present study is carried out with the following objectives:

Objective 1: To understand the sustainability issues and measures adopted by aviation industry post covid scenario

Objective 2: To analyze the applicability of triple bottom line measures adopted by aviation industry

METHODOLOGY

This study employs a qualitative approach to research. This study's approach was determined to be a comprehensive literature review. Understanding the level of business sustainability in the aviation sector and identifying the application of sustainability metrics in the aviation industry were the initial objectives of this study, which utilized a literature analysis. The literature review included the study and assessment of research papers, book chapters, conference papers, articles, reports, and other publications in the field of sustainability in aviation management. For the literature review, several combinations of complimentary search phrases are entered into online databases such as Scopus, Google Scholar, and Science Direct, IEEE, PubMed, Emerald insights, T&F, Wiley and MDPI etc. For a focused and well-coordinated literature review (from 2006 to 2022), there is a three-stage review method. The review of digital databases and deliberately develop three stages entailing planning, conducting, and reporting the review (Sabharwal & Miah, 2022). The three-stage literature review is mentioned below:

Figure 1. Three Step Literature Review

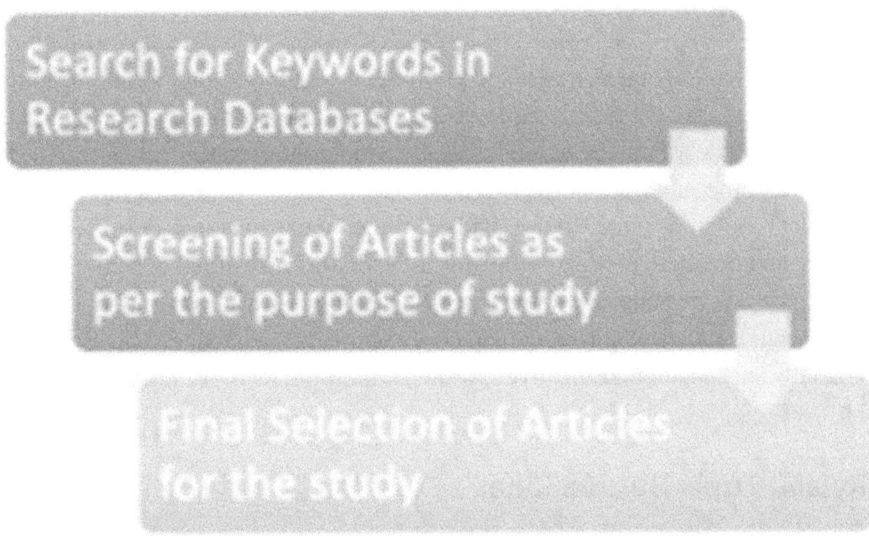

DISCUSSION

This section presents a detailed discussion on issues impacting sustainability in the aviation industry, sustainability through a triple bottom line, and how overall aviation sustainability can be achieved by adopting the triple bottom line approach focusing on economic sustainability, social sustainability, and environmental sustainability.

Aviation Industry: Issues Impacting Sustainability

The growth prospects of an economy are positively correlated with the expansion of the aviation market. The global aviation sector is facing a dismal decline in revenue. According to the International Air Transport Association (IATA), the global airline industry's income is expected to decrease by around USD 113 billion. Every 100 instances of COVID-19 will result in a decline in passenger demand, which will cost global airlines roughly USD 113 billion. The aviation industry has increased commercial activity by permitting speedier commutes, opening up previously inaccessible regions, promoting tourism, providing jobs, bolstering other business sectors, and improving commerce. These aviation sector products enhance the quality of life and contribute to global economic expansion. Statistics demonstrate that aviation has become a vital component of the global economy because it facilitates increased access to labor, expertise, investment, and other economic activity (Singh et al., 2022). The aviation sector has terrible effects on the environment and human health as a result of its usage of fossil fuels, toxic pollutants, and the employment of enormous and varied workforces. These problems have been exacerbated by the expansion of the aviation business, both in terms of passenger and freight transportation, because this demand growth entails an increase in fuel consumption, hazardous waste, and noise. Because the industry is global, these bad effects don't just happen in the countries where the companies are based. They also happen in other countries, which leads to global climate change (Karaman, Kilic, & Uyar, 2018).

Sustainability

The objective of conventional business models is to produce wealth for shareholders, sometimes at the expense of other stakeholders. Sustainable firms are changing the corporate ecosystem by developing business models that provide value for all stakeholders, including employees, shareholders, supplier chains, civil society, and the environment. Embedded sustainability measures have a favorable effect on corporate success (Whelan & Fink, 2016). The continued existence and viability of airlines need a shift in their strategy and business models to bolster their financial resilience. Major variables impacting airline performance are overcapacity, severe competition, and high operating expenses. Optimal resource use, collaboration rather than rivalry, and cost minimization appear to be realistic solutions for sustaining a financially successful take-off over hard terrain in the face of the current COVID dilemma (Agrawal, 2021).

Sustainability and Triple Bottom Line

The origin of sustainability may date back to over 130 years ago from an idea known as spaceship earth. Used interchangeably, TBL and sustainability are related constructs; however, there seems to be an inconsistent use of sustainability (Alhaddi, 2015). John Elkington's Triple Bottom Line Theory has altered how corporations, organizations, and governments evaluate the sustainability and effectiveness of initiatives and programs. Beyond the cornerstone of assessing sustainability on three fronts — people, planet, and profits — the triple bottom line's adaptability enables firms to implement the idea in a manner that meets their requirements (Slaper & Hall, 2011). The Triple Bottom Line is an approach focused on three crucial dimensions: environmental quality, social equality, and economic prosperity. Similar to more known marketing research, these dimensions may be seen as the foundation for market-focused resources, competencies, and a competitive advantage over market rivals. (Arowoshegbe & Emmanuel, 2016) suggested that social dimension includes the company's impact on its employees and the social system within its community. When looking at the environmental dimension, companies need to look at the qualitative and quantitative effects they are having on their local, national, and international resources. The last, but certainly not the least, economic dimension includes the company's financial performance, the flow of capital, and their economic involvement in society. As per the study by (Miller, 2020) beyond helping companies capitalize on a growing market for sustainable goods, embracing sustainable business strategies can be highly attractive to investors. According to Sustainable Business Strategy, evidence has increasingly shown that firms with promising environmental, social, and governance (ESG) metrics tend to produce superior financial returns. As a result, more investors have begun focusing on ESG metrics when making investment decisions. (Correia, 2019) In his study found the key role that sustainability plays in organizations it is important that they incorporate sustainability into their strategies and accesses the issues with their implementation. (Willard, 2012)demonstrated that a sustainable behavior can contribute to a company profits by increasing revenue and employee productivity, reduce energy, water and materials expenses, reduce turnover and decrease strategic and operational risks. (Ali, Kaur, & Khan, 2022) with their study drew attention to green operations for energy and resource conservations, promotes the role of sustainable work culture, sustainable strategies, and policies for their role in encouraging sustainability performance outcomes.

Improving Aviation Business Sustainability Through The Triple Bottom Line

Aviation sustainability looks for ways to lessen the negative effects that air transportation has on both society and the environment. Its goal is to lessen aviation's impact on climate change through revolutionary innovation and new practices.

Future aircraft will need to be more efficient, have lower direct operating costs, carry higher payloads, and continue to fly from existing airports to preserve the aviation business. These regulations will make the aviation sector more economically and ecologically sustainable. Southwest Airlines' implementation of Triple Bottom Line set it apart from other airlines and propelled it to the top of the US airline industry. In response to the COVID-19 outbreak, Southwest Airlines is implementing a novel methodology for managing its ability to stand apart from the competition. In the next few weeks, the company's capacity reductions will be less severe than those of its competitors, a decision justified by the realization that it is simpler to remove resources from the system than to restore supply to its timetable. The airline is attempting to keep as many destinations as possible, and the majority of these destinations remain unaltered at this time. However, Southwest insists that if demand remains low, it will allocate additional resources to its operations. Southwest is also cooperating with other airlines to ensure that passengers feel comfortable traveling in the next weeks and months, including limiting airplane reservations so that passengers can exercise social distancing (Triple Bottom Line: Evidence from the Aviation Sector, 2020). But at some point, airlines will have to find a balance between being cut off from society and making sure that the cost of flying stays the same (COVID-19: Southwest Airlines: Nuanced Crisis Management, CAA, 2020).

Environmental Sustainability

The aviation sector can improve its sustainability by employing various effective substitute technologies that have a lower environmental impact (Bertoni et al., 2015). Numerous studies indicate that superior aircraft engines, alternative fuels, air-to-air refueling, well-organized flight routes, ecologically friendly impulsion systems, and fuel savings can all contribute to aviation sustainability (Agarwal, 2010; Elhmoud & Kutty, 2020; ICAO, 2012; Warwick & Norris, 2011).

The research by Somerville et al. (2016) showed the multiple benefits of box-wing technology, including the fact that its deployment will result in considerable fuel savings for aircraft. Reduced aircraft emissions are a byproduct of reduced fuel consumption; as a result, environmental sustainability in the aviation sector will be enhanced. The multiple benefits of box-wing technology, including the fact that its deployment will result in considerable fuel savings for aircraft (Ragbir et al., 2016). Reduced aircraft emissions are a byproduct of reduced fuel consumption; as a result, environmental sustainability in the aviation sector will be enhanced. The research by Ragbir et al. (2021) supported prior studies on environmental sustainability that indicated consumers' willingness to pay more for planes powered by biofuels and constructed with sustainable materials.

To be sustainable, firms must adopt new approaches and methods for routine tasks such as designing and sourcing aircraft and necessary components. The industry must be ready for the transition in procurement by moving away from buying decisions from the lowest pricing suppliers to those producing items fulfilling longer-term sustainability standards. A CO2-friendly component may cost more upfront, but it offers possibilities for recycling that cut down on carbon emissions (Galer, 2022). Additionally, designers will need to engage much more strongly with their suppliers, sharing information to acquire the finest quality items that will reduce rework and subsequent CO2 emissions, while airplane makers explore sustainable fuel choices and materials. Moreover, the question is not only how to build a more environmentally friendly system for new aircraft fleets, but also how to improve the efficiency of the "existing fleets". It has been evident that aviation sustainability and efficiency can be greatly improved with changes to the design of an aircraft, fuel efficiency, engines, and materials.

Hence, aircraft manufacturers must look for ways to improve their current processes while transferring more data up and down the value chain to develop and build the next generation of aircraft. This will help them operate their businesses in a carbon-neutral manner. Moreover, it is important to note that a sustainable aviation business connects data from each aspect of engineering, sourcing, logistics, manufacturing, sales, and funds (Galer, 2022). Finally, to meet the sustainable goals, the aviation sector will need to consider a variety of strategies, including improved efficiency, technological innovation, and the use of Sustainable Aviation Fuel (SAF).

Social Sustainability

The aviation industry has created millions of jobs worldwide and contributed significantly to improving human life. All countries can now travel freely thanks to air travel. It provides access to international marketplaces while bridging cultures, nations, and people. Air travel provides a means of inclusion for isolated communities when it is the only mode of transportation, connecting developing nations to wealthier nations. Most people on the globe can enjoy the cultural and recreational opportunities that come with international travel because of the low cost of air travel. In turn, this makes the world more accessible to tourism, which greatly raises living standards and reduces poverty in many countries (Elhmoud & Kutty, 2020). In addition to serving as a means of travel for tourists, aviation also serves the crucial purpose of enabling the quick transportation of medical supplies, emergency aid relief, and organs for the purpose of transplantation.

The airline sector also makes use of its operations to support charitable groups and give back to the community at large. Each year, every major airline has made significant financial contributions to philanthropic endeavors and charity. The main carriers typically support both national organizations and local charities in the hub cities where they operate. Additionally, most airlines that provide a mile incentive program permit the donation of miles for free travel to charitable organizations (Elhmoud & Kutty, 2020).

The "Social Life Cycle Assessment" is a frequently employed instrument for evaluating sustainability in the social domain (Barke et al., 2019). The stakeholder implications resulting from their actions throughout the life cycle are accounted for by the social impacts that exist. Reporting on corporate social responsibility (CSR) is another well-known assessment technique. By taking appropriate legal, moral, and charitable measures, the airline sector is required by the CSR to protect and advance the welfare of both the passenger community and the industry as a whole (Anttila & Kretzschmar, 2010). The findings of the study (Ragbir et al., 2021) indicated that affect and emotions, especially the amount of happiness and concern for sustainability in aviation, explain why an individual would be prepared to pay extra to support sustainable practices. As sustainable projects continue to be adopted on a global scale, the aviation sector must incorporate sustainable practices into airports, planes, and their entire operations. Also, consumers are becoming increasingly concerned with human resource strategies that help to promote social sustainability. With more options, disruptive changes in technology, and changes in consumption and lifestyles that go along with them, the airline industry will need to stay flexible and open to new trends and changing dynamics to stay competitive.

Economic Sustainability

Economic competence, economic growth, and financial viability comprise economic sustainability. In addition, methodologies for the estimation of costs and performance can be used to evaluate economic viability (Elhmoud & Kutty, 2020). For achieving economic sustainability in the aviation industry, the usage of SAF is one of the most significant ways. Amounting to the goals of the aviation sector to meet passenger needs and make a profit, the use of SAFs is influenced by the price of aviation fuel and regulatory incentives. A sizable return on investment may encourage bio-refineries to establish SAF production facilities to meet the expanding demand (Elhmoud & Kutty, 2020).

On the other side, the development of more fuel-efficient aircraft is a promising way to increase the aviation sector's economic sustainability. The latest models from leading airline manufacturers have been released, and they have the potential to cut current fuel use by 20% (DuPont, 2012). Even if the new airplanes have a little high price tag, the economic benefits for all parties concerned are substantial in both the short and long terms. In the end, new fuel-efficient aircraft save money on fuel costs as well as on operating expenses, maintenance expenses, and taxes. These cost reductions will be felt by both customers and airline companies. The recent increase in ticket prices will be lessened by more fuel-efficient aircraft and lower costs. Increased fuel prices are directly responsible for higher ticket prices. Additionally, as the demand for crude oil falls globally due to the aircraft industry's reduced fuel use, the price per barrel will also decline. New fuel-efficient aircraft will ultimately have a big financial impact on the entire world community. The financial savings will be felt by households, small businesses, and aircraft manufacturers.

Understanding the links between manufacturing, operations, and decommissioning in the airline industry necessitated the development of a decision-support model based on the LCC strategy (Thokala, Scanlan, & Chipperfield, 2010). The worldwide standard for the integration of financial reports with corporate sustainability reporting as a single report will contribute to speeding up the process (Koç & Durmaz, 2015).

CONCLUSION

The aviation sector has played a significant role in the global development process. The necessity for sustainability in this industry has been completely acknowledged by industry leaders in a more integrated way. The purpose of this article was to undertake a review of the literature covering the need for sustainability and incorporating methods for all aspects of sustainability (economic, social, and environmental) for sustainability in the aviation industry. Post-COVID-19, the airlines must define and prioritize their responses to the current situation and commercial objectives. Inventiveness in technology is already pushing past limits in several areas, which is good for the whole world. The finest illustration of this is the fact that social distancing measures have sparked a technological revolution; virtual connections and gatherings now aim to expand internationally.

The present study focuses on achieving sustainability in the aviation business post-COVID era through a Triple Bottom Line approach. The study discusses the factors for achieving overall business sustainability such as economic sustainability, social sustainability, and environmental sustainability. The study shows that there is a high requirement for sustainability in the aviation sector especially post-COVID era as the industry has already been highly affected by the pandemic. Several interventions are required at different facets of the aviation supply chain to overcome the challenges faced by the current aviation industry. The study suggests that the aviation sector must adapt to more fuel-efficient aircraft for its operations. This would help to sustainably improve the environment by drastically reducing carbon emissions. Similarly, fuel-efficient airplanes use less fuel and have a positive economic impact on sustainability. Also, the airline sector needs to be more socially conscious, and CSR is crucial to achieving sustainability in all areas.

To meet sustainable goals, the aviation sector will need to consider a variety of strategies, including improved efficiency, technological innovation, and the use of sustainable aviation fuel. The study also suggests that by adopting a Triple Bottom Line approach, the aviation industry not only can minimize the cost but also minimize the environmental burden through their sustainable operations, which will lead to improved performance and a better future for the aviation industry. As aviation is a robust business that has recovered from past shocks, the road to the Triple Bottom Line would accelerate necessary reform. Governments, businesses, and global institutions should work together to make sure that innovative solutions can be put into place quickly. This will help the world recover from the pandemic more quickly and effectively, while also leaving a legacy of sustainability for future generations to enjoy. The methods discussed in this study encourage decision-making to promote sustainable operations in the aviation sector, thus enabling mitigate the existing issues and challenges.

The present study is based on secondary literature and limited to the analysis of available literature in the field of aviation sustainability. This study offers several research avenues in the field of aviation sustainability where the present study can be used as a guide to conduct an empirical study covering the primary data. Moreover, individual sustainability aspect such as economic aspect or social aspect or environmental aspect can also be studied with reference to a particular nation.

REFERENCES

Abraham, A. V. (2022, April 1). *Frost Perspective: Frost & Sullivan*. Retrieved from Frost & Sullivan. https://www.frost.com/frost-perspectives/impact-of-the-russia-ukraine-crisis-on-aviation/

Agrawal, A. (2021). Sustainability of airlines in India with Covid-19: Challenges ahead and possible way-outs. *Journal of Revenue and Pricing Management, 20*(4), 457–472. doi:10.105741272-020-00257-z

Alhaddi, H. (2015). Triple Bottom Line and Sustainability: A Literature Review. *Business and Management Studies*, 6-11.

Ali, S. S., Kaur, R., & Khan, S. (2022). Evaluating sustainability initiatives in warehouse for measuring sustainability performance: An emerging economy perspective. *Annals of Operations Research*, 1–18. doi:10.100710479-021-04454-w PMID:36091933

Anttila, T., & Kretzschmar, A. (2010). *Application of CSR programs in the airline industry*. Semantic Scholar.

Arowoshegbe, A. O., & Emmanuel, U. (2016). Sustainability And Triple Bottom Line: An Overview Of Two Interrelated Concepts. *Igbinedion University Journal of Accounting*, 88-116.

Barke, A., Thies, C., Melo, S. P., Cerdas, F., Herrmann, C., & Spengler, T. S. (2020). Socio-economic life cycle assessment of future aircraft systems. *Procedia CIRP*, *90*, 262–267. doi:10.1016/j.procir.2020.01.096

Bouwer, J., Krishnan, V., Saxon, S., & Tufft, C. (2022). *Taking stock of the pandemic's impact on global aviation*. McKinsey & Company.

Bouwer, J. K., Saxon, S., & Tufft, C. (2022, March 31). *Industry: mckinsey*. Mckinsey. https://www.mckinsey.com/industries/travel-logistics-and-infrastructure/our-insights/taking-stock-of-the-pandemics-impact-on-global-aviation

Connected Aviation Industry. (2021, May 18). Blog. *Connected Aviation Today*. https://connectedaviationtoday.com/sustainability-in-aviation-reclaims-attention-airlines-airports-ramp-up-recovery-plans/

Correia, M. S. (2019). Sustainability: An Overview of the Triple Bottom Line and Sustainability Implementation. *International Journal of Strategic Engineering*, *2*(1), 29–38. doi:10.4018/IJoSE.2019010103

Elhmoud, E. R., & Kutty, A. A. (2020). Sustainability Assessment in Aviation Industry: A MiniReview on the Tools, Models and Methods of Assessment. *International Conference on Industrial Engineering and Operations Management* (pp. 1228-1238). IEOM Society International.

Karaman, A. S., Kilic, M., & Uyar, A. (2018). Sustainability reporting in the aviation industry: worldwide evidence. *Sustainability Accounting, Management and Policy Journal*, 362-391.

Koç, S., & Durmaz, V. (2015). Airport Corporate Sustainability: An Analysis of Indicators Reported in the Sustainability Practices. *International Conference on Leadership, Technology and Innovation Management* (pp. 158-170). Procedia - Social and Behavioral Sciences.

Triple Bottom Line: Evidence from Aviation Sector. (2020). *International Journal of Business Ethics and Governance (IJBEG)*, 71-80.

Markatos, D. N., & Pantelakis, S. G. (2022). Assessment of the Impact of Material Selection on Aviation Sustainability, from a Circular Economy Perspective. *Aerospace (Basel, Switzerland)*, *9*(2), 2–14. doi:10.3390/aerospace9020052

Miller, K. (2020, December 8). Blog: Harward Business School. *Harward Business School*. https://online.hbs.edu/blog/post/what-is-the-triple-bottom-line

Ragbir, N. K., Rice, S., Winter, S. R., & Choy, E. C. (2021). Emotions and caring mediate the relationship between knowledge of sustainability and willingness to pay for greener aviation. *Technology in Society*, *64*, 1–7. doi:10.1016/j.techsoc.2020.101491

Sabharwal, R., & Miah, S. J. (2022). An intelligent literature review: Adopting inductive approach to define machine learning applications in the clinical domain. *Journal of Big Data*, *9*(1), 1–18. doi:10.118640537-022-00605-3

Singh, J., Rana, S., Hamid, A. B., & Gupta, P. (2022). Who should hold the baton of aviation sustainability? *Social Responsibility Journal*.

Slaper, T. F., & Hall, T. J. (2011). The Triple Bottom Line: What Is It and How Does It Work? *Indiana Business Review*, 1-8.

Somerville, A., Marino, M., Baxter, G., & Graham, W. (2016). Understanding Box Wing Aircraft: Essential Technology To Improve Sustainability In The Aviation Industry. *Aviation*, 129–136.

Song, K. H., Choi, S., & Han, I. H. (2020). Competitiveness Evaluation Methodology for Aviation Industry Sustainability Using Network DEA. *Sustainability*, *12*(24), 1–16. doi:10.3390u122410323

Suk, M., & Kim, W. (2021). COVID-19 and the airline industry: Crisis management and resilience. *Tourism Review*, *76*(4), 984–998. doi:10.1108/TR-07-2020-0348

Thokala, P., Scanlan, J., & Chipperfield, A. (2010). Life cycle cost modelling as an aircraft design support tool. *Journal of Aerospace Engineering*, 477–488.

Whelan, T., & Fink, C. (2016, October 21). Sustainable Business Practices. *Harvard Business Review*. https://hbr.org/2016/10/the-comprehensive-business-case-for-sustainability

Willard, B. (2012). The New Sustainability Advantage: Seven Business Case Benefits of a Triple Bottom Line. Gabriola Island: New Society Publishers.

Chapter 13
The Effective Occupational Commitment of the Employees in the Aviation Industry and the Suggestions for the Future

Esra Zeynel
Isparta University of Applied Science, Turkey

ABSTRACT

The technological change in the industrial field, which started with Industry 1.0, called the "Industrial Revolution," is now moving towards a new era called Industry 4.0. With the pandemic process that started in 2020, remote working systems have become widespread and remote working has been carried out mostly with digital tools. In the continuation of the process in which the pandemic was experienced very intensely, a process accepted as the new normal took place This change and development in the industrial field also affects the workforce behaviorally. This research examines the occupational commitment levels of the employee in the aviation industry during the post-pandemic period and offers suggestions for the future. Participants consist of airport employees operating in Turkey. According to the results of the research, the occupational commitment levels of the participants was high.

INTRODUCTION

The industrial change that started with Industry 1.0 is now advancing to a new and different period which is called Industry 4.0. This change and development in the industrial field have affected the organizations in almost every aspect. It also affects the workforce psychologically and behaviorally. During the Classical Management Approach, which emerged with the developments brought by the industrial revolution, prioritized efficiency and rationality, it neglected employee psychology. However, the Human Relations School that emerged with the famous Hawthorne Studies and the different studies carried out together reveal morale and motivation in the workplace as an important source of productivity. Afterwards, with the Neoclassical Management Approach, behavioral elements in the workplace have taken their place

DOI: 10.4018/978-1-6684-6835-7.ch013

in the literature as an important subject of study (Tutar, 2016, p. 11, 19). Organizational Behavior science has developed with the knowledge of these subjects. Therefore, positive attitudes and behaviors of employees are considered important in terms of workplace productivity and industrial development. For this reason, the fact that organizational leaders are aware of how these changes in organizations have an impact on the workforce plays an important role in determining organizational policies. Because human resources are the most important value of organizations. The higher the occupational commitment of the workforce of the organizations, the higher the productivity and efficiency of work. Organizational outputs are obtained more effectively.

Digitalization, which has become the most current issue and focus of recent times, transforms organizational structures. Many businesses and their employees around the world are trying to adapt to this rapid transformation. With this rapid change and development in information technologies, businesses need to develop different perspectives and adapt to change in order to maintain their sustainable competitive power. This change causes shifts in working systems. For all these reasons, the future expectations of the workforce are affected, and uncertainty creates anxiety and worry. Negative feelings and thoughts also affect the thoughts of employees about their profession. The pandemic process, which started in 2020, accelerated the change in working systems and caused digitalization to enter organizational environments rapidly. Remote working systems were quickly implemented for the isolation of individuals in the workplace. Remote work is provided by technological methods. Simultaneously, the use of digital systems for employees and customers has accelerated. In addition, its importance for the future has increased. This change in organizational environments has also affected the attitudes and behaviors of employees. This study deals with the change in the aviation industry in terms of organizational behavior. Understanding how employees' occupational commitment is affected will be a guide for organizational policies to be implemented in the future.

The aviation industry has been one of the most affected by the pandemic. Employees were affected by these processes. In this process, the negative impact of the sector on the economy increased the uncertainty about the future. They also experienced changes in their working systems. For this reason, employees were worried about how their working systems would be in the future (IATA, 2020; ILO, 2020). This research was carried out in the post-pandemic period, which is seen as the new normalization process, when the restrictions and bans of the pandemic were lifted. The aim of this study is to measure the occupational commitment of employees in the aviation industry during the post-pandemic period. Employees were asked about their working methods in the post-pandemic period. It has been tested whether the occupational commitment of the employees has a difference according to the working method. In addition, it was measured whether or not their occupational commitment differed according to some sociodemographic variables such as gender, age and job characteristics like department.

AVIATION SECTOR AND COVID-19 PANDEMIC

The coronavirus epidemic, which emerged in Wuhan, China in December, 2019 and affected many countries in a short time, was described as a COVID-19 Pandemic by the World Health Organization on March 11, 2020 (World Health Organization [WHO], 2020). By 2020, the modern world faced with an epidemic on a scale it never experienced before (Akça, 2020). Comprehensive measures were taken by governments against the epidemic. The borders of the countries were closed and international travel was taken under control. Decreased air traffic with these restrictions in the aviation sector caused flights to be stopped (International Labour Organization [ILO], 2020). With these measures, passenger demand decreased significantly. The worldwide suspension of flights has deeply affected the aviation industry. In this process, all stakeholders of the aviation industry and personnel working in the aviation industry were adversely affected (International Air Transport Association [IATA], 2020). The disease was felt globally. The pandemic's impact on air transport progressed to a low point in April, where passenger traffic dropped 94% from the previous year and much of the industry stalled from February. As a result of governments' travel restrictions from outside their national borders to prevent the importation of the COVID-19 virus, international air travel has been much more affected than domestic air travel or global air cargo. International Toll Mileage started to decline in February, down 98% in March and April compared to the previous year. Global IDPs remained almost 89% lower through September than a year ago. Recovery in the industry has been very slow. At the height of the crisis in April 2020, 90% of the aviation business disappeared. The industry lost $137.7 billion in 2020 and $42.1 billion in 2021 (IATA, 2020).

After the initial impact of the pandemic and for most of 2020 and early 2021, international air travel was negatively impacted by travel restrictions. During this period, passengers preferred less restricted domestic travel. In mid-2021, there was a strong demand for international travel, especially for leisure and short-haul travel, as governments began to lift restrictions on international travel. In contrast, domestic revenue passenger kilometers (PMIs) remained flat as Asian markets such as China, which initially recovered quickly, were hit by renewed virus outbreaks. (IATA,2022).

Due to the declaration of the pandemic on March 11, 2020 and the rapid progression of the COVID-19 disease, the government of the Republic of Turkey also took extensive measures. The Turkish government, like other states, closed the borders and stopped flights. These measures also had a huge impact on the Turkish aviation industry. In this process and afterwards, passenger demands and thus the number of flights decreased, as individuals did not want to enter places where the risk of contamination is high, such as airplane cabins, and did not want to be in crowded environments such as airports.

Air flight traffic information for 2020 and 2020 is shown in the figures below.

Figure 1. Daily Traffic Variation- States (2020)
Supporting European Aviation (Eurocontrol)- Dashboard

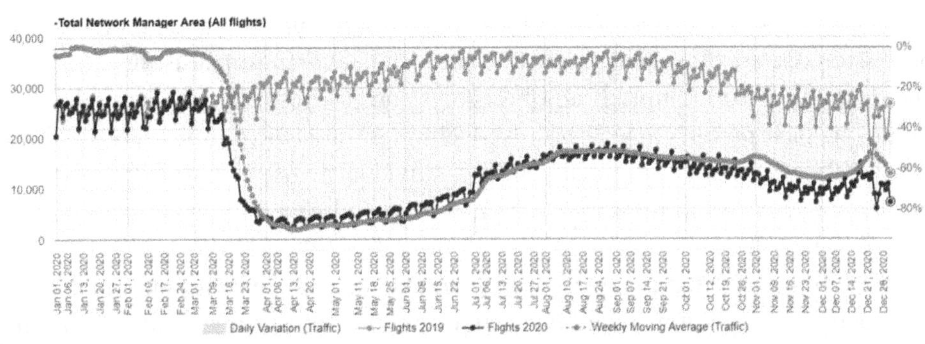

In fact, the importance of aviation was made absolutely clear by the pandemic restrictions. People recognized that their quality of life deteriorated and economies suffered. Recently, again, with increasing numbers and increasing excitement, it seems that people's desire to travel, connect with each other and see the world has increased again. By the end of 2023, most regions are expected to reach or exceed pre-pandemic demand levels. The emerging value of life aviation to the challenges facing the world as a result of the COVID-19 pandemic. Impacts are most evident in the travel and hospitality sectors. Millions lost their jobs and global gross domestic product was hit hard. Aviation is a very important driving force in these areas (IATA, 2022).

Figure 2. Daily Traffic Variation- States (2022)
Supporting European Aviation (Eurocontrol)- Dashboard

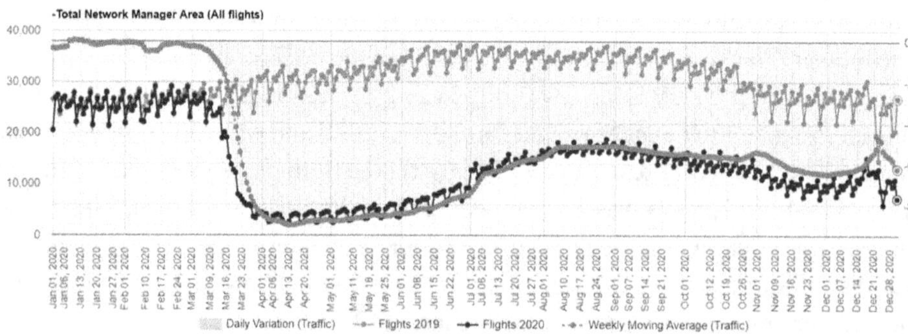

AVIATION SECTOR AND OCCUPATIONAL COMMITMENT

It is stated that the occupational commitment of individuals to their profession has significant effects on business results. Possibly, this effect is greater than organizational commitment and work engagement. Individuals lead and are guided by their professional expectations rather than their organizational expectations (Cohen, 2000). Bagraim (2003) defined the concept as a psychological state that shapes the relationship of the employee with his profession and emphasized that it has strong effects on whether the individual will continue his profession or not. Occupational commitment; It is the psychological bond that an individual develops with his profession, which is based on an emotional reaction to a profession (Lee et al., 2000). Occupational commitment reflects the individual motivation to work in a preferred profession (Carson & Bedeian, 1994). Occupational commitment is the emotional bond that an individual develops between himself and his profession (Meyer et al., 1993). Individuals who are highly committed to their profession have positive feelings towards their profession (Blau, 2003; Turner & Chelladurai, 2005). In addition, it is claimed that individuals who are committed to their profession show more research activities to improve their profession and generally experience conflict between work and family (Balay, 2000).

Meyer et al. (1993) stated that occupational commitment has three basic sub-dimensions: affective commitment, continuity commitment and normative commitment. Irving, et al. (1997) supported the measurement reliability and validity with their research on government employees in Canada, and Snape & Redman (2003) in the United Kingdom, with their research specific to human resource management experts. Meyer et al. (1993) argued that each type of commitment in the three-dimensional occupational commitment model is an important variable on the intention to leave the profession and the decision to continue the profession. While the dimension of affective commitment to the profession in the professional commitment model explains the strong desire of the individual to continue the profession, the dimension of continuance commitment to the profession explains the high cost of leaving the profession based on relatively less emotional basis. Normative commitment dimension is attributed to the feeling of obligation to continue the profession (Blau, 2001).

Affective commitment is explained as the individual's participation in the profession and the feeling of an emotional connection between the individual and the profession. According to this type of commitment, it is stated that individuals continue their profession at their own will, become members of professional associations, and follow professional developments closely by attending professional meetings. Affective professional commitment expresses a stronger type of attachment than professional continuance commitment or professional normative commitment, as it reflects individuals' own desires to continue their profession (Meyer & Herscovitch, 2001). In this type of commitment, individuals can exert more effort when necessary because they see themselves as a part of the profession (Bagraim, 2003). Affective commitment is the most important type of commitment in terms of the individual's identification with his profession and devotion to his profession and its values (Mowday, 1998). Affective commitment begins with the person's choice of profession and is shaped by their current professional experience (Weng & McElroy, 2012). Having positive experiences that will increase the level of job satisfaction is an important factor that improves affective occupational commitment (Meyer et al., 1993; Lee et al., 2000). Occupational commitment, personal well-being and satisfaction; organizationally, willingness to stay in the organization and productive behaviors; professionally, it provides positive results such as professional participation, development and continuing the profession (Meyer & Espinoza, 2016).

In the results of the research conducted by Yüksek, et al. (2022), it has been shown that COVID-19 anxiety has a significant effect on well-being and affective and normative commitment. There are also studies measuring the effects of stress and burnout caused by COVID-19 on occupational commitment (Öztürk & Uslu, 2021). There are studies on flexible working systems and affective commitment. For example, Bilkay (2022) conducted a research on the mediating role of affective commitment in the effect of flexible working arrangements on workaholism during the pandemic period.

Due to the high competition, manager expectations, customer demands, time pressure and excessive workload in the aviation industry, the employees of the sector are physically and psychologically worn out. These difficulties can cause negative consequences such as stress and burnout. Therefore, organizational commitment and occupational commitment employees in the aviation industry can be cultivated from these difficult conditions. For this reason, in the literature, besides occupational commitment studies on aviation sector employees, organizational commitment studies are encountered. For example, Seçgin (2016) conducted a study in Eskişehir aviation sector to determine the effects of positive organizational behavior on organizational commitment. Arisoy & Yumak (2020) conducted a research on organizational commitment civil aviation cabin crews Korkmaz (2020) conducted a research on the relationship between human resources management functions and organizational commitment in the aviation industry. Dursun (2021) a study conducted on effect of work wtress and burnout perceptions of aviation sector employees on organizational commitment. In addition, a strong negative relationship was found between organizational commitment and turnover intention), while a positive and strong relationship was observed between perceived organizational support and organizational commitment (Satardien, et al., 2019). The research finding is that there is significant influence of job satisfaction and normative commitment (a component of organizational commitment) on turnover intention among aviation maintenance personnel. Anak Edwin, A. (2020). According to a study, there is a significant relationship between executive support and organizational commitment (Turk, et al, 2021). Li, et al.,

(2021) stated that organizational trust, organizational identification, organizational commitment, and safety operation behavior (SOB) were significantly correlated with each other. Moreover, airline pilots' organizational identification and organizational commitment mediated the relationship between their organizational trust and SOB in aviation.

METHODOLOGY

Design

The study is based on quantitative research methods. The data of the study were obtained from the people working at the three big airports operating in Turkey according to the convenience sampling method. A questionnaire was applied to the participants. This questionnaire consists of a occupational commitment scale to determine affective occupational commitment, together with a section covering the sociodemographic characteristics of the participants and the way they work on the pandemic and post-pandemic.

In this study, data were collected from individuals working at three airports operating in İstanbul and İzmir, Türkiye. Questionnaires were directed to people working in different statuses and positions. Questionnaires were directed to people working in different statuses and positions. Those who took the time to participate in the survey voluntarily participated. 252 questionnaires were returned. However, outlier data and missing data were examined in detail. As a result, it was decided that 248 data were suitable for analysis. The occupational commitment levels and average answers of the participants were determined by descriptive analysis. With the ANOVA test, it was tested that the occupational commitment levels of the individuals who participated in the survey were not significantly different according to the age factor and working time. The TUKEY test was conducted to determine whether there is a difference in the occupational commitment of the employees according to the level of education

Hypotheses

The hypotheses formed according to the purpose of the study are as follows.

H1: Occupational commitment differs by gender
H2: Occupational commitment differs by age
H3: Occupational commitment differs by department
H4: Occupational commitment differs by working method in post pandemic era
H5: Occupational commitment differs by organizational support

Measures

To measure the professional commitment levels of the participants, the Turkish version of the scale developed by Meyer et al., (1993) and the Turkish validity scale was used by Tak & Çiftçioğlu (2009) was used. The scale is three-dimensional and based on a 5-point Likert scale.In this study, it was especially desired to measure the level of emotional commitment. Analyzes were made on affective occupational commitment. Participants were also asked about their working methods during the pandemic and post-pandemic periods and organizational support. In addition, the participants were asked questions about their sociodemographic information.

RESULTS

Descriptives Statistics

Information about the participants is shown in the table below. According to the findings, 53.2% (132) of the 248 participants were male and 46.8% (116) were female. The departments involved in the study include Cabin Services (41, 16.5%), Passenger Services (35, 14.1%), Operation/Ground Services (78, 31.5%), Aircraft Maintenance/Technical (71, 28.6%), Human Resources (7, 2.8%), Occupational health and Safety (8, 3.2%), Purchasing/Administrative Affairs (2, 0.8%). And 6 (2.4%) people who do not belong to these groups are grouped as other. (Table 1).

The subject of the research is to examine the effects of the post-pandemic period on employees. During the pandemic process, remote working methods were applied. In some of the organizations, these working methods continued to be applied, albeit partially, during the post-pandemic period. Therefore, employees were asked about the working method in the pandemic era. Moreover, they were asked about the working method in the first period of the post-pandemic period in which the research was conducted. In addition, employees were asked about the supports that are provided by the organization. Accordingly, the second table includes information about the working method and the findings of the organizational support (Table 2).

Confirmatory Factor Analysis

Confirmatory factor analysis was performed to test the construct validity of the occupational commitment scale. Amos 26 program was used. Analysis findings and path diagram are as follows. It was observed that there was a relationship between the 1st and 2nd questions in the scale. In addition, it was observed that there was a relationship between 4 and 5 items. A connection was established between these substances according to the modification indices.

The Table 3 showing the goodness of fit criteria of the model and the values of goodness of fit related to the findings of the study was created by the author using the sources of Hair et al., 2010, Tabachnick & Fidell, 2007, Kline, 2011, Byrne, 2012. According to the findings of the model, while some values show good fit, some values are between acceptable values. When all values are evaluated together, the model has an acceptable level of validity. An appropriate model is expected to be meaningless ($p > .05$), but in large samples χ^2 is usually significant ($p < .05$) (Tabachnick & Fidell, 2007). Therefore, p-value and the ratio of χ^2 to the degrees of freedom should be consider together. According to findings of the study (χ^2/df) = 3.8 was acceptable value. The Goodness of fit index (GFI) = .96, The normed fit index (NFI) = .95 and Comparative fit index (CFI) = .96 values values indicated that the model had good fit ((Hair et al., 2010). However, Hu & Bentler (1999) stated that a CFI value of .95 and above with a SRMR value of .08 and below together indicate acceptable fit (Kline, 2011). MacCallum et al. 1996 stated that RMSEA values ranging from .08 to .10 showed mediocre fit.

Table 1. Sample characteristics

Gender		N	%
	Male	132	53.2
	Female	116	46.8
	Total	248	100.0
Age			
	19-24	54	21.8
	25-34	104	41.9
	35-44	68	27.4
	45-54	20	8.1
	55+	2	.8
	Total	248	100.0
Education			
	High School Graduate Degree	32	12.9
	Bachelor's Degree	166	66.9
	Master/Doctoral Degree	50	20.2
	Total	248	100.0
Department			
	Cabin Services	41	16.5
	Passenger Services	35	14.1
	Operation/Ground Services	78	31.5
	Aircraft Maintenance/Technical	71	28.6
	Human Resources	7	2.8
	Occupational health and Safety	8	3.2
	Purchasing/Administrative Affairs	2	.8
	Other	6	2.4
	Total	248	100.0
Sector			
	Public	28	11.3
	Special	220	88.7
	Total	248	100.0
Managerial Position			
	Non-manager	157	63.3
	Lower/middle level manager	70	28.2
	Top manager	21	8.5
	Total	248	100.0

Table 2. Characteristics of organizational policies related to work

	Working Method in the Pandemic Era	N	%
	I was compulsorily working from home, I was working at the workplace before	88	35.5
	I chose to work from home, I could work from my workplace if I wished	16	6.5
	I've been working from home, but I've been working from home at certain times before	8	3.2
	I've been working from home but I've been working entirely from home before	6	2.4
	I was working at the workplace	130	52.4
	Total	248	100.0
Working Method in the Post Pandemic Era			
	I was compulsorily working from home, I was working at the workplace before	4	1.6
	I chose to work from home, I could work from my workplace if I wished	8	3.2
	I work from home, I have worked from home for certain times before	4	1.6
	I work from home, but I've already worked from home before	7	2.8
	I work at workplace	225	90.7
	Total	248	100.0
Organizational Support			
	No support provided	29	11.7
	Office tools, Pc/Notebook, Telephone/Telephone line	19	7.7
	Business processes software/technical support	11	4.4
	Educational support	17	6.9
	Pc/Notebook, Telephone/Telephone line, Business processes software/technical support	8	3.2
	Office tools, Pc/Notebook, Telephone/Telephone line, Educational support	27	10.9
	Business processes software/technical support, Educational support	14	5.6
	Office tools, Pc/Notebook, Telephone/Telephone line, Business processes software/technical support, Educational support	123	49.6
	Total	248	100.0

Table 3. Model fit summary

	Observed Values of the Model	Goodness of Fit Indices	Acceptable Indices
P	.000 ($p < .05$)	$p > .05$ (insignificant)	
CMIN/df (χ^2/df)	3.8 (26,597/7)	CMIN/df < 3	3 ≤ CMIN/df < 5
GFI	.96	.95 < GFI ≤ 1.00	.90 < GFI ≤ .95
AGFI	.89	.90 < AGFI ≤ 1.00	.85 < AGFI ≤ .90
NFI	.95	.95 < NFI ≤ 1.00	.90 < NFI ≤ .95
CFI	.96	.95 ≤ CFI ≤ 1.00	.90 < CFI < .95
RMSEA	.10	0 ≤ RMSEA < 05	.05 ≤ RMSEA < 08
RMR	.08	< .05	.05≤ RMR ≤ .08

CMIN (χ^2): Chi-squared test indicates the difference between observed and expected covariance matrices df: degree of freedom GFI: Goodness of fit index AGFI: Adjusted goodness of fit index NFI: Normed fit index CFI: Comparative fit index RMSEA: The root mean square error of approximation RMR: Root mean square residual

Figure 3. Path Diagram

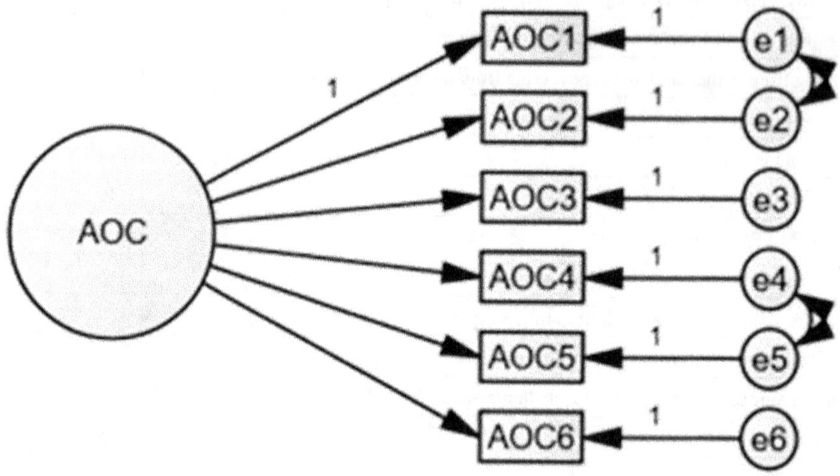

Descriptive Statistics of the Occupational Commitment

It was seen that the occupational commitment levels of the participants were at a level that could be considered high. It is seen that the average of the affective commitment levels of the employees in the aviation sector is 3.94 (Table 4). Considering that the scale used has a rating between 1 and 5, a level close to 4 can be considered as a high level of occupational commitment.

Table 4. The mean of the affective occupational commitment

	Affective Occupational Commitment
N	248
Mean	3.9496
S.E. of Mean	.05731
S.D	.90248
Minimum	1.00
Maximum	5.00

Difference Tests

Difference Between Affective Occupational Commitment and Gender

Whether the occupational commitment levels of the employees in the aviation sector differ according to gender was investigated through the data obtained from 248 people. According to the Independent samples t test results, there was no significant difference regarding the occupational commitment of the employees (t=1,097 ; *p*< .05). The average of the opinions of female employees regarding their occupational commitment levels (Mean= 4, 008; S.D = .86720). It was higher than the average of male employees' opinions about their occupational commitment levels (Mean=3, 8822 ; S.D = .94025). However, there was no significant difference between the professional commitment levels of female and male employees (Table 5). H1 hypothesis was rejected.

According to Özmen et al. (2005), it was found that there was no significant difference in the occupational commitment of female compared to their male colleagues in the examination of the relationship between gender and organizational commitment.

Table 5. Difference between affective occupational commitment and gender using ındependent t- test

							Equal variances not assumed		
	Gender	**N**	**Mean**	**S.D**	**F**	**Sig.**	**T**	**df**	**Sig. (2-tailed)**
AOC	Male	132	4.0088	.86720	2.099	.149	1.097	235.615	.274
	Female	116	3.8822	.94025					

**p*<.05

Difference Between Affective Occupational Commitment and Age

According to the Levene test (*p*=.153, p >.05), the variances were equal. According to the Anova test findings, the significance ratio (Sig.) was greater than .05 (*p* >.05) and there was no significant difference between the opinions of the employees regarding their occupational commitment according to the age. H2 hypothesis was rejected.

Table 6. Difference between affective occupational commitment and age using ANOVA

	N	**Mean**	**S.D**		**Sum of Squares**	**Df**	**F**	**Sig.**
19-24	54	3.8765	1.01724	Between Groups	1.801	4	.549	.700
25-34	104	4.0192	.87926					
35-44	68	3.8799	.88372	Within Groups	199.374	243		
45-54	20	4.0667	.80641					
55+	2	3.5000	.47140	Total	201.176	247		
Total	248	3.9496	.90248					

Difference Between Affective Occupational Commitment and Department

According to the Levene test (*p*=.002, *p*<.05), the variances were not equal, so Welch (*p*= .005) and *Brown-Forsythe* (*p*=.004) test results were checked. According to the results of analysis of variance, there was a significant difference in the occupational commitment levels of the workforce working in different departments (F (4.353), $p < .05$). H3 hypothesis was accepted (Table 7).

Table 7. Difference between affective occupational commitment and department using ANOVA

Affective Occupational Commitment											
	N	**Mean**	**S.D.**	**Min**	**Max**		**Sum of Squares**	**df**	**Mean Square**	**F**	**Sig.**
Cabin Services	41	4.0732	.86124	2.17	5.00	Between Groups	22.666	7	3.238	4.353	.000 Welch (*p*= .005 *Brown-Forsythe* (*p*=.004)
Passenger Services	35	3.2810	.76885	1.83	5.00						
Operation/Ground Services	78	4.0662	.94294	1.33	5.00						
Aircraft Maintenance/ Technical	71	4.1221	.78273	2.67	5.00						
Human Resources	7	3.7381	1.48404	1.00	5.00	Within Groups	178.509	240	.744		
Purchasing/Administrative Affairs	8	4.1042	.61681	3.33	5.00						
Occupational health and Safety	2	4.4167	.82496	3.83	5.00						
Other	6	3.3333	.40825	2.83	4.00						
Total	248	3.9496	.90248	1.00	5.00	Total	201.176	247			

*$p < .05$

Tamhane test was used in the post hoc test, since the variances were not equal according to the Levene test (p= .002). According to the results of the test, which shows which groups the differences originate from, there is a significant difference between the occupational differences of passenger service employees differ according to the occupational commitment levels of Cabin Services (Mean= -.79222, S.E = .19847), Operation/Ground Services (Mean= -.78529, S.E = .17546) and Aircraft Maintenance/ Technical employees (Mean= -.84111, S.E = .17812). (Table 8).

Difference Between Affective Occupational Commitment and Working Method in Post Pandemic Era

According to the Levene test (p= .108, $p > .05$), the variances were equal. According to the Anova test findings, there was a difference in the occupational commitment levels of the participants, according to Working Method in Post Pandemic Era (F (2.428); $p < .05$). There was a significant difference in the occupational commitment and working method in the post pandemic era (F (2.428), $p < .05$). (Table 9). H4 hypothesis was accepted.

LSD method was used in Post Hoc test because sample size were not equal of the groups. According to the results of the test, there was a significant difference between the occupational differences among those who answered "I have worked from home for certain times before" and "I was compulsively working from home, I was working at the workplace before" (Mean= 1.41667, S.E =.63090) ; I chose to work from home, I could work from my workplace if I wished (Mean=-1.25000*, S.E = .54638) ; I work at workplace (Mean= -1.01056, S.E = .45006). (Table 10). It is an important result.

Table 8. Multiple comparisons of the department

(I) Department	(J) Department	Mean Difference (I-J)	S.E.	Sig.
Cabin Services	Passenger Services	.79222*	.19847	.002
	Operation/Ground Services	.00693	.16636	1.000
	Aircraft Maintenance/Technical	-.04889	.16917	1.000
	Human Resources	.33508	.35270	.981
	Occupational health and Safety	-.03100	.33334	1.000
	Purchasing/Administrative Affairs	-.34350	.62453	.999
	Other	.73984	.37697	.509
Passenger Services	Cabin Services	-.79222*	.19847	.002
	Operation/Ground Services	-.78529*	.17546	.000
	Aircraft Maintenance/Technical	-.84111*	.17812	.000
	Human Resources	-.45714	.35708	.905
	Operation/Ground Services	-.82321	.33797	.229
	Purchasing/Administrative Affairs	-1.13571	.62701	.613
	Other	-.05238	.38107	1.000
Operation/Ground Services	Cabin Services	-.00693	.16636	1.000
	Passenger Services	.78529*	.17546	.000
	Aircraft Maintenance/Technical	-.05583	.14146	1.000
	Human Resources	.32814	.34028	.979
	Occupational health and Safety	-.03793	.32017	1.000
	Purchasing/Administrative Affairs	-.35043	.61760	.999
	Other	.73291	.36538	.480
Aircraft Maintenance/Technical	Cabin Services	.04889	.16917	1.000
	Passenger Services	.84111*	.17812	.000
	Operation/Ground Services	.05583	.14146	1.000
	Human Resources	.38397	.34166	.951
	Occupational health and Safety	.01790	.32164	1.000
	Purchasing/Administrative Affairs	-.29460	.61836	1.000
	Other	.78873	.36666	.385
Human Resources	Cabin Services	-.33508	.35270	.981
	Passenger Services	.45714	.35708	.905
	Operation/Ground Services	-.32814	.34028	.979
	Aircraft Maintenance/Technical	-.38397	.34166	.951
	Occupational health and Safety	-.36607	.44635	.992
	Purchasing/Administrative Affairs	-.67857	.69148	.977
	Other	.40476	.47981	.990

continues on following page

Table 8. Continued

(I) Department	(J) Department	Mean Difference (I-J)	S.E.	Sig.
Purchasing/Administrative Affairs	Cabin Services	.03100	.33334	1.000
	Passenger Services	.82321	.33797	.229
	Operation/Ground Services	.03793	.32017	1.000
	Aircraft Maintenance/Technical	-.01790	.32164	1.000
	Human Resources	.36607	.44635	.992
	Purchasing/Administrative Affairs	-.31250	.68181	1.000
	Other	.77083	.46577	.716
Occupational health and Safety	Cabin Services	.34350	.62453	.999
	Passenger Services	1.13571	.62701	.613
	Operation/Ground Services	.35043	.61760	.999
	Aircraft Maintenance/Technical	.29460	.61836	1.000
	Human Resources	.67857	.69148	.977
	Purchasing/Administrative Affairs	.31250	.68181	1.000
	Other	1.08333	.70417	.786
Other	Cabin Services	-.73984	.37697	.509
	Passenger Services	.05238	.38107	1.000
	Operation/Ground Services	-.73291	.36538	.480
	Aircraft Maintenance/Technical	-.78873	.36666	.385
	Human Resources	-.40476	.47981	.990
	Occupational health and Safety	-.77083	.46577	.716
	Purchasing/Administrative Affairs	-1.08333	.70417	.786

$^*p < .05$

Table 9. Difference between affective occupational commitment and working method in post pandemic era using ANOVA

	N	Mean	S.D.		Sum of Squares	Df	Mean Square	F	Sig.
I was compulsorily working from home, I was working at the workplace before	4	4.3750	1.03972	Between Groups	7.731	4	1.933	2.428	.049
I chose to work from home, I could work from my workplace if I wished	8	4.2083	.97080						
I work from home, I have worked from home for certain times before	4	2.9583	1.57159	Within Groups	193.445	243	.796		
I work from home, but I've already worked from home before	7	3.3571	.44544						
I work at workplace	225	3.9689	.88389	Total	201.176	247			
Total	248	3.9496	.90248						

Table 10. Multiple comparisons of the working method in the post pandemic era

(I) Working Method in the Post Pandemic.Era	(J) Working Method in the Post Pandemic Era	Mean Difference (I-J)	S.E.	Sig.
I was compulsorily working from home, I was working at the workplace before	I chose to work from home, I could work from my workplace if I wished	.16667	.54638	.761
	I work from home, I have worked from home for certain times before	1.41667*	.63090	.026
	I work from home, but I've already worked from home before	1.01786	.55923	.070
	I work at workplace	.40611	.45006	.368
I chose to work from home, I could work from my workplace if I wished	I was compulsorily working from home, I was working at the workplace before	-.16667	.54638	.761
	I work from home, I have worked from home for certain times before	1.25000*	.54638	.023
	I work from home, but I've already worked from home before	.85119	.46177	.067
	I work at workplace	.23944	.32101	.456
I work from home, I have worked from home for certain times before	I was compulsorily working from home, I was working at the workplace before	-1.41667*	.63090	.026
	I chose to work from home, I could work from my workplace if I wished	-1.25000*	.54638	.023
	I work from home, but I've already worked from home before	-.39881	.55923	.476
	I work at workplace	-1.01056*	.45006	.026
I work from home, but I've already worked from home before	I was compulsorily working from home, I was working at the workplace before	-1.01786	.55923	.070
	I chose to work from home, I could work from my workplace if I wished	-.85119	.46177	.067
	I work from home, I have worked from home for certain times before	.39881	.55923	.476
	I work at workplace	-.61175	.34244	.075
I work at workplace	I was compulsorily working from home, I was working at the workplace before	-.40611	.45006	.368
	I chose to work from home, I could work from my workplace if I wished	-.23944	.32101	.456
	I work from home, I have worked from home for certain times before	1.01056*	.45006	.026
	I work from home, but I've already worked from home before	.61175	.34244	.075

*p < .05

Difference Between Occupational Commitment and Organizational Support

According to the Levene test (p= .031, $p < .05$), so Welch (p= .007) and *Brown-Forsythe* (p=.007) test results were checked. the variances were not equal. According to the results of the analysis of variance, there was a significant difference in the occupational commitment levels of the employees according to the organizational support (F (3.537), $p < .05$). (Table 11). H5 hypothesis was accepted.

Tamhane method was used in Post Hoc test because variances were not equal (Levene statistics, p=.31). According to the results, there was a significant difference between the occupational differences of those who had the support of "office tools, pc/notebook, telephone/telephone line, business processes software/technical, training" (Mean= .65685; S.E = .17994) and those who had not received any support (Table 12).

Table 11. Difference between affective occupational commitment and organizational support using ANOVA

	N	Mean	S.D.		Sum of Squares	df	Mean Square	F	Sig.
No support provided	29	3.5057	.88918	Between Groups	8.814	7	2.688	3.537	.001 Welch (*p*= .007) *Brown-Forsythe* (*p*=.007)
Office tools, PC/Notebook, Telephone/Telephone line	19	3.5614	1.10862						
Business processes software/technical support	11	3.5758	.63842						
Training support	17	4.1275	.83026						
PC/Notebook, Phone/Hotline Business processes software/technical support	8	3.6042	.93409						
PC/Notebook, Phone/Telephone line, Business processes software/technical support, Training support	27	4.0370	.82215	Within the Groups	182.361	240	.760		
Business processes software/technical support, Training support	14	3.6310	1.32074						
Office tools Pc/Notebook, Phone/Telephone line, Business processes software/technical support, Training support	123	4.1626	.79257						
Total	248	3.9496	.90248	Total	201.176	247			

Table 12. Multiple comparisons of the organizational support

(I)Organizational Support	(J) Organizational Support	Mean Difference (I-J)	S.E.	Sig.
No support provided	Office tools, Pc/Notebook, Telephone/Telephone line	-.05566	.25728	1.000
	Business processes software/technical support	-.07001	.30867	1.000
	Training support	-.62170	.26627	.279
	PC/Notebook, Telephone/Telephone line, Business processes software/technical support	-.09842	.34811	1.000
	PC/Notebook, Telephone/Telephone line, Training support	-.53129	.23312	.310
	Business processes software/technical support, Training support	-.12521	.28368	1.000
	Office tools, PC/Notebook, Telephone/Telephone line, Business processes software/technical support, Training support	-.65685*	.17994	.008

continues on following page

Table 12. Continued

(I)Organizational Support	(J) Organizational Support	Mean Difference (I-J)	S.E.	Sig.
Office tools, Pc / Notebook, Telephone / Telephone line	No support provided	.05566	.25728	1.000
	Business processes software/technical support	-.01435	.33025	1.000
	Training support	-.56605	.29101	.521
	PC/Notebook, Telephone/Telephone line, Business processes software/technical support	-.04276	.36738	1.000
	PC/Notebook, Telephone/Telephone line, Training support	-.47563	.26102	.606
	Business processes software/technical support, Training support	-.06955	.30703	1.000
	Office tools, Pc/Notebook, Telephone/Telephone line, Business processes software/technical support, Training support	-.60120	.21487	.100
Business processes software / technical support	No support provided	.07001	.30867	1.000
	Office tools, PC/Notebook, Telephone/Telephone line	.01435	.33025	1.000
	Training support	-.55169	.33730	.728
	PC/Notebook, Telephone/Telephone line, Business processes software/technical support	-.02841	.40504	1.000
	Pc/Notebook, Telephone/Telephone line, Training support	-.46128	.31180	.818
	Business processes software/technical support Training support	-.05519	.35121	1.000
	Office tools, Pc/Notebook, Telephone/Telephone line, Business processes software/technical support, Training support	-.58684	.27432	.393
Training support	No support provided	.62170	.26627	.279
	Office tools, Pc/Notebook, Telephone/Telephone line	.56605	.29101	.521
	Business processes software/technical support	.55169	.33730	.728
	Pc/Notebook, Telephone/Telephone line, Business processes software/ technical support	.52328	.37373	.857
	Pc/Notebook, Telephone/Telephone line, Training support	.09041	.26989	1.000
	Business processes software/technical support, Training support	.49650	.31460	.763
	Office tools, Pc/Notebook, Telephone/Telephone line, Business processes software/technical support, Training support	-.03515	.22555	1.000
Pc/Notebook, Telephone/ Telephone line, Business processes software/technical support	No support provided	.09842	.34811	1.000
	Office tools Pc/Notebook, Telephone/Telephone line	.04276	.36738	1.000
	Business processes software/technical support	.02841	.40504	1.000
	Training support	-.52328	.37373	.857
	Office tools Pc/Notebook, Telephone/Telephone line	-.43287	.35089	.921
	Business processes software/technical support,Training support	-.02679	.38633	1.000
	Pc/Notebook, Telephone/Telephone line, Business processes software/ technical support, Training support	-.55843	.31805	.650

continues on following page

Table 12. Continued

(I)Organizational Support	(J) Organizational Support	Mean Difference (I-J)	S.E.	Sig.
Pc/ Notebook, Telephone / Telephone line, Training support	No support provided	.53129	.23312	.310
	Office tools, Pc/Notebook, Phone/Telephone line	.47563	.26102	.606
	Business processes software/technical support	.46128	.31180	.818
	Training support	-.09041	.26989	1.000
	Office tools, Pc/Notebook, Phone/Telephone line, Business processes software/technical support	.43287	.35089	.921
	Business processes software/technical support, Training support	.40608	.28708	.850
	Office tools Pc/Notebook, Telephone/Telephone line, Business processes software/ technical support, Training support	-.12556	.18526	.998
Business processes software / technical support, Training support	No support provided	.12521	.28368	1.000
	Office tools, Pc/Notebook, Phone/Telephone line,	.06955	.30703	1.000
	Business processes software/technical support	.05519	.35121	1.000
	Training support	-.49650	.31460	.763
	Pc/Notebook, Telephone/Telephone line, Business processes software/ technical support	.02679	.38633	1.000
	Pc/Notebook, Telephone/Telephone line, Training support	-.40608	.28708	.850
	Office tools, Pc/Notebook, Telephone/Telephone line, Business processes software/technical support, Training support	-.53165	.24587	.378
Office tools Pc / Notebook, Telephone / Telephone line, Business processes software / technical support, Training support	No support provided	.65685*	.17994	.008
	Office tools, Pc/Notebook, Telephone/Telephone line	.60120	.21487	.100
	Business processes software/technical support	.58684	.27432	.393
	Training support	.03515	.22555	1.000
	Pc/Notebook, Phone/Telephone line, Business processes software/ technical support	.55843	.31805	.650
	Pc/Notebook, Telephone/Telephone line, Training support	.12556	.18526	.998
	Business processes software/technical support, Training support	.53165	.24587	.378

$*p < .05$

CONCLUSION

The COVID-19 epidemic, which was defined as a pandemic as of March 2020, brought about a very different period. Communication between people has decreased at an unprecedented level. Quarantine was implemented, the borders of countries were closed to international travel, working systems were changed and remote working methods were applied. These measures and restrictions have affected businesses in many ways. The aviation sector was one of the most affected sectors. This crisis actually brings with it a transformation. It is important for the aviation industry to take the right decisions and make preparations for the future. Those working in the aviation industry experienced great stress during the pandemic period. Employees' occupational commitment was impacted by health concerns, uncertainty brought about by the change in work systems, and the industry's financial impact.

This research was conducted during the post-pandemic period. In this period, it is a period called the new normal, in which measures and restrictions are removed. Employees began to return to work. After the difficulties of the pandemic, it was aimed to determine how the new normal employees' job commitment levels were affected. The findings of the research show that the occupational commitment of those employees in the aviation industry in this period is at a level that can be considered high (Mean=3.9). The reason for this may be that the uncertainty experienced in the pandemic has made individuals more aware of the importance of their profession. On the other hand, returning to the workplace after being away from the workplace may have increased their positive attitudes towards their profession. In future studies, the opinions of the employees on this subject can be determined in more detail with in-depth interviews. A study measuring occupational commitment during the pandemic also showed that occupational commitment increased. The reason for this may be that people cling to their professions against the negativities and uncertainties experienced. It was investigated whether the professional commitment of the employees was affected by different working methods (such as remote working / working at the workplace) during the post-pandemic period. Accordingly, it was found that the levels of professional commitment differed. In a study conducted by Bilkay (2022), it was found that flexible working order affects the affective organizational commitment of aviation workers and affective organizational commitment regulates the relationship between flexible working order and workaholism. In addition, as a result of the tests performed according to demographic variables, it was seen that occupational commitment did not change according to gender and age. According to Derin et al. (2017), there is no significant difference between the mean professional commitment of nurses according to gender, age and education.

The emerging value of life aviation to the challenges facing the world as a result of the COVID-19 pandemic. Impacts are most evident in the travel and hospitality sectors. Millions lost their jobs and global gross domestic product was hit hard. Aviation is a very important driving force in these areas. The wishes and needs of the employees should be given importance. Policies that increase morale and motivation should be developed. Investigating the emotional commitment of employees will give an idea for decision makers about working arrangements. Measures should be taken to protect employees against difficulties and crises. Employees should be valued. It should not be forgotten that their support is important in crises. It is important to appreciate, to give confidence, to feel organizational support in the face of difficulties. It is necessary to create an effective organizational communication climate that strengthens the communication of employees with managers. As a result, the dedication shown by aviation employees in times of crisis and even during the pandemic is the result of their emotional commitment to their profession in the aviation industry and the aviation company they work for. Qualified human resources and organizational behavior policies in the aviation sector should be sustainable and increase this commitment.

Employees are still concerned about the contagion of epidemics and diseases. It will be important that the hygiene rules are sustainable in organizational environments and that precautions such as masks and vaccinations are taken into account. Vaccines, masks and hygiene measures should be given importance. States should follow policies that support the aviation industry.Sustainability policies should be strengthened.Policies that are balanced on gender, that care about female employees and bring them to the sector should be followed. Organizations in the aviation sector must adapt to the changing forms of business life in order to protect the workforce based on international labor standards. Higher levels of protection and enhanced measures should be implemented to further mitigate the effects of the crisis.

REFERENCES

Akca, M. (2020). COVID-19'un havacılık sektörüne etkisi [The impact of COVID-19 on the aviation industry]. *Avrasya Sosyal ve Ekonomi Araştırmaları Dergisi, 7 (4),* 45-64. https://dergipark.org.tr/tr/pub/asead/issue/54055/723434

Arisoy, B., & Yumak, E. M. B. (2020). Örgütsel bağlılık sivil havacılık kabin ekipleri üzerine bir araştırma [A research on organizational commitment civil aviation cabin crews]. *International Journal of New Trends in Social Sciences*, *4*(2), 94–104. doi:10.18844/ijntss.v4i2.5572

Bagraim, J. J. (2003). The dimensionality of professional commitment. *SA Journal of Industrial Psychology*, *29*(2), 6–9. doi:10.4102ajip.v29i2.104

Balay, R. (2000). *Yönetici ve öğretmenlerde örgütsel bağlılık* [Organizational commitment in administrators and teachers.]. Nobel Yayın Dağıtım.

Bilkay, S. (2022). Pandemi Dönemindeki Esnek Çalışma Düzenlemelerinin İşkolikliğe Etkisinde Duygusal Bağlılığın Aracılık Rolünün Araştırılması. [Investigating the Mediating Role of Emotional Attachment in the Impact of Flexible Working Arrangements on Workaholism in the Pandemic Period.]. *Pearson Journal*, *7*(18), 156–169. doi:10.46872/pj.496

Blau, G. (2003). Testing for a four-dimensional structure of occupational commitment. *Journal of Occupational and Organizational Psychology*, *76*(4), 469–488. doi:10.1348/096317903322591596

Blau, G. J. (2001). On assessing the construct validity of two multidimensional constructs occupational commitment and occupational entrenchmen. *Human Resource Management Review*, *11*(3), 279–298. doi:10.1016/S1053-4822(00)00052-8

Byrne, B. M. (2012). *Structural Equation Modeling with Mplus: Basic Concepts, Applications, and Programming. NY.* Routledge.

Cohen, A. (2000). The relationship between commitment forms and work outcomes: A comprasion of three models. *Human Relations*, *5*(3), 387–417. doi:10.1177/0018726700533005

Derin, N., Şimşek Ilkım, N., & Yayan, H. (2017). Hemşirelerde mesleki bağlılığın mesleki prestijle açıklanması.[Explanation of professional commitment in nurses with professional prestige.] *Hacettepe Üniversitesi Hemşirelik Fakültesi Dergisi. 4*(3), 24-37. dergipart.org.tr

Dursun, D. E. (2021). Effect of work stress and burnout perceptions of aviation sector employees on organizational commitment. *Journal of Contemporary Issues in Business and Government*, *27*(3), 412–417. doi:10.47750/cibg.2021.27.03.056

Edwin, A. A. (2020). The factor that influence turnover intention among aviation maintenance personnel in kota kinabalu international airport sabah. [MJBE]. *Malaysian Journal of Business and Economics*, *7*(2), 73. doi:10.51200/mjbe.vi.2880

Hair, J. F., Black, W. C., Babin, B. J., & Anderson, R. E. (2010). *Multivariate Data Analysis* (7th ed.). Pearson.

International Air Transport Association (IATA) (2020). *Annual review 2020.* IATA.

International Air Transport Association (IATA) (2022). *Annual review 2022*. IATA.

International Labour Organization (ILO) (2020). *Sectoral brief, 09/04/2020*. ILO.

Irving, P. G., Coleman, D. F., & Cooper, C. L. (1997). Further assessments of a three- component model of occupational commitment: Generalizability and differences across occupations. *The Journal of Applied Psychology*, *82*(3), 444–452. doi:10.1037/0021-9010.82.3.444

Kline, R. B. (2011). *Principles and Practice of Structural Equation Modeling*. Guilford Press.

Korkmaz, S. (2020). *Havacılık sektöründe insan kaynakları yönetim fonksiyonları ile örgütsel bağlılık ilişkisi üzerine bir araştırma: TGS yer hizmetleri işletmesi örneği. [A study on the relationship between human resources management functions and organizational commitment in the aviation sector: The case of TGS ground handling operations.]* [Unpublished master dissertation, İstanbul Gelişim Üniversitesi, Lisansüstü Eğitim Enstitüsü, İstanbul, Türkiye]. acikerisim.gelisim.edu.tr

Lee, K., Carswell, J., & Allen, N. (2000). A meta-analytic review of occupational commitment: Relations with person and work-related variables. *The Journal of Applied Psychology*, *85*(5), 799–811. doi:10.1037/0021-9010.85.5.799 PMID:11055151

Li, Y., Liu, Z., Qin, K., Cui, J., Zeng, X., Ji, M., & Li, Y. (2021). Organizational trust and safety operation behavior in airline pilots: The mediating effects of organizational identification and organizational commitment. *Journal of Air Transport Management*, *92*, 102018. doi:10.1016/j.jairtraman.2021.102018

MacCallum, R. C., Browne, M. W., & Sugawara, H. M. (1996). Power analysis and determination of sample size for covariance structure modeling. *Psychological Methods*, *1*(2), 130–149. doi:10.1037/1082-989X.1.2.130

Meyer, J. P., & Allen, N. J. (1991). A three component conceptualization of organizational commitmen. *Human Resource Management Review*, *1*(1), 61–89. doi:10.1016/1053-4822(91)90011-Z

Meyer, J. P., Allen, N. J., & Smith, C. A. (1993). Commitment to organizations and occupations: Extension and test of a three component conceptualization. *The Journal of Applied Psychology*, *78*(4), 538–551. doi:10.1037/0021-9010.78.4.538

Meyer, J. P., & Espinoza, J. A. (2016). Occupational commitment. In J. P. Meyer (Ed.), *Handbook of employee commitment* (pp. 135–149). Edward Elgar Publishing. doi:10.4337/9781784711740.00021

Meyer, J. P., & Herscovitch, L. (2001). Commitment in the workplace toward a general model. *Human Resource Management Review*, *11*(3), 299–326. doi:10.1016/S1053-4822(00)00053-X

Mowday, R. T. (1998). Reflections on the study and relevance of organizational commitment. *Human Resource Management Review*, *8*(4), 387–401. doi:10.1016/S1053-4822(99)00006-6

Özmen, Ö. T., Özer, P. S., & Saatçioğlu, Ö. Y. (2005). Akademisyenlerde örgütsel ve mesleki bağliliğin incelenmesine ilişkin bir örnek araştirma. [An exemplary study of organizational and professional commitment in academics.] *Dokuz Eylül Üniversitesi İşletme Fakültesi Dergisi, 6 (2)* 1-14. https://dergipark.org.tr/en/pub/ifede/issue/4603/62878

Öztürk, H., & Uslu, T. (2021). The effect of stress and burnout levels of employees on their occupational commitment during the covid-19 pandemic process: Aviation industry example: Covid-19 pandemisi sürecinde çalışanların stres ve tükenmişlik düzeylerinin mesleki bağlılıklarına etkisi: Havacılık sektörü örneği. *Journal of Human Sciences, 18*(4), 728–736. doi:10.14687/jhs.v18i4.6246

Satardien, M., Jano, R., & Mahembe, B. (2019). The relationship between perceived organisational support, organisational commitment, and turnover intention among employees in a selected organisation in the aviation industry. *SA Journal of Human Resource Management/SA Tydskrif vir Menslikehulpbronbestuur, 17*(0), a1123. doi:10.4102/sajhrm.v17i0.1123

Seçgin, S. (2016). *Pozitif örgütsel davranışın örgütsel bağlılık üzerindeki etkilerini belirlemeye yönelik Eskişehir havacılık sektöründe yapılan bir araştırma [A research conducted in Eskişehir aviation sector to determine the effects of positive organizational behavior on organizational commitment.].* [Unpublished master dissertation, Anadolu Üniversitesi, Sosyal Bilimler Enstitüsü, Eskişehir, Türkiye]. earsiv.anadolu.edu.tr

Snape, E., & Redman, T. (2003). An evaluation of a three-component model of occupational commitment: Dimensionality and consequences among United Kingdom human resource management specialists. *The Journal of Applied Psychology, 88*(1), 152–159. doi:10.1037/0021-9010.88.1.152 PMID:12675402

Supporting European Aviation (Eurokontrol) (2020). Dashboard, Daily traffic variation – states. Eurocontrol.int.

Tabachnick, B. G., & Fidell, L. S. (2007). *Using multivariate statistics* (5th ed.). Allyn & Bacon/Pearson Education.

Tabachnick, B. G., & Fidell, L. S. (2012). Using multivariate statistics. 6h edition, Person Education, Boston. Tak, B. & Çiftcioglu Aydemir B. (2009). Testing three dimentional occupational commitment scale on Turkish sample: An empirical investigation. *Journal of Business Faculty, 10*(1), 35–54.

Türk, A., Cevher, M. F., & Mizrak, K. C. (2021). The effect of informal relations and executive support on organizational commitment in the aviation cector. *International Journal of Innovative Science and Research Technology, 6*(2), 243–253. http://bitly.ws/9nMw

Tutar, H. (2016). *Örgütsel davranış.1.baskı* [Organizational behavior.1st edition.]. Detay Yayıncılık.

Weng, Q., & McElroy, J. C. (2012). Organizational career growth, affective occupational commitment and turnover ıntentions. *Journal of Vocational Behavior, 80*(2), 56–265. doi:10.1016/j.jvb.2012.01.014

World Health Organization (WHO). (2020). WHO. https://www.who.int/

Yüksek, H., Çelik, M., & Keser, A. (2022). The mediator role of well-being in the effect of COVID-19 anxiety on occupational commitment: Research in the aviation sector. *International Journal of Occupational Safety and Ergonomics*, 1–17. doi:10.1080/10803548.2022.2057011 PMID:35322753

KEY TERMS AND DEFINITIONS

Aviation: It is a term that covers travels with airplanes, the aircraft industry, sector employees, businesses in the sector.

COVID-19 Pandemic: It is the term that expresses the epidemic defined as a pandemic by the World Health Organization on March 11, 2020, with the SARS-Cov-2 virus becoming widespread and effective all over the world.

Eurocontrol: It is a European civil military organization that supports European aviation.

IATA: It stands for International Air Transport Association. This trade association represents approximately 300 airlines or 83% of total air traffic.

ILO: It refers to a specialized agency of the United Nations. It stands for International Labor Organization.

Occupational Commitment: It is the bond that enables an individual to choose her/his profession and stay in it.

Remote Working: It is one of the flexible working systems and is the term that refers to working from outside the workplace.

Compilation of References

Aaker, D. A. (1996). Measuring Brand Equity Across Products and Markets. *California Management Review*, *38*(3), 102–120. Advance online publication. doi:10.2307/41165845

Abate, M., Christidis, P., & Purwanto, A. J. (2020). Government support to airlines in the aftermath of the COVID-19 pandemic. *Journal of Air Transport Management*, *89*, 1–15. doi:10.1016/j.jairtraman.2020.101931 PMID:32952317

Abeyratne, R. I. R. (1995). Proposals and Guidelines for the Carriage of Elderly and Disabled Persons by Air. *Journal of Travel Research*, *33*(3), 52–59. doi:10.1177/004728759503300310

Abraham, A. V. (2022, April 1). *Frost Perspective: Frost & Sullivan*. Retrieved from Frost & Sullivan. https://www.frost.com/frost-perspectives/impact-of-the-russia-ukraine-crisis-on-aviation/

Abrantes, J. (2016). *O contributo das companhias aéreas de baixo custo para o desenvolvimento dos hostels nas cidades de Lisboa e Porto [The contribution of low cost carriers for the development of hostels in the cities of Lisbon and Oporto]*. [PhD thesis, Universidade de Lisboa, Instituto de Geografia e Ordenamento do Território, Lisboa, Portugal].

Abu-Rayash, A., & Dincer, I. (2020). Analysis of mobility trends during the COVID-19 coronavirus pandemic: Exploring the impacts on global aviation and travel in selected cities. *Energy Research & Social Science*, *68*, 101693. doi:10.1016/j.erss.2020.101693 PMID:32839706

Acar, N., & Nur, R. (2018). *Evaluation of the Airport Services According to Handicap of Passengers*. Government Accountability Office.

Adeleke, A. Q., Bahaudin, A. Y., Kamaruddeen, A. M., Bamgbade, J. A., Salimon, M. G., Khan, M. W. A., & Sorooshian, S. (2018). The Influence of Organizational External Factors on Construction Risk Management among Nigerian Construction Companies. *Safety and Health at Work*, *9*(1), 115–124. doi:10.1016/j.shaw.2017.05.004 PMID:30363069

Adiguzel, Z., Ozcinar, F. M., & Karadal, H. (2020). Does Servant Leadership Moderate the Link Between Strategic Human Resource Management on Rule Breaking and Job Satisfaction? *European Research on Management and Business Economics*, *26*(2), 103–110. doi:10.1016/j.iedeen.2020.04.002

Adler, N., & Golany, B. (2001). *Evaluation of deregulated airline networks using data envelopment analysis combined with principal component analysis with an application to Western Europe*. www.elsevier.com/locate/dsw

Aeromedia. (2012). Nuovo volo Torino-Lisbona operato dalla TAP Portugal/New Turin-Lisbon Service Operated by TAP Portugal [New Turin-Lisbon flight operated by TAP Portugal/New Turin-Lisbon Service Operated by TAP Portugal.]. *Aero Media*. http://www.aeromedia.it/taptolis.html

Agrawal, S., De Smet, A., Lacroix, S., & Reich, A. (2020). *To emerge stronger from the COVID-19 crisis, companies should start reskilling their workforces now*. McKinsey. https://www.mckinsey.com/business-functions/organization/our-insights/to-emerge-stronger-from-the-covid-19-crisis-companies-should-startreskilling-their-workforces-now

Agrawal, A. (2021). Sustainability of airlines in India with Covid-19: Challenges ahead and possible way-outs. *Journal of Revenue and Pricing Management*, *20*(4), 457–472. doi:10.105741272-020-00257-z

Agrawal, R. K., & Swaroop, P. (2009). Effect of Employer Brand Image on Application Intentions of B-School Undergraduates. *Vision: The Journal of Business Perspective*, *13*(3), 41–49. doi:10.1177/097226290901300304

Ahammad, F. M., Glaister, W. K., & Gomes, E. (2020). Strategic Agility and Human Resource Management. *Human Resource Management Review*, *30*(100700), 1–3.

Ahmad, N., & Qahmash, A. (2021). Smartism: Implementation and assessment of interpretive structural modeling. *Sustainability*, *13*(16), 1–27. doi:10.3390u13168801

Ahmady, A. G., Mehrpour, M., & Nikooravesh, A. (2016). Organizational Structure. *Social And Behavioral Sciences.*, *230*, 455–462.

Air Nostrum_IBERIA Regional. (2022). Air Nostrum. http://www.airnostrum.es/home.asp?idioma=2

Air Transport Action Group (ATAG). (2020). *Aviation Benefits Beyond Borders*. ATAG. https://aviationbenefits.org/environmental-efficiency/ climate-action/waypoint-2050/

Airbus. (2004). Getting to Grips with Fuel Economy. *Airbus*, *4*, pp. 82.

Aires, L. (2015). Paradigma qualitativo e práticas de investigação educacional [Qualitative paradigm and educational research practices.]. Research Gate.

Airport Research Center. (2009). *Study on the Impact of Directive 96/67/EC on Ground Handling Services 1996-2007 Final Report Content*. Airport Consultants. www.airport-consultants.com

Ajzen, I. (1991). The Theory of Planned Behavior. *Organizational Behavior and Human Decision Processes*, *50*(2), 179–211. doi:10.1016/0749-5978(91)90020-T

Akbar, Y. H., & Kisilowski, M. (2020). To bargain or not to bargain: Airlines, legitimacy and nonmarket strategy in a COVID-19 world. *Journal of Air Transport Management*, *88*, 101867. doi:10.1016/j.jairtraman.2020.101867 PMID:32834693

Akca, M. (2020). COVID-19'un havacılık sektörüne etkisi [The impact of COVID-19 on the aviation industry]. *Avrasya Sosyal ve Ekonomi Araştırmaları Dergisi, 7 (4),* 45-64. https://dergipark.org.tr/tr/pub/asead/issue/54055/723434

Akıncı, Z. (2011). *Konaklama İşletmelerinde Kriz Sürecinde İnsan Kaynakları Yönetiminin Rolü Ve Önemi* [The Role and Importance of Human Resources Management in the Crisis Process in Hospitality Enterprises.]. Süleyman Demirel Üniversitesi Vizyoner Dergisi.

Alamdari, F. (2002). Regional development in airlines and travel agents relationship. *Journal of Air Transport Management*, *8*(5), 339–348. doi:10.1016/S0969-6997(02)00014-5

Albayrak, T., Aksoy, Ş., & Caber, M. (2013). The effect of environmental concern and scepticism on green purchase behaviour. *Marketing Intelligence & Planning*, *31*(1), 27–39. doi:10.1108/02634501311292902

Albers, S., & Rundshagen, V. (2020). European airlines¢ strategic responses to the COVID-19 pandemic (January-May, 2020). *Journal of Air Transport Management*, *87*, 101863. doi:10.1016/j.jairtraman.2020.101863 PMID:32834690

Alhaddi, H. (2015). Triple Bottom Line and Sustainability: A Literature Review. *Business and Management Studies*, 6-11.

Al-Harbi, K. M. A. S. (2001). Application of the AHP in project management. *International Journal of Project Management*, *19*(1), 19–27. doi:10.1016/S0263-7863(99)00038-1

Alherbawi, M., McKay, G., Mackey, H. R., & Al-Ansari, T. (2021). A novel integrated pathway for Jet Biofuel production from whole energy crops: A Jatropha curcas case study. *Energy Conversion and Management, 229*, 113662. doi:10.1016/j.enconman.2020.113662

Ali, S. S., Kaur, R., & Khan, S. (2022). Evaluating sustainability initiatives in warehouse for measuring sustainability performance: An emerging economy perspective. *Annals of Operations Research*, 1–18. doi:10.100710479-021-04454-w PMID:36091933

Allen, M. R., Dube, O. P., Solecki, W., Aragón-Durand, W., Cramer, S., Humphreys, M., & Mulugetta, R. (2019). Framing and Context. In: Global Warming of 1.5 C. An IPCC Special Report on the impacts of global warming of 1.5 C above pre-industrial levels and related global greenhouse gas emission pathways, in the context of strengthening the global response t.

Almeida, C., Costa, V., & Abrantes, J. (2020). Airline Business Models and Tourism Sector. In L. Carvalho, L. Calisto, & N. Gustavo (Eds.), *Strategic Business Models to Support Demand, Supply, and Destination Management in the Tourism and Hospitality Industry* (pp. 216–239). IGI Global., doi:10.4018/978-1-5225-9936-4.ch012

Alves, C. F., & Barbot, C. (2007). Do low cost carriers have different corporate governance models? *Journal of Air Transport Management, 13*(2), 116–120. doi:10.1016/j.jairtraman.2006.11.001

Aly, N. M. (2017). A review on utilization of textile composites in transportation towards sustainability. [). IOP Publishing.]. *IOP Conference Series. Materials Science and Engineering, 254*(4), 042002. doi:10.1088/1757-899X/254/4/042002

Alzaydi, Z. M., Al-Hajla, A., Nguyen, B., & Jayawardhena, C. (2018). A review of service quality and service delivery. *Business Process Management Journal, 24*(1), 295–328. doi:10.1108/BPMJ-09-2016-0185

Amaeshi, F., Lamba, S., & Choudhary, N. (2013). Impact Of Hrm Practices On Organizational Commitment Of Employees. *International Journal of Advancements in Research & Technology, 2.*

Amankwah-Amoah, J. (2020). Note: Mayday, Mayday, Mayday! Responding to environmental shocks: Insights on global airlines' responses to COVID-19. *Transportation Research Part E: Logistics and Transportation Review, 143.* doi:10.1016/j.tre.2020.102098

Amankwah-Amoah, J. (2016). Ebola and global airline business: An integrated framework of companies' responses to adverse environmental shock. *Thunderbird International Business Review, 58*(5), 385–397. doi:10.1002/tie.21789

Amankwah-Amoah, J. (2021). COVID-19 pandemic and innovation activities in the global airline industry: A review. *Environment International, 156*, 1–7. doi:10.1016/j.envint.2021.106719 PMID:34166874

Amankwak-Amoah, J., Khan, Z., & Osabutey, E. L. C. (2021). COVID-19 and Business Renewal: Lessons and Insights from the Global Airline Industry. *International Business Review, 30*, 01802. PMID:36568574

Ambler, T., & Barrow, S. (1996). The employer brand. *Journal of Brand Management, 4*(3), 185–206. doi:10.1057/bm.1996.42

ANAC. (2022). *Monitorização Mensal do Tráfego Aéreo nas Infraestruturas Aeroportuárias Nacionais. [Monthly Air Traffic Monitoring in National Airport Infrastructures].* Autoridade Nacional de Aviação Civil. https://www.anac.pt/vPT/Generico/PublicacoesINAC/newslettersdetrafego/Paginas/NewslettersdeTrafego.aspx

Ancell, D., & Graham, A. (2016). A Framework for Evaluating the European Airline Costs of Disabled Persons and Persons with Reduced Mobility. *Journal of Air Transport Management, 50*, 41–44. doi:10.1016/j.jairtraman.2015.10.001

Andersson, K., Carr, F., Feron, E., & Hall, W. D. (2000). *Analysis and Modeling of Ground Operations at Hub Airports.* MIT.

Anttila, T., & Kretzschmar, A. (2010). *Application of CSR programs in the airline industry.* Semantic Scholar.

Appelbaum, H. S., & Fewster, M. B. (2003). Global Aviation Human Resource Management: Contemporary Compensation and Benefits Practices. *Management Research News, 26*(7), 59–71. doi:10.1108/01409170310783592

Arachchige, B. J. H., & Robertson, A. (2011). Business Student Perceptions of a Preferred Employer: A Study Identifying Determinants of Employer Branding. *The IUP Journal of Brand Management, 8*(3), 25–46.

Arisoy, B., & Yumak, E. M. B. (2020). Örgütsel bağlılık sivil havacılık kabin ekipleri üzerine bir araştırma [A research on organizational commitment civil aviation cabin crews]. *International Journal of New Trends in Social Sciences, 4*(2), 94–104. doi:10.18844/ijntss.v4i2.5572

Armstrong, B., Reynolds, C., Bridge, G., Oakden, L., Wang, C., Panzone, L., Schmidt Rivera, X., Kause, A., Ffoulkes, C., Krawczyk, C., Miller, G., & Serjeant, S. (2021). How Does Citizen Science Compare to Online Survey Panels? A Comparison of Food Knowledge and Perceptions Between the Zooniverse, Prolific and Qualtrics UK Panels. *Frontiers in Sustainable Food Systems, 4*, 1–11. doi:10.3389/fsufs.2020.575021

Arowoshegbe, A. O., & Emmanuel, U. (2016). Sustainability And Triple Bottom Line: An Overview Of Two Interrelated Concepts. *Igbinedion University Journal of Accounting*, 88-116.

Arrhenius, S. (1896). On the influence of carbonic acid in the air upon the temperature on the ground. *Philosophical Magazine, 4*, 237–276.

Artar, O., & Türkay, U. İ. (2021). Havacılık sektöründe havalimanlarının dijital dönüşümü. [Digital transformation of airports in the aviation industry.] *Working Paper Series, 2*(1), 86-97.

Ashford, N. (2013). *Airport operations.* McGraw-Hill.

Assaf, A. G., & Josiassen, A. (2012). European vs. U.S. airlines: Performance comparison in a dynamic market. *Tourism Management, 33*(2), 317–326. doi:10.1016/j.tourman.2011.03.012

ATAG. (2020). *Aviation: Benefits Beyond Borders.* Air Transport Action Group.

ATAG. (2020). *Facts & Figures.* https://www.atag.org/facts-figures.html

Atems, B., & Yimga, J. (2021). Quantifying the impact of the COVID-19 pandemic on US airline stock prices. *Journal of Air Transport Management, 97*, 102141. doi:10.1016/j.jairtraman.2021.102141 PMID:34566280

Aviation Image Library. (2018). *Tupolev Tu-155.* AirTeamImages. https://www.airteamimages.com/tpolev-tu-155_CCCP-85035_tupolev-design-bureau_354061.html

Aviation Voice. (2020). China Offers Airlines Financial Support amid COVID-19 Outbreak. *Aviation Voice.* https://aviationvoice.com/china-offers-airlines-financial-support-amid-covid-19- outbreak-2-202003061421/

Avila, D., & Sherry, L. (2016, April). Method for analysis of Ice Super Saturated Regions (ISSR) in the US airspace. In 2016 Integrated Communications Navigation and Surveillance (ICNS) (pp. 10B2-1). IEEE.

Avila, D., Sherry, L., & Thompson, T. (2019). Reducing global warming by airline contrail avoidance: A case study of annual benefits for the contiguous United States. *Transportation Research Interdisciplinary Perspectives, 2*, 100033. doi:10.1016/j.trip.2019.100033

Azeem, M., Ahmed, M., Haider, S., & Sajjad, M. (2021). Expanding Competitive Advantage Through Organizational Culture, Knowledge sharing and Organizational Innovation. *Technology in Society, 66*, 101635. doi:10.1016/j.techsoc.2021.101635

Backhaus, K., & Tikoo, S. (2004). Conceptualizing and researching employer branding. *Career Development International, 9*(5), 501–517. doi:10.1108/13620430410550754

Baggia, A., & Rodič, B. (2010). *Dynamic Airport Ground Crew Scheduling.* Research Gate. https://www.researchgate.net/publication/256789547

Bagraim, J. J. (2003). The dimensionality of professional commitment. *SA Journal of Industrial Psychology, 29*(2), 6–9. doi:10.4102ajip.v29i2.104

Bai, W., Feng, Y., Yue, Y., & Feng, L. (2017). Organizational Structure. *Cross-Functional Integration and Performance of New Product Development Team., 174*, 621–629.

Bakir, M., Akan, S., & Atalik, O. (2020). An evaluation for Long-Haul Low-Cost Carriers sing User-Generated Content: The impact of perceived service quality on Value for Money. In C. Ramos, C. R. Almeida, & P. O. Fernandes (Eds.), *Handbook of Research on Social Media Applications for the Tourism and Hospitality Sector* (pp. 231–251). IGI Global., doi:10.4018/978-1-7998-1947-9.ch014

Balay, R. (2000). *Yönetici ve öğretmenlerde örgütsel bağlılık* [Organizational commitment in administrators and teachers.]. Nobel Yayın Dağıtım.

Balk, A. D. (2007). *Safety of ground handling*. NLR. www.nlr.nl

Bangkokbiznews. (2021). dœ̄nthāng pī mai thūa lōk pūan! yoklœ̄k kwā sī phan thīeo bin - khāt nakbin - lūkrư̄a [In Thai: A New Year Travelling Chaos: Over four Thousand Flights were cancelled due to the lack of pilots and crew]. *Bangkok Biz News.* https://www.bangkokbiznews.com/news/892592

Bangkokbiznews. (2021). IATA chī lūkčhāng ‘ sāi kān bin ‘ sūan yai sǭ tokngān nai sipsǭng dư̄an [In Thai: IATA says most airline workers are likely to lose their jobs in 12 months]. *Bangkok Biz News.* https://www.bangkokbiznews.com/news/892592

Barbot, C., Roberto Frias, R., Costa, Á., & Sochirca, E. (2008). *Airlines performance in the new market context: a comparative productivity and efficiency analysis.* MIT.

Barilo, N. (2015). *Safety Considerations for Hydrogen and Fuel Cell Applications.* Hydrogen Tools, Portal. https://h2tools.org/sites/default/files/Safety_Considerations_for_Hydrogen_and_Fuel_Cell_Applications.pdf

Barke, A., Thies, C., Melo, S. P., Cerdas, F., Herrmann, C., & Spengler, T. S. (2020). Socio-economic life cycle assessment of future aircraft systems. *Procedia CIRP, 90*, 262–267. doi:10.1016/j.procir.2020.01.096

Barrett, S. (1997). The implications of the Ireland-UK airline deregulation for an EU internal market. *Journal of Air Transport Management, 3*(2), 67–73. doi:10.1016/S0969-6997(97)00003-3

Barrett, S. D. (2004). How do the demands for airport services differ between full-service carriers and low-cost carriers? *Journal of Air Transport Management, 10*(1), 33–39. doi:10.1016/j.jairtraman.2003.10.006

Bauer, L. B., Bloch, D., & Merkert, R. (2020). Ultra Long-Haul: An emerging business model accelerated by COVID-19. *Journal of Air Transport Management, 89*, 101901. doi:10.1016/j.jairtraman.2020.101901 PMID:32839647

Baumeister, S., Nyrhinen, J., Kemppainen, T., & Wilska, T.-A. (2022). Does airlines' eco-friendliness matter? Customer satisfaction towards an environmentally responsible airline. *Transport Policy, 128*, 89–97. doi:10.1016/j.tranpol.2022.09.016

Baumeister, S., Zeng, C., & Hoffendahl, A. (2022). The effect of an eco-label on the booking decisions of air passengers. *Transport Policy, 124*, 175–182. doi:10.1016/j.tranpol.2020.07.009

Bayraktaroğlu, S. (2011). *İnsan Kaynakları Yönetimi. [Human Resources Management.]* Sakarya: Sakarya Yayıncılık.

BCG. (2020). What 12,000 Employees Have to Say About the Future of Remote Work. *BCG*. https://www.bcg.com/publications/2020/valuable-productivity-gains-covid-19

Bellis, M. (2019). History of Airships and Balloons. *ThoughtCo, Life Long Learning*. https://www.thoughtco.com/history-of-airships-and-balloons-1991241

Benligiray, S. (2020). İnsan Kaynakları Yönetimine Giriş. İnsan Kaynakları Yönetimi (s. 15). [Introduction to Human Resources Management. Human Resource Management (p. 15).] içinde Eskişehir: AÖF.

Benraïss-Noailles, L., & Viot, C. (2021). Employer brand equity effects on employees well-being and loyalty. *Journal of Business Research*, *126*, 605–613. doi:10.1016/j.jbusres.2020.02.002

Bergmann, K. (2022). JP 9102 Bidders strengthen their industry teams. *Asia-Pacific Defence Reporter (2002)*, *48*(6), 30-32.

Berry, L. L., Hensel, J. S., & Burke, M. C. (1976). Improving retailer capability for effective consumerism response. *Journal of Retailing*, *52*(3), 3–14.

Berthon, P., Ewing, M., & Hah, L. L. (2005). Captivating company: Dimensions of attractiveness in employer branding. *International Journal of Advertising*, *24*(2), 151–172. doi:10.1080/02650487.2005.11072912

BertilloJ. B.SalandoJ. (n.d.) Human Resource Management Practices in Airline Industry: The British Airways Global Perspective. Available at: SSRN: doi:10.2139/ssrn.2292797

Bethune, G. (1999). *From Worst to First (1. Baskı b.)*. Wiley.

Bican, P. M., & Brem, A. (2020). Digital business model, digital transformation, digital entrepreneurship: Is there a sustainable "digital"? *Sustainability*, *12*(13), 5239. doi:10.3390u12135239

Bick, A., Blandin, A., & Mertens, K. (2020). *Work from home after the COVID-19 Outbreak*. Federal Reserve Bank of Dallas.

Bigby, C., Johnson, H., O'Halloran, R., Douglas, J., West, D., & Bould, E. (2019). Communication access on trains: A qualitative exploration of the perspectives of passengers with communication disabilities. *Disability and Rehabilitation*, *41*(2), 125–132. doi:10.1080/09638288.2017.1380721 PMID:28949244

Bilkay, S. (2022). *Pandemi Dönemindeki Esnek Çalışma Düzenlemelerinin İşkolikliğe Etkisinde Duygusal Bağlılığın Aracılık Rolünün Araştırılması [Investigating the Mediating Role of Emotional Attachment in the Impact of Flexible Working Arrangements on Workaholism in the Pandemic Period.*]. Pearson Journal Of Socıal Scıences & Humanıtıes.

Bilkay, S. (2022). Pandemi Dönemindeki Esnek Çalışma Düzenlemelerinin İşkolikliğe Etkisinde Duygusal Bağlılığın Aracılık Rolünün Araştırılması. [Investigating the Mediating Role of Emotional Attachment in the Impact of Flexible Working Arrangements on Workaholism in the Pandemic Period.]. *Pearson Journal*, *7*(18), 156–169. doi:10.46872/pj.496

Bingöl, D. (2013). *İnsan Kaynakları Yönetimi [Human Resources Management.*]. Beta Yayınevi.

Biswas, A. (2017). A consumption value-gap analysis for sustainable consumption. *Environmental Science and Pollution Research International*, *24*(8), 7714–7725. doi:10.100711356-016-8355-9 PMID:28124271

Biswas, M. K., & Suar, D. (2016). Antecedents and Consequences of Employer Branding. *Journal of Business Ethics*, *136*(1), 57–72. doi:10.100710551-014-2502-3

Bitzan, J., & Peoples, J. (2016). A comparative analysis of cost change for low-cost, full-service, and other carriers in the US airline industry. *Research in Transportation Economics*, *56*, 25–41. doi:10.1016/j.retrec.2016.07.003

Blau, G. (2003). Testing for a four-dimensional structure of occupational commitment. *Journal of Occupational and Organizational Psychology*, *76*(4), 469–488. doi:10.1348/096317903322591596

Blau, G. J. (2001). On assessing the construct validity of two multidimensional constructs occupational commitment and occupational entrenchmen. *Human Resource Management Review*, *11*(3), 279–298. doi:10.1016/S1053-4822(00)00052-8

Blettner, D. P. (2011). Adaptation of allocation of resources and attention in response to external shocks: The case of Southwest Airlines. *Management Research Review*, *34*(3), 284–293. doi:10.1108/01409171111116303

Bloomberg, J. (2018). Digitization, digitalization, and digital transformation: confuse them at your peril. *Forbes.*

Bloomberg. (2020). Covid 19 impact: Jobs are being wiped out at airlines worldwide, and there's worse to come. *HRWorld.* https://hr.economictimes.indiatimes.com/news/industry/covid-19-impact-jobs-are-being-wiped-out-at-airlines-worldwide-and-theres-worse-to-come/77142768?redirect=1

Bodell, L. (2022). *Despite a good 4th Quarter easyJet expects a Loss for 2022.* Simple Flying. https://simpleflying.com/easyjet-2022-loss-despite-good-q4/

Boeing. (2022). *787 Dreamliner by Design: Advanced Composite Use.* Boeing. https://www.boeing.com/commercial/787/by-design/#/advanced-composite-use

Bogaisky, J. (2019). Startups bet hydrogen Fuel Cells are ready for takeoff in aviation. *Forbes Magazine.* https://www.forbes.com/sites/jeremybogaisky/2019/08/14/zeroavia-alakai-hydrogen-fuel-cell/#2cc402b918aa

Bogdan, R., & Biklen, S. (1994). Investigação Qualitativa e Educação [Qualitative Research and Education.]. Scientific Research Publishing.

Bolisani, E., Scarso, E., Ipsen, C., Kirchner, K., & Hansen, J. P. (2020). Working from home during COVID-19 pandemic: Lessons learned and issues. *Management & Marketing. Challenges for the Knowledge Society*, *15*(1), 458–476. doi:10.2478/mmcks-2020-0027

Bondarouk, T., & Brewster, C. (2016). Conceptualising the future of HRM and technology research. *International Journal of Human Resource Management*, *27*(21), 2652–2671. doi:10.1080/09585192.2016.1232296

Bouranta, N., Chitiris, L., & Paravantis, J. (2009). The relationship between internal and external service quality. *International Journal of Contemporary Hospitality Management*, *21*(3), 275–293. doi:10.1108/09596110910948297

Bourgeois-Bougrine, S. (2020). The illusion of Aircrews' Fatigue Risk Control. *Transportation Research Interdisciplinary Perspectives.*, *4*, 100104. doi:10.1016/j.trip.2020.100104

Bouwer, J. K., Saxon, S., & Tufft, C. (2022, March 31). *Industry: mckinsey.* Mckinsey. https://www.mckinsey.com/industries/travel-logistics-and-infrastructure/our-insights/taking-stock-of-the-pandemics-impact-on-global-aviation

Bouwer, J., Saxon, S., & Wittkamp, N. (2021). Back to the future? Airline sector poised for change post-COVID-19. In McKinsey & Company.

Bouwer, J., Krishnan, V., Saxon, S., & Tufft, C. (2022). *Taking stock of the pandemic's impact on global aviation.* McKinsey & Company.

Boyer, G. (2002). English Poor Laws. *Eh.Net Encyclopedia.* https://eh.net/encyclopedia/english-poor-laws/

Bräuer, T., Voigt, C., Sauer, D., Kaufmann, S., Hahn, V., Scheibe, M., Schlager, H., Huber, F., Le Clercq, P., Moore, R. H., & Anderson, B. E. (2021). Reduced ice number concentrations in contrails from low-aromatic biofuel blends. *Atmospheric Chemistry and Physics*, *21*(22), 16817–16826. doi:10.5194/acp-21-16817-2021

Braverman, M. (2003). Managing the Human Impact of Crisis. *Risk Management*, 475–488.

Brewer, G. D. (1919). *Hydrogen aircraft technology*. CRC Press.

Brewer, G. D. (1983). An Assessment of the Safety of Hydrogen-Fueled Aircraft. *Journal of Aircraft*, *20*(11), 935–939. doi:10.2514/3.48195

Brightmore, D. (2020). KLM is recycling plastic bottles to make 3D printed tools. *Manufacturing Digital*. https://manufacturingdigital.com: https://manufacturingdigital.com/technology/klm-recycling-plastic-bottles-make-3d-printed-tools

Brochado, A., Rita, P., Oliveira, C., & Oliveira, F. (2019). Airline passengers' perceptions of service quality: Themes in online reviews. *International Journal of Contemporary Hospitality Management*, *31*(2), 855–873. doi:10.1108/IJCHM-09-2017-0572

Brown, S. P., & Lam, S. K. (2008). A Meta-Analysis of Relationships Linking Employee Satisfaction to Customer Responses. *Journal of Retailing*, *84*(3), 243–255. doi:10.1016/j.jretai.2008.06.001

Brown, T. A. (2015). *Confirmatory factor analysis for applied research*. The Guilford Press.

Budd, L., & Ison, S. (2020). Supporting the Needs of Special Assistance (Including PRM) Passengers: An International Survey of Disabled Air Passenger Rights Legislation. *Journal of Air Transport Management*, *87*, 101851. doi:10.1016/j.jairtraman.2020.101851

Budd, L., Ison, S., & Adrienne, N. (2020). European airline response to the COVID-19 pandemic – Contraction, consolidation and future considerations for airline business and management. *Research in Transportation Business & Management*, *37*, 100578. Advance online publication. doi:10.1016/j.rtbm.2020.100578

Buhusayen, B., Seetand, P.-S., & Coetzer, A. (2020). Turnaround Management of Airport Service ProvidersOperating during COVID-19 Restrictions. *Sustainability*, 1–24.

BukhariS. (2011). What is Comparative Study? SSRN. doi:10.2139/ssrn.1962328

Burbidge, R. (2018). Adapting aviation to a changing climate: Key priorities for action. *Journal of Air Transport Management*, *71*, 167–174. doi:10.1016/j.jairtraman.2018.04.004

Bureau of Information Office of the Permanent Secretary. (2022, June 16). *Meet & Greet Thailand Moving Together*. Bureau of Information.www.pr.moph.go.th

Bureau of Transportation Statistics. (2022, June 06). *U.S. Airlines' April 2022 Fuel Consumption Down 9.5% from Pre-Pandemic 2019; Aviation Fuel Cost per Gallon Hits Record High*. BTS. www.bts.gov.us

Button, K. (1998). Opening U.S. Skies to Global Airline Competition – Executive Summary. Center for Trade Policy Studies, CATO Institute.

Button, K. (2001). Deregulation and liberalisation of European air transport markets. *Innovation (Abingdon)*, *14*(3), 255–275. doi:10.1080/13511610120102619

Buyck, C. (2009). Iberian Success: Air Nostrum Thrives as an Independent Regional With Strong Ties to Iberia. *ATW: Air Transport World*, *46*(6). https://trid.trb.org/view/890798

Byrne, B. M. (2012). *Structural Equation Modeling with Mplus: Basic Concepts, Applications, and Programming. NY.* Routledge.

Cable, D. M., & Turban, D. B. (2001). Establishing the dimensions, sources and value of job seekers' employer knowledge during recruitment. *Research in Personnel and Human Resources Management, 20*, 115–163. doi:10.1016/S0742-7301(01)20002-4

Çakır, E., & Ulukan, Z. (2022). Digitalization on Aviation 4.0: Designing A Scikit-Fuzzy Control System for In-Flight Catering Customer Satisfaction. In *Intelligent and Fuzzy Techniques in Aviation 4.0* (pp. 123–146). Springer. doi:10.1007/978-3-030-75067-1_6

Caleb, A., & Kunycky, A. (2019). *Hydrogen as a Renewable Energy Carrier for Commercial Aircraft.* ResearchGate. https://www.researchgate.net/publication/336551230_Hydrogen_as_a_Renewable_Energy_Carrier_for_Commercial_Aircraft

Camilleri, M. (2018). *Travel Marketing, Tourism Economics and the Airline Product.* Springler International., doi:10.1007/978-3-319-49849-2

Campos, A. (2022a, July 14th). Adiar o aeroporto até 2027 vai custar ao país 6,8 mil milhões, diz CTP. Esperar por Alcochete resultará numa perda de 21,4 mil milhões. [Postponing the airport until 2027 will cost the country 6.8 billion, says CTP. Waiting for Alcochete will result in a loss of 21.4 billion]. *Expresso.* https://expresso.pt/economia/2022-09-29-Costa-espera-privatizacao-da-TAP-nos-proximos-12-meses-e-admite-que-Estado-possa-perder-dinheiro-d4e66e83

Campos, A. (2022b, September 20th). Costa espera privatização da TAP nos próximos 12 meses e admite que Estado possa perder dinheiro. [Costa expects privatization of TAP in the next 12 months and admits that the State could lose money]. *Expresso.* https://expresso.pt/economia/2022-09-29-Costa-espera-privatizacao-da-TAP-nos-proximos-12-meses-e-admite-que-Estado-possa-perder-dinheiro-d4e66e83

CAPA. (2020). *Aviation 2020 to 2030. A COVID reality check for airlines.* Centre for Aviation. https://centreforaviation.com/analysis/reports/aviation-2020-to-2030-a-covid-reality-check-for-airlines-545006

CAPA. (2022). *Europe's low cost carriers continue to drive region's capacity recovery.* Centre for Aviation. https://centreforaviation.com/analysis/reports/europes-low-cost-carriers-continue-to-drive-region-capacity-recovery-604992

Carlota, V. (2019). Diehl Aviation worked with Airbus to create 3D printed modules for Qatar Airways. *3D Natives.* https://www.3dnatives.com/en/qatar-airways-090420194/

Carlucci, D. (2020). The Novel Coronavirus May Damage Aviation More Than SARS. *MarshMcLennan.* https://www.brinknews.com/the-novel-coronavirus-may-damage-aviation-more-than-sars/

Carmona, R. (2008). *Literature review of the passenger airline business models.* Semantic Scholar.

Carmona-Benítez, R., & Lodewijks, G. (2008). *Literature review of the passenger airline business models.* TRAIL Research School.

CDC. (2022). *CDC Museum COVID-19 Timeline.* David J. Sencer CDC Museum. https://www.cdc.gov/museum/timeline/covid19.html# Charumilind,

Chang, Y., & Chen, L. (2007). *Solve The Vehicle Routing Problem With Time Windows Via A Genetic Algorithm.* AIMS. www.AIMSciences.org

Chang, Y. C., & Chen, C. F. (2011). Identifying Mobility Service Needs for Disabled Air Passengers. *Tourism Management, 32*(5), 1214–1217. doi:10.1016/j.tourman.2010.11.001

Charumilind, S., Craven, M., Lamb, J., Sabow, A., Singhal, S., & Wilson, M. (2022). *When will the COVID-19 pandemic end?* Retrieved November 20, 2022, from https://www.mckinsey.com/industries/healthcare-systems-and-services/our-insights/when-will-the-covid-19-pandemic-end

Chatman, J. A., & Jehn, K. A. (1994). Assessing the Relationship between Industry Characteristics and Organizational Culture: How Different Can You Be? *Academy of Management Journal*, *37*(3), 522–553. doi:10.2307/256699

Chen, C.-F. (2006). Job satisfaction, organizational commitment, and flight attendants' turnover intentions: A note. *Journal of Air Transport Management*, *12*(5), 274–276. doi:10.1016/j.jairtraman.2006.05.001

Chen, Y. S. (2008). The driver of green innovation and green image - Green core competence. *Journal of Business Ethics*, *81*(3), 531–543. doi:10.100710551-007-9522-1

Chen, Y. S., & Chang, C. H. (2012). Enhance green purchase intentions: The roles of green perceived value, green perceived risk, and green trust. *Management Decision*, *50*(3), 502–520. doi:10.1108/00251741211216250

Chesbrough, H. (2007). Business model innovation: It's not just about technology anymore. *Strategy and Leadership*, *35*(6), 12–17. doi:10.1108/10878570710833714

Chiou, Y. C., & Chen, Y. H. (2010). Factors influencing the intentions of passengers regarding full service and low cost carriers: A note. *Journal of Air Transport Management*, *16*(4), 226–228. doi:10.1016/j.jairtraman.2009.11.005

Chiscano, M. C., & Darcy, S. (2022). An accessible and inclusive public transportation management response to COVID-19 through a co-creation process with people with disability. The case of Metro Barcelona. *Research in Transportation Business & Management*, *45*, 100880. doi:10.1016/j.rtbm.2022.100880

Chokshi, N., & Krauss, C. (2022, April 15). Fuel Prices Send Airfares Higher, but Travelers Seem Ready to Pay. *The New York Times*. www.nytimes.com

Civil Aviation Authority of Thailand- CAAT. (2021). *CAAT NEWS 28/2564*. CAAT. www.caat.or.th

Clausen, T. (2010). Airport Ground Staff Scheduling. Technical University of Denmark. In DTU Management (Issue 2). APA.

CNN. (2002). EasyJet selects Airbus over Boeing. *CNN.com*. http://edition.cnn.com/2002/BUSINESS/10/14/airbus.easyjet/

Codes—Airline and Location Codes Search. (n.d.). IATA. https://www.iata.org/en/publications/directories/code-search/

Cohen, A. (2000). The relationship between commitment forms and work outcomes: A comprasion of three models. *Human Relations*, *5*(3), 387–417. doi:10.1177/0018726700533005

Connected Aviation Industry. (2021, May 18). Blog. *Connected Aviation Today*. https://connectedaviationtoday.com/sustainability-in-aviation-reclaims-attention-airlines-airports-ramp-up-recovery-plans/

Coombs, T. W. (2015). *Ongoing Crisis Communication: Planning, Managing, and Responding* (4th ed.). SAGE.

Cooper, C., & Dewe, P. J. (2008). *Stress: A brief history*. John Wiley & Sons.

Correia, M. S. (2019). Sustainability: An Overview of the Triple Bottom Line and Sustainability Implementation. *International Journal of Strategic Engineering*, *2*(1), 29–38. doi:10.4018/IJoSE.2019010103

Cowling, P. I., Colledge, N. J., Dahal, K. P., & Remde, S. M. (2006). *The trade off between diversity and quality for multiobjective workforce scheduling*. In Evolutionary computationin combinatorial optimization. Proceedings of the 6th EuropeanConference (EvoCOP 2006), Budapest, Hungary.

Craven, M., Liu, L., Mysore, M., & Wilson, M. (2020). *COVID-19: Briefing note*. McKinsey&Company.

Creaton, S. (2003, April 12th). Ryanair saves €3.8m on Buzz deal. *The Irish Times*. https://www.irishtimes.com/business/ryanair-saves-3-8m-on-buzz-deal-1.355488

Creaton, S. (2007). *Ryanair. The full story of the controversial low-cost airline.* Aurum Press.

Crovini, C., Ossola, G., & Britzelmaier, B. (2021). How to Reconsider Risk Management in SMEs? And Advanced, Reasoned, and Organized Literature Review. *European Management Journal, 39*(1), 118–134. doi:10.1016/j.emj.2020.11.002

CSIRO. (2020). *Opportunities for hydrogen in commercial aviation.* Main Report, Energy and Resources, CSIRO. https://www.csiro.au/en/Do-business/Futures/Reports/Energy-and-Resources/hydrogen-commercial-aviation

Cunnington, G. R., & Parmley, R. T. (1980). Aerodynamic Surface Cooling for Laminar Flow Control for Hydrogen-Fuelled Subsonic Aircraft. *SAE Transactions*. doi:10.4271/801155

Curley, A., Dichter, A., Krishnan, V., Riedel, R., & Saxon, S. (2020). *Coronavirus: Airlines brace for severe turbulence.* McKinsey & Company.

D'Angelo, C. D., Gazzaroli, D., Corvino, C., & Gozzoli, C. (2022). Changes and Challenges in Human Resources Management: An Analysis of Human Resources Roles in a Bank Context (after COVID-19). *Sustainability, 14*(8), 1–14. doi:10.3390u14084847

da Rocha, P. M., Costa, H. G., & da Silva, G. B. (2022). Gaps, Trends and Challenges in Assessing Quality of Service at Airport Terminals: A Systematic Review and Bibliometric Analysis. *Sustainability, 14*(7), 1–17. doi:10.3390u14073796 PMID:36090804

Dabirian, A., Berthon, P., & Kietzmann, J. (2019). Enticing the IT crowd: Employer branding in the information economy. *Journal of Business and Industrial Marketing, 34*(7), 1403–1409. doi:10.1108/JBIM-11-2018-0333

Dabirian, A., Paschen, J., & Kietzmann, J. (2019). Employer Branding: Understanding Employer Attractiveness of IT Companies. *IT Professional, 21*(1), 82–89. doi:10.1109/MITP.2018.2876980

Daft, J., & Albers, S. (2014). *An Empirical Analysis of Airline Business Model Convergence Working Paper Series.* Department Of Business Policy And Logistics.

Dai, L., Eden, L., & Beamish, P. W. (2017). Caught in the crossfire: Dimensions of vulnerability and foreign multinationals' exit from war-afflicted countries. *Strategic Management Journal, 38*(7), 1478–1498. doi:10.1002mj.2599

Daly, M., & Delaney, L. (2013). The scarring effect of unemployment throughout adulthood on psychological distress at age 50: Estimates controlling for early adulthood distress and childhood psychological factors. *Social Science & Medicine, 80*, 19–23. doi:10.1016/j.socscimed.2012.12.008 PMID:23415587

Dangi, N., Gupta, S. K., & Narula, S. A. (2020). Consumer buying behaviour and purchase intention of organic food: A conceptual framework. *Management of Environmental Quality, 31*(6), 1515–1530. doi:10.1108/MEQ-01-2020-0014

Davies, P., Tidey, A., & Ortahamamcilar, B. (2020, August 18). Covid-19'un Avrupa ekonomisine etkileri: Hangi firma ne kadar işçi çıkardı? [The effects of Covid-19 on the European economy: Which company laid off how many workers?] *Euronews*. https://tr.euronews.com/: https://tr.euronews.com/2020/08/18/covid-19-un-avrupa-ekonomisine-etkileri-hangi-firma-ne-kadar-isci-c-kard

de Moor, S. (2020). A Fuel Efficiency Masterclass: Part Two of Three. *Aircraft Analytics*. https://aircraft-analytics.com/insights/a-fuel-efficiency-masterclass-part-two-of-three

Demek, C. K., Raschke, I. R., Janvrin, J. D., & Dilla, N. W. (2018). Do Organizations Use a Formalized Risk Management Process to Address Social Media Risk? *International Journal of Accounting Information Systems, 28*, 31–44. doi:10.1016/j.accinf.2017.12.004

Denison, D. R. (1990). *Corporate Culture and Organizational Effectiveness.* Wiley.

Derin, N., Şimşek Ilkım, N., & Yayan, H. (2017). Hemşirelerde mesleki bağlılığın mesleki prestijle açıklanması.[Explanation of professional commitment in nurses with professional prestige.] *Hacettepe Üniversitesi Hemşirelik Fakültesi Dergisi. 4*(3), 24-37. dergipart.org.tr

Deshaies, B. (1997). *Metodologia da Investigação em Ciências Humanas* [Research Methodology in Human Sciences.]. Scribd.

Desilver, D. (2022, June 15). *In the U.S. and around the World, Inflation is High and getting Higher*. Pew Research Center. www.pweresearch.org

Devanna, M. A., Fombrun, C., & Tichy, N. (1981). Human resources management: A strategic perspective. *Organizational Dynamics, 9*(3), 51–67. doi:10.1016/0090-2616(81)90038-3

Deveci, M., Çiftçi, M. E., Akyurt, İ. Z., & Gonzale, E. D. (2022). Impact of COVID-19 pandemic on the Turkish civil aviation industry. *Sustainable Operations and Computer*, 93-102.

Deveci, M., Çiftçi, M. E., Akyurt, İ. Z., & Gonzalez, E. D. R. S. (2022). Impact of COVID-19 pandemic on the Turkish civil aviation industry. *Sustainable Operations and Computers, 3*, 93–102. doi:10.1016/j.susoc.2021.11.002

DGCA. (2015), Instruction for Airline Passengers with Disabilities or Restricted Mobility. *SHT Barrier Free.* https://web.shgm.gov.tr/documents/sivilhavacilik/files/mevzuat/sektorel/talimatlar/sht-engelsız.pdf

Diaconu, L. (2012). The Evolution of the European Low-cost Airlines' Business Models. Ryanair Case Study. *Procedia: Social and Behavioral Sciences, 62*, 342–346. doi:10.1016/j.sbspro.2012.09.054

Diário de Notícias. (2021, December 4th). "TAP parece ter desistido desta parte do país", acusa Turismo do Norte. ["TAP seems to have given up on this part of the country", accuses Turismo do Norte]. *Diário de Notícias.* https://www.dn.pt/dinheiro/tap-parece-ter-desistido-desta-parte-do-pais-acusa-turismo-do-norte-14379732.html

Díez, I. L. (2021). *Análisis económico-financiero de Vueling Airlines S.A.*[Economic-financial analysis of Vueling Airlines S.A.]

Dobruszkes, F. (2006). An analysis of European low-cost airlines and their networks. *Journal of Transport Geography, 14*(4), 249–264. doi:10.1016/j.jtrangeo.2005.08.005

Dobruszkes, F. (2009). New Europe, new low-cost air services. *Journal of Transport Geography, 17*(6), 423–432. doi:10.1016/j.jtrangeo.2009.05.005

Dockery, M., & Bawa, S. (2020). Working from Home in the COVID-19 Lockdown. *BCEC, 19*, 1–5.

Doğan, A. (2011). Elektronik İnsan Kaynakları Yönetimi Ve Fonksiyonları. [Electronic Human Resources Management and Functions.] *Journal of Internet Applications and Management*, 51-80.

Doganis, R. (2006). *The airline business*. Routledge.

Dogan, V. (2018). The Use and Misuse of Structural Equation Modeling in Turkish Marketing Academia: Problems and Solutions. *The Journal of American Science, 16*(32), 201–230.

Dombey, O. (2004). The Effects Of Sars On The Chinese Tourism Industry. *Journal of Vacation Marketing, 10*(1), 4–10. doi:10.1177/135676670301000101

Douglas, S. P., & Craig, C. S. (2007). Collaborative and iterative translation: An alternative approach to back translation. *Journal of International Marketing, 15*(1), 30–43. doi:10.1509/jimk.15.1.030

Drăgan, O. (2021). Sustainable Aviation Needs Airports to Become Complex Hydrogen-Based Ecosystems. *Autoevolution.com*. [Photograph]. Retrieved from https://s1.cdn.autoevolution.com/images/news/sustainable-aviation-needs-airports-to-become-complex-hydrogen-based-ecosystems-163720_1.jpg

Dube, K., Nhamo, G., & Chikodzi, D. (2021). COVID-19 pandemic and prospects for recovery of the global aviation industry. *Journal of Air Transport Management*, *92*, 102022. doi:10.1016/j.jairtraman.2021.102022 PMID:36567961

Dudek, E., Krzykowska-Piotrowska, K., & Siergiejczyk, M. (2020). Risk Management in (Air Transport with Exemplary Risk Analysis Based on the Tolerability Matrix. *Transport Problems*, *15*(2), 143–156. doi:10.21307/tp-2020-027

Durst, S., Hinteregger, C., & Zieba, M. (2019). The Linkage Between Knowledge Risk Management and Organizational Performance. *Journal of Business Research*, *105*, 1–10. doi:10.1016/j.jbusres.2019.08.002

Dursun, D. E. (2021). Effect of work stress and burnout perceptions of aviation sector employees on organizational commitment. *Journal of Contemporary Issues in Business and Government*, *27*(3), 412–417. doi:10.47750/cibg.2021.27.03.056

EASA (European Union Aviation Safety Agency). (2019). European Aviation. *Environment Reporter*, *2019*, 112.

easyJet (2017). *easyJet announces 'Worldwide by easyJet'*. easyJet. https://mediacentre.easyjet.com/story/4000/easyjet-announces-worldwide-by-easyjet

easyJet (n.d.). *Our Journey*. easyJet. https://corporate.easyjet.com/about/our-journey

EasyJet. (2022). *Reports and Presentations*. EasyJet. https://corporate.easyjet.com/investors/reports-and-presentations/2022

Edwin, A. A. (2020). The factor that influence turnover intention among aviation maintenance personnel in kota kinabalu international airport sabah. [MJBE]. *Malaysian Journal of Business and Economics*, *7*(2), 73. doi:10.51200/mjbe.vi.2880

Eger, L., Komárková, L., Egerová, D., & Mičík, M. (2021). The effect of COVID-19 on consumer shopping behaviour: Generational cohort perspective. *Journal of Retailing and Consumer Services*, *61*, 102542. doi:10.1016/j.jretconser.2021.102542

Ehteshami, S. M. M., & Chan, S. H. (2014). The role of hydrogen and fuel cells to store renewable energy in the future energy network–potentials and challenges. *Energy Policy*, *73*, 103–109. doi:10.1016/j.enpol.2014.04.046

Eichhorst, W., Marx, P., Rinne, U., & Brunner, J. (2022). *Job retention schemes during COVID-19: A review of policy responses*. International Labour Organization publications.

Ek, Å., & Akselsson, R. (2007). Aviation on the Ground: Safety Culture in a Ground Handling Company. *The International Journal of Aviation Psychology*, *17*(1), 59–76. doi:10.1080/10508410709336937

Ekutu, A. C., Edeh, O. F., Alamina, P. U., Fern, S. Y., Kumari, P., & Eder, E. J. (2020). Effects of Organizational Structure on Employee Resilience. *Journal of Business School.*, *3*(2), 75–85.

Elfirdoussi, S., Lachgar, M., Kabaili, H., Rochdi, A., Goujdami, D., & El Firdoussi, L. (2020). Assessing distance learning in higher education during the COVID-19 pandemic. *Education Research International*, *2020*, 2020. doi:10.1155/2020/8890633

Elhmoud, E. R., & Kutty, A. A. (2020). Sustainability Assessment in Aviation Industry: A MiniReview on the Tools, Models and Methods of Assessment. *International Conference on Industrial Engineering and Operations Management* (pp. 1228-1238). IEOM Society International.

Elkhweldi, A. M., & Elmabrouk, Kh. S. (2015). *Aviation Risk Management Strategies*. In the 2015 International Conference on Industrial Engineering and Operations Management, Dubai, United Arab Emirates.

ElMarady, A. A., & Rahouma, K. H. (2022). The Impact of COVID-19 on the Cybersecurity in Civil Aviation: Review and Analysis. In *2022 International Telecommunications Conference (ITC-Egypt)* (pp. 1-6). IEEE.

Engau, C., & Hoffmann, V. H. (2011). Strategizing in an unpredictable climate: Exploring corporate strategies to cope with regulatory uncertainty. *Long Range Planning*, *44*(1), 42–63. doi:10.1016/j.lrp.2010.11.003

Engau, C., Hoffmann, V. H., & Busch, T. (2011). Airlines' flexibility in facing regulatory uncertainty: To anticipate or adapt? *California Management Review*, *54*(1), 107–125. doi:10.1525/cmr.2011.54.1.107

Epstein, A. H., & O'Flarity, S. M. (2019). Considerations for reducing aviation's CO2 with aircraft electric propulsion. *Journal of Propulsion and Power*, *35*(3), 572–582. doi:10.2514/1.B37015

Erdağ, T. (2021). Havayolu Taşımacılığı Sektörü Gelişim Evrelerinin PEST Analizi ile Değerlendirilmesi:5 Dönem+Covid-19 Pandemi Süreci Dönemi Kapsamında Bir İnceleme. [Evaluation of the Development Stages of the Air Transport Sector by PEST Analysis: 5 Semesters + A Review within the Scope of the Covid-19 Pandemic Process Period.]. *The International Journal of Social Sciences (Islamabad)*, 422–461.

Ernst, A. T., Jiang, H., Krishnamoorthy, M., & Sier, D. (2004). Staff scheduling and rostering: A review of applications, methods and models. *European Journal of Operational Research*, *153*(1), 3–27. doi:10.1016/S0377-2217(03)00095-X

Eroğlu, U. (2001). *İnsan Kaynakları Yönetiminde Yeni Yönelimleri Hazırlayan İki Güç: Sanal Organizasyonlar Ve Stratejik Tabanlı Düşünce* [*Two Forces Preparing New Directions in Human Resource Management: Virtual Organizations and Strategically Based Thinking.*]. ISGUC The Journal of Industrial Relations and Human Resources.

Ettenreich, R. (1919). Wolkenbildung über einer Feuersbrunst und an Flugzeugabgasen. [Cloud formation over a conflagration and aircraft exhaust fumes.]. *Meteorologische Zeitschrift (Braunschweig)*, *36*, 355–356.

EU Air Transport Liberalisation Process, Impacts and Future Considerations . (2015). International Transport Forum Discussion Papers N. 2015/04; International Transport Forum Discussion Papers, *vol. 2015/04*. doi:10.1787/2223439X

Eurocontrol. (2020). *ANS Performance dashboard.* Eurocontrol. www.ansperformance.eu

European Commission, (2020). *Future of Transport: Update on the economic impacts of COVID-19. The* European Commission's science and knowledge service.

Europen Comission. (1996). *DIRECTIV 96/67/CE.* EC. https://eur-lex.europa.eu/legal-content/PT/ALL/?uri=CELEX:31996L0067

Fageda, X. & Flores-Fillol, R. (2012). Air services on thin routes: Regional versus low-cost airlines. *Elsevier Enhanced Reader.* doi:10.1016/j.regsciurbeco.2012.03.005

Fageda, X., Suau-Sanchez, P., & Mason, K. J. (2015). The evolving low-cost business model: Network implications of fare bundling and connecting flights in Europe. *Journal of Air Transport Management*, *42*, 289–296. doi:10.1016/j.jairtraman.2014.12.002

Farine, F. K., Aditia, R., & Rahmadana, F. M. (2021). An Empirical Study of Leardership, Organizational Culture, Conflict, and Work Ethic in Determining Work Performance in Indonesia's Education Authority. *Heliyon*, *7*(7), e07698. doi:10.1016/j.heliyon.2021.e07698 PMID:34386638

Farooq, M. S., Salam, M., Fayolle, A., Jaafar, N., & Ayupp, K. (2018). Impact of service quality on customer satisfaction in Malaysia airlines: A PLS-SEM approach. *Journal of Air Transport Management*, *67*, 169–180. doi:10.1016/j.jairtraman.2017.12.008

Fernandes, A. A. R., & Solimun. (2017). Moderating effects orientation and innovation strategy on the effect of uncertainty on the performance of business environment. *International Journal of Law and Management, 59*(6), 1211–1219. doi:10.1108/IJLMA-10-2016-0088

Fethi, M. D., Jackson, P. M., & Weyman-Jones, T. G. (2001). European Airlines: a Stochastic DEA study of efficiency with market. *NC-ND 4.0 REPOSITORY RECORD*. https://hdl.handle.net/2134/420

Fioriti, M., Vercella, V. & Viola, N. (2018). Cost-Estimating Model for Aircraft Maintenance. *Aerospace Research Centre* (*Vol. 55*, no. 4).

Flight Safety Foundation. (n.d.). *Ground Accident Prevention (GAP)*. Ground Accident Prevention (GAPhttps://flightsafety.org/toolkits-resources/past-safety-initiatives/ground-accident-prevention-gap/

Forbes, S., & Lederman, M. (2007). The Role of Regional Airlines in the U.S. Airline Industry. *Advances in Airline Economics, 2. Full article: Impact of COVID-19 on different business models of European airlines*. https://www.tandfonline.com/doi/full/10.1080/13683500.2021.1960284

Forgas, S., Moliner, M. A., Sánchez, J., & Palau, R. (2010). Antecedents of airline passenger loyalty: Low-cost versus traditional airlines. *Journal of Air Transport Management, 16*(4), 229–233. doi:10.1016/j.jairtraman.2010.01.001

Fornell, C., & Larcker, D. F. (1981). Evaluating Structural Equation Models with Unobservable Variables and Measurement Error. *JMR, Journal of Marketing Research, 18*(1), 39–50. doi:10.1177/002224378101800104

Francis, G., Fidato, A., & Humphreys, I. (2003). Airport–airline interaction: The impact of low-cost carriers on two European airports. *Journal of Air Transport Management, 9*(4), 206–273. doi:10.1016/ S0969-6997(03)00004-8

Frassinelli, M. (2012). The Hindenburg 75 years later: Memories time cannot erase. *New Jersey Real-Time News*. https://www.nj.com/news/2012/05/the_hindenberg_75_years_later.html

Frow, P., & Payne, A. (2011). A stakeholder perspective of the value proposition concept. *European Journal of Marketing, 45*(1/2), 223–240. doi:10.1108/03090561111095676

Frye, A. (2010). *Air Passenger Rights for Disabled People*. Research Gate.

Gan, X., Chang, R., Zuo, J., Wen, T., & Zillante, G. (2018). Barriers to the transition towards off-site construction in China: An Interpretive structural modeling approach. *Journal of Cleaner Production, 197*, 8–18. doi:10.1016/j.jclepro.2018.06.184

Gaskin, J. (2021a). "CFA". *Gaskination's StatWiki*. http://statwiki.gaskination.com/

Gaskin, J. (2021b). "EFA". *Gaskination's StatWiki*. http://statwiki.gaskination.com/

Gaskin, J. (2021c). "Plugins". *Gaskination's StatWiki*. http://statwiki.gaskination.com

Gavine, A. (2019). British Airways is exploring 3D printing for cabin efficiencies. *Aircraft Interiors International*. https://www.aircraftinteriorsinternational.com/news/airline-news/british-airways-is-exploring-3d-printing-for-cabin-efficiencies.html

Genç, M., & Işıktaş, S. (2022). Covıd-19 Pandemi Sürecinin, Türk Havacılık Sektörü Çalışanlarına Ve Müşterilerine Olan Etkileri: Swot Analizi. [The Effects of the Covid-19 Pandemic Process on Turkish Aviation Sector Employees and Customers: Swot Analysis.] *International Journal of Arts & Social Studies*, 44-65.

George, W. R. (1990). Internal marketing and organizational behavior: A partnership in developing customer-conscious employees at every level. *Journal of Business Research, 20*(1), 63–70. doi:10.1016/0148-2963(90)90043-D

Georgiadis, P. (2022, August 8). *Ryanair chief warns fares will rise for 5 years because flying is 'too cheap'*. Financial Times. 8 August https://www.ft.com/content/32108696-9ef5-49ad-8a07-fa30375702bf

Gilal, N. G., Zhang, J., & Gilal, F. G. (2018). The four-factor model of product design: Scale development and validation. *Journal of Product and Brand Management*, *27*(6), 684–700. doi:10.1108/JPBM-11-2017-1659

Gillen, D., & Lall, A. (1997). Developing Measures Of Airport Productivity And Performance: An Application Of Data Envelope Analysis. *Science Direct.*

Gillen, D., & Lall, A. (2004). Competitive advantage of low-cost carriers: Some implications for airports. *Journal of Air Transport Management*, *10*(1), 41–50. doi:10.1016/j.jairtraman.2003.10.009

Gillespie, M. (2020). *Roadmap to runway: Getting hydrogen off the ground. CSIROscope*. Blog.

Gittell, J. H., Cameron, K., Lim, S., & Rivas, V. (2006). Relationships, layoffs, and organizational resilience: Airline industry responses to September 11. *The Journal of Applied Behavioral Science*, *42*(3), 300–329. doi:10.1177/0021886306286466

Gjersing, L., Caplehorn, J. R., & Clausen, T. (2010). Cross-cultural adaptation of research instruments: Language, setting, time and statistical considerations. *BMC Medical Research Methodology*, *10*(1), 13. doi:10.1186/1471-2288-10-13 PMID:20144247

GNAL. (1972). *Estudo da Localização do Novo Aeroporto de Lisboa* [Study for the Location of the New Lisbon Airport]. Ministério das Comunicações.

Goll, I., & Rasheed, A. A. (2011). The effects of 9/11/2001 on business strategy variability in the US air carrier industry. *Management Decision*, *49*(6), 948–961. doi:10.1108/00251741111143621

Good, D. H., Roller, L.-H., & Sickles, R. C. (1995). Airline efficiency differences between Europe and the US: Implications for the pace of EC integration and domestic regulation. In European Journal of Operational Research, 80.

Goodbody, W. (2021, August 9th). *Ryanair to cease flying to and from Northern Ireland.* Rté.ie. https://www.rte.ie/news/business/2021/0824/1242608-ryanair-to-cease-flying-to-and-from-northern-ireland/

Gössling, S. (2020). Risks, resilience, and pathways to sustainable aviation: A COVID-19 perspective. *Journal of Air Transport Management*, *89*, 101933. doi:10.1016/j.jairtraman.2020.101933 PMID:32952322

Gössling, S., Humpe, A., & Bausch, T. (2020). Does 'flight shame' affect social norms? Changing perspectives on the desirability of air travel in Germany. *Journal of Cleaner Production*, *266*, 122015. doi:10.1016/j.jclepro.2020.122015

Gounaris, S. (2008). The notion of internal market orientation and employee job satisfaction: Some preliminary evidence. *Journal of Services Marketing*, *22*(1), 68–90. doi:10.1108/08876040810851978

Gratton, G. B., Williams, P. D., Padhra, A., & Rapsomanikis, S. (2022). Reviewing the impacts of climate change on air transport operations. *Aeronautical Journal*, *126*(1295), 209–221. doi:10.1017/aer.2021.109

Gratton, G., Padhra, A., Rapsomanikis, S., & Williams, P. D. (2020). The impacts of climate change on Greek airports. *Climatic Change*, *160*(2), 219–231. doi:10.100710584-019-02634-z

Gray, J., & Rumpe, B. (2015). Models for digitalization. *Software & Systems Modeling*, *14*(4), 1319–1320. doi:10.100710270-015-0494-9

Greene, W. E., Walls, G. D., & Schrest, L. J. (1994). Internal Marketing: The Key to External Marketing Success. *Journal of Services Marketing*, *8*(4), 5–13. doi:10.1108/08876049410070682

Gudmundsson, S. V., Cattaneo, M., & Redondi, R. (2021). Forecasting temporal world recovery in air transport markets in the presence of large economic shocks: The case of COVID-19. *Journal of Air Transport Management*, *91*, 102007. doi:10.1016/j.jairtraman.2020.102007 PMID:36568736

Guerrero, J. E., Sanguineti, M., & Wittkowski, K. (2020). Variable cant angle winglets for improvement of aircraft flight performance. *Meccanica*, *55*(10), 1917–1947. doi:10.100711012-020-01230-1

Gummesson, E. (2007). Exit services marketing - enter service marketing. *Journal of Customer Behaviour*, *6*(2), 113–141. doi:10.1362/147539207X223357

Gupta, B. (2011). A comparative study of organizational strategy and culture across industry. *Benchmarking*, *18*(4), 510–528. doi:10.1108/14635771111147614

Gupta, S., & Dhingra, S. (2022). Modeling the key factors influencing the adoption of mobile financial services: An interpretive structural modeling approach. *Journal of Financial Services Marketing*, *27*(2), 96–110. doi:10.105741264-021-00101-4

Gürbüz, S. (2019). *Structural Equation Modelling with AMOS*. Seçkin Yayıncılık.

Güreş, N., Arslan, S., Yüksel, C., & Yılmaz, H. (2019). Mobil Uygulamaya Sahip Havayolu İşletmelerinin Yolculara Yönelik Hizmetlerinin Araştırılması [Investigation of passenger services of airlines with mobile applications]. *Çukurova Üniversitesi Sosyal Bilimler Enstitüsü Dergisi, 28*(3), 252-265.

Gürüz, D., & Yaylacı, Ö. G. (2004). *İletişimci gözüyle insan kaynakları yönetimi* [*Human resources management through the eyes of a communicator.*]. MediaCat Kitapları.

Hagmann, C., Semeijn, J., & Vellenga, D. B. (2015). Exploring the green image of airlines: Passenger perceptions and airline choice. *Journal of Air Transport Management*, *43*, 37–45. doi:10.1016/j.jairtraman.2015.01.003

Hair, J. F., Black, W. C., Babin, B. J., & Anderson, R. E. (2019). *Multivariate Data Analysis* (8th ed.). Cengage Learning.

Hair, J. F., Risher, J. J., Sarstedt, M., & Ringle, C. M. (2019). When to use and how to report the results of PLS-SEM. *European Business Review*, *31*(1), 2–24. doi:10.1108/EBR-11-2018-0203

Halowell, R. (1996). Southwest Airlines: A Case Study Linking Employee Needs Satisfaction And Organizational Capabilities to Competitive Management. *Human Resource Management*, *35*(4), 513–534. doi:10.1002/(SICI)1099-050X(199624)35:4<513::AID-HRM5>3.0.CO;2-Z

Hamouche, S. (2021). Human Resource Management and the COVID-19 Crisis: Implications, Challenges, Opportunities, and Future Organizational Directions. *Journal of Management & Organization*, 1–16. Advance online publication. doi:10.1017/jmo.2021.15

Hardiman, J. (2021, August 9). *Why Does easyJet Have Swiss And Austrian Subsidiaries?* Simple Flying. 9 August https://simpleflying.com/easyjet-swiss-austrian-subsidiaries/

Hardman, S., Chandan, A., & Steinberger-Wickens, R. (2015). Fuel cell added value for early market applications. *Journal of Power Sources*, *287*, 297–306. doi:10.1016/j.jpowsour.2015.04.056

Harris, C. L., & Ogbonna, E. (2001). Strategic Human Resource Management, Market Orientation, and Organizational Performance. *Journal of Business Research*, *51*(2), 157–166. doi:10.1016/S0148-2963(99)00057-0

Harvey, G., & Turnbull, P. (2020). Ricardo flies Ryanair: Strategic human resource management and competitive advantage in a Single European Aviation Market. *Human Resource Management Journal*, *30*(4), 553–565. doi:10.1111/1748-8583.12315

Hayne, C., & Free, C. (2014). Hybridized Professional Groups and Institutional Work: COSO and the Rise of Enterprise Risk Management. *Accounting, Organizations and Society, 39*(5), 309–330. doi:10.1016/j.aos.2014.05.002

Hayward, J. (2022, january 13). *What Is The IATA Summer And Winter Schedule?* Simple Flying. https://simpleflying.com/iata-summer-winter-schedule/

Heathrow. (2018). *Heathrow Carbon Footprint 2017*, p. 5. Heathrow Airport Ltd.

Heiets, I., & Yibing, X. İ. E. (2021). The Impact of the COVID-19 Pandemic on the Aviation Industry. *Journal of Aviation, 5*(2), 111–126. doi:10.30518/jav.933296

Henderson, J. C., & Sadi, M. A. (2000). The asian economic crisis and the aviation industry: Impacts and response strategies. *Transport Reviews, 20*(3), 347–367. doi:10.1080/014416400412841

Herpertz, S., Nizielski, S., Hock, M., & Schütz, A. (2016). The relevance of emotional intelligence in personnel selection for high emotional labor jobs. *PLoS One, 11*(4), e0154432. doi:10.1371/journal.pone.0154432 PMID:27124201

Homann, Q. (2019). Aviation, Fuel Cell & Hydrogen Energy Association. https://www.fchea.org/in-transition/2019/11/25/aviation

Hopkins, B., & Markham, J. (2003). *E-Hr: Using Intranets to Improve the Effectiveness of Your People*. Gower Publishing Limited.

Horcajuelo, G. (2020). Airbus, France Discuss State Support Amid COVID-19-CEO. *Teletrader*. https://www.teletrader.com/airbus-france-discuss-state-support-amid-covid-19 ceo/news/details/52014346?ts=1595118512588

Hosen, M., Ogbeibu, S., Giridharan, B., Cham, T.-H., Lim, W. M., & Paul, J. (2021). Individual motivation and social media influence on student knowledge sharing and learning performance: Evidence from an emerging economy. *Computers & Education, 172*, 104262. doi:10.1016/j.compedu.2021.104262

Hu, L., & Bentler, P. M. (1999). Cutoff criteria for fit indexes in covariance structure analysis: Conventional criteria versus new alternatives. *Structural Equation Modeling, 6*(1), 1–55. doi:10.1080/10705519909540118

Hunter, L. (2006). Low Cost Airlines. Business Model and Employment Relations. *European Management Journal, 24*(5), 315–321. doi:10.1016/j.emj.2006.08.001

Hwang, J., & Lyu, S. O. (2020). Relationships among green image, consumer attitudes, desire, and customer citizenship behavior in the airline industry. *International Journal of Sustainable Transportation, 14*(6), 437–447. doi:10.1080/15568318.2019.1573280

Hydrogen-powered Aviation. (2020). A fact-based study of hydrogen technology, economics, and climate impacts by 2050. *Fuel Cells and Hydrogen Joint Undertakings*. Retrieved from https://www.euractiv.com/wp-content/uploads/sites/2/2020/06/20200507_Hydrogen-Powered-Aviation-report_FINAL-web-ID-8706035.pdf

IAG. (2022). *International Airlines Group*. Nuestras Marcas - Vueling.

IATA. (2008). *Airlines to Lose US $ 5.2 Billion in 2008 - Slowing Demand and High Oil to*. IATA.

IATA. (2015). *ACMG Report*. Airline Airline Cost Management Group. IATA.

IATA. (2018). *Airport Service Level Agreement (SLA)*, 3. IATA.https://www.iata.org/policy/infrastructure/Documents/airport-service-level-agreement.pdf

IATA. (2019). *Economic Performance of the Airline Industry. 2019 End Year Report*. IATA.

IATA. (2020). *Passenger air traffic dropped 90% at COVID-19 peak.* Airlines IATA. https://airlines.iata.org/analysis/passenger-air-traffic-dropped-90-at-covid-19-peak

IATA. (2021). *Net-Zero Carbon Emissions by 2050.* IATA. https://www.iata.org/en/pressroom/2021-releases/2021-10-04-03/

IATA. (2021, 04 05). COVID-19 Travel Regulations Map (powered by Timatic). *Travel Center.* https://www.iatatravelcentre.com/world.php

IATA. (2022). *Airline Industry Economic Performance – June 2022.* International Air Transport Association. https://www.iata.org/en/iata-repository/publications/economic-reports/airline-industry-economic-performance---june-2022---data-tables/

IATA. (2022). iata.org. https://www.iata.org/en/publications/economics/fuel-monitor/

ICAO. (2011). Identify stakeholders map. ICAO. https://www.icao.int/apac/meetings/2011_fpl_am_tf4_seminar/ıdentify%20stakeholders.pdf

ICAO. (2018). *Manual on the regulation of International Air Transport* (3rd ed.). International Civil Aviation Organization.

ICAO. (2019). *ICAO Environmental Report 2019: Destination Green – The Next Chapter.* ICAO.

ICAO. (2020). Effects of Novel Coronavirus (COVID-19) on Civil Aviation: Economic Impact Analysis. In International Civil Aviation Organization (ICAO), Montréal, Canada. IACO. https://www.icao.int/sustainability/Documents/COVID-19/ICAO Coronavirus 2020 04 02 Econ Impact.pdf

ICAO. (2021). *Presentation of 2020 Air Transport Statistical Results.* ICAO. https://www.icao.int/annual-report-2020/Documents/ARC_2020_Air Transport Statistics_final_sched.pdf

ICAO. (2022). *2021 global air passenger totals show improvement from 2020, but still only half pre-pandemic levels.* ICAO. https://www.icao.int/Newsroom/Pages/2021-global-air-passenger-totals-show-improvement.aspx

ICAO. (2022). Effects of Novel Coronavirus (COVID-19) on Civil Aviation: Economic Impact Analysis. IACO. https://www.icao.int/sustainability/Documents/COVID-19/ICAO_Coronavirus_Econ_Impact.pd

ICAO. (2022). *Effects of Novel Coronavirus (COVID-19) on Civil Aviation: Economic Impact Analysis. Montreal.* ICAO. -

ICAO. (2022). Eurocontrol Comprehensive Aviation Assessment. ICAO. https://www.eurocontrol.int/publication/eurocontrol-comprehensive-aviation-assessment

ICAO. (2022). *ICAO Aircraft Engine Emissions Databank.* ICAO, Doc 9646. https://www.easa.europa.eu/domains/environment/icao-aircraft-engine-emissions-databank

ICAO. (2022). Passenger with disablities. ICAO. https://store.icao.int/en/manual-on-access-to-air-transport-by-persons-with-disabilities-doc-9984

IFC. (2021). *The impact of COVID-19 on Airports: An Analysis.* IFC.

Imam, H. (2020). New Normal: An HRM Perspective of Post COVID-19 in the Airline Industry. Preprints. 2020120723.

İnal, İ. H., Akdemir, A., & Cihan, S. (2021). Pandemi sonrası oluşan VUCA ortamının çalışan insan kaynakları kaygı düzeyi ve verimliliği üzerine etkisi [The effect of the post-pandemic VUCA environment on employee human resources anxiety level and productivity]. *Yönetim Bilimleri Dergisi, 19*(40), 347–374.

İnan, T. T. (2018). Yer Hizmetleri Yönetiminin Ramp Hizmetleri Sürecinde Performans Ölçüm Sistemi (PSM) ve Kurumsal Karne (BSC) Stratejilerinin İncelenmesi [Investigation of Performance Measurement System (PSM) and Corporate Scorecard (BSC) Strategies in the Ramp Services Process of Ground Handling Management.]. *Journal of Aviation*. doi:10.30518/jav.401511

Indeed (2021). *What Is Comparative Analysis and How Is It Used?* Indeed. https://www.indeed.com/career-advice/career-development/comparative-analysis

Inglada, V., Rey, B., Rodríguez-Alvarez, A., & Coto-Millan, P. (2006). Liberalisation and efficiency in international air transport. *Transportation Research Part A, Policy and Practice*, *40*(2), 95–105. doi:10.1016/j.tra.2005.04.006

International Air Transport Association (IATA) (2020). *Annual review 2020.* IATA.

International Air Transport Association (IATA) (2022). *Annual review 2022.* IATA.

International Labour Office (ILO). (2003). *The impact of the restructuring of civil aviation on employment and social practices.* Submission to the Worldwide Air Transport Conference: Challenges and Opportunities of Liberalization, Geneva, ILO

International Labour Organization (ILO) (2020). *Sectoral brief, 09/04/2020.* ILO.

IPCC. (2021). Summary for Policymakers. In *Climate Change 2021: The Physical Science Basis. Contribution of Working Group I to the Sixth Assessment Report of the Intergovernmental Panel on Climate Change* [Masson-Delmotte, V., P. Zhai, A. Pirani, S.L. Connors, C. Péan, S. Berger, N. Caud, Y. Chen, L. Goldfarb, M.I. Gomis, M. Huang, K. Leitzell, E. Lonnoy, J.B.R. Matthews, T.K. Maycock, T. Waterfield, O. Yelekçi, R. Yu, and B. Zhou (eds.)]. (pp. 3–32). Cambridge University Press. doi:10.1017/9781009157896.001

Irving, P. G., Coleman, D. F., & Cooper, C. L. (1997). Further assessments of a three- component model of occupational commitment: Generalizability and differences across occupations. *The Journal of Applied Psychology*, *82*(3), 444–452. doi:10.1037/0021-9010.82.3.444

ItalianoT. club. (2007). *Portogallo.* Touring Editore.

James, L. (2017), *Importance of Airport Stakeholder Outreach.* Aviation Pros. https://www.aviationpros.com/airports/blog/12388063/importance-of-airport-stakeholder-outreach

Jean, D., & Lohmann, G. (2016). Revisiting the airline business model spectrum: The influence of post global financial crisis and airline mergers in the US (2011–2013). *Research in Transportation Business & Management*, *21*, 76–83. doi:10.1016/j.rtbm.2016.06.002

Jiang, T., & Iles, P. (2011). Employer-brand equity, organizational attractiveness and talent management in the Zhejiang private sector, China. *Journal of Technology Management in China*, *6*(1), 97–110. doi:10.1108/17468771111105686

Jin, J., Li, S., Chen, Z., & Wang, L. (2022). Innovating, exit or both? Strategic responses to crisis revisited from resource redeployment perspective: evidence from China. *International Journal of Emerging Markets.*

Johns Hopkins University. (2022). COVID-19 Dashboard by the Center for Systems Science and Engineering (CSSE). JHU. https://coronavirus.jhu. edu/map.html.

Josephs, L. (2020, March 16). *US airlines seek more than $50 billion in government assistance as coronavirus roils business.* CNBC. https://www.cnbc.com/2020/03/16/us-airlines-seek-more-than-50-billion-in-aid-as-coronavirus-roils-business.html

Kampitak, T., & Yoopetch, C. (2018). The Study of Employee Retention in Thailand's Airlines Industry. *Dusit Thani College Journal.*, *12*(2), 193–208.

Kankaew, K. (2016). Thai Airways International: The economy Crisis Resolutions. *Actual Problems of Economic.*, *2*(176), 261–265.

Kankaew, K. (Ed.). (2022). *Global Air Transport Management and Reshaping Business Models for the New Era*. IGI Global Publishing. doi:10.4018/978-1-6684-4615-7

Kankaew, K., Ekachat, T., Rojanard, W., Nisara, P., Kumgwol, K., Sitikarn, B., & Charernit, K. (2021). Coping with the Changes that Challenge Business and Education Sectors in Thailand 4.0. *International Journal of Asian Business and Information Management*, *13*(2), 1–11. doi:10.4018/IJABIM.287588

Kankaew, K., & Treruttanaset, P. (2020). The Organizational Culture Affecting Job Performance of Newly Hired Employees: A Case Study of the Customs Bureau at Bangkok Suvarnabhumi International Airport, Thailand. In A. Bejaoui (Ed.), *Corporate Leadership and Its Role in Shaping Organizational Culture and Performance*. IGI Global Publishing.

Kannan, G., Pokharel, S., & Kumar, P. S. (2009). A hybrid approach using ISM and fuzzy TOPSIS for the selection of reverse logistics provider. *Resources, Conservation and Recycling*, *54*(1), 28–36. doi:10.1016/j.resconrec.2009.06.004

Karaman, A. S., Kilic, M., & Uyar, A. (2018). Sustainability reporting in the aviation industry: worldwide evidence. *Sustainability Accounting, Management and Policy Journal*, 362-391.

Katou, A. A. (2008). Measuring the impact of HRM on organisational performance. *Journal of Industrial Engineering and Management*, *1*(2), 119–142. doi:10.3926/jiem.2008.v1n2.p119-142

Kelly, J. (2021). Airlines Lost Over 400,00 Workers—United Airlines Announced Another 14,000 Jobs May Be Lost. *Forbes*. https://www.forbes.com/sites/jackkelly/2021/02/01/airlines-lost-over-40000-workers-united-airlines-announced-another-14000-jobs-may-be-lost/?sh=74443cd024b3

Khandelwal, B., Karakurt, A., Sekaran, P. R., Sethi, V., & Singh, R. (2013). Hydrogen powered aircraft: The future of air transport. *Progress in Aerospace Sciences*, *60*, 45–59. doi:10.1016/j.paerosci.2012.12.002

Khan, W., Akhtar, A., Ansari, S. A., & Dhamija, A. (2020). Enablers of halal food purchase among Muslim consumers in an emerging economy: An interpretive structural modeling approach. *British Food Journal*, *122*(7), 2273–2287. doi:10.1108/BFJ-08-2018-0528

Kharina, A., & Rutherford, D. (2015). Fuel Efficiency Trends for New Commercial Jet Aircraft: 1960 to 2014. *The International Council on Clean Transportation, White Paper*, p. 27.

Kierzkowski, A., & Kisiel, T. (2017). *The Human Factor in the Passenger Boarding Process at the Airport*. Science Direct.

Kilpatrick, J. (2020). COVID-19 Managing supply chain risk and disruption. *Deloitte*. https://www.deloitte.com/global/en/services/risk-advisory/analysis/covid-19-managing-supply-chain-risk-and-disruption.html

Kim, M., & Sohn, J. (2022). Passenger, Airline, and Policy Responses to the COVID-19 Crisis: The Case of South Korea. *Journal of Air Transport Management*, *98*, 102144. doi:10.1016/j.jairtraman.2021.102144 PMID:34539103

Kim, S., Wong, A. K. F., Han, H., & Yeung, M. W. (2022). Airline employees' stress amidst the COVID-19 pandemic and its job-related consequences. *Asia Pacific Journal of Tourism Research*, *27*(1), 30–47. doi:10.1080/10941665.2021.1998158

Kissel, P., & Büttgen, M. (2015). Using social media to communicate employer brand identity: The impact on corporate image and employer attractiveness. *Journal of Brand Management*, *22*(9), 755–777. doi:10.1057/bm.2015.42

Kline, R. B. (2011). *Principles and Practice of Structural Equation Modeling*. Guilford Press.

Koç, S., & Durmaz, V. (2015). Airport Corporate Sustainability: An Analysis of Indicators Reported in the Sustainability Practices. *International Conference on Leadership, Technology and Innovation Management* (pp. 158-170). Procedia - Social and Behavioral Sciences.

Kökény, L., Kenesei, Z., & Neszveda, G. (2022). Impact of COVID-19 on different business models of European airlines. *Current Issues in Tourism, 25*(3), 458–474. doi:10.1080/13683500.2021.1960284

Kollmuss, A., & Agyeman, J. (2002). Mind the Gap: Why do people act environmentally and what are the barriers to pro-environmental behavior? *Environmental Education Research, 8*(3), 239–260. doi:10.1080/13504620220145401

Kolukısa, A. (2011). *Evaluating Aircraft Turnaround Process in the Framework of Airport Design and Airline Behaviour.*

Korkmaz, S. (2020). *Havacılık sektöründe insan kaynakları yönetim fonksiyonları ile örgütsel bağlılık ilişkisi üzerine bir araştırma: TGS yer hizmetleri işletmesi örneği. [A study on the relationship between human resources management functions and organizational commitment in the aviation sector: The case of TGS ground handling operations.]* [Unpublished master dissertation, İstanbul Gelişim Üniversitesi, Lisansüstü Eğitim Enstitüsü, İstanbul, Türkiye]. acikerisim.gelisim.edu.tr

Koudis, G. S., Hu, S. J., Majumdar, A., Jones, R., & Stettler, M. E. (2017). Airport emissions reductions from reduced thrust takeoff operations. *Transportation Research Part D, Transport and Environment, 52*, 15–28. doi:10.1016/j.trd.2017.02.004

Kropp, B. (2021). Trends that will shape work in 2021 and beyond. *Harvard Business Review*. www.hbr.org

Küçükönal, H., & Korul, V. (2002). *Havayolu İşletmelerinde İnsan Kaynakları Yönetimi [Human Resources Management in Airline Operations.*]. Sosyal Bilimler Dergisi.

Kuisma, N. (2018). *Digitalization and Its Impact on Commercial Aviation.* MDPI.

Kurtcebe, M. (2020). Tıme And Money in Dıgıtalızatıon of Industry: A Socıologıcal Research on Aviation in Turkey. *And Differentiation II*, 24.

Kurulgan, M. (2020). *İnsan Kaynakları Planlaması. İnsan Kaynakları Yönetimi (s. 75). [Human Resources Planning. Human Resource Management (p. 75).]* İçinde Eskişehir: Anadolu Üniversitesi.

Lakshmanan, R. (2022). Air Transport Safety and Security. In K. Kankaew (Ed.), *Global Air Transport Management and Reshaping Business Models for the New Era.* IGI Global Publishing. doi:10.4018/978-1-6684-4615-7.ch004

Lamb, T. L., Winter, S. R., Rice, S., Ruskin, K. J., & Vaughn, A. (2020). Factors that predict passengers willingness to fly during and after the COVID-19 pandemic. *Journal of Air Transport Management, 89*, 101897. doi:10.1016/j.jairtraman.2020.101897 PMID:32837029

Land, K. C., Lovell, K., & Thore, S. (1993). *Chance-constrained data envelopment analysis New efficiency and productivity indexes.* Research Gate. https://www.researchgate.net/publication/313058771

Lappas, I., & Kourousis, K. I. (2016). Anticipating the need for new skills for the future aerospace and aviation professionals. *Journal of Aerospace Technology and Management, 8*(2), 232–241. doi:10.5028/jatm.v8i2.616

Lauder, R. (1969). The Goal Is: Mobility! Environmental and Transportation Barriers Encountered by The Disabled. *National Citizens Conference on Rehabilitation of the Disabled and Disadvantaged.* OhioLINK.

Lawton, T., & Solomko, S. (2005). When being the lowest cost is not enough: Building a successful lowfare airline business model in Asia. *Journal of Air Transport Management, 11*(6), 355–362. doi:. jairtraman.2005.03.001 doi:10.1016/j

Lawton, T., Rajwani, T., & O´Kane, C. (2011). Strategic Reorientation and Business Turnaround: The Case of Global Legacy Airlines. *Journal of Strategy and Managment*, 215–237. doi:10.1108/17554251111152252

Learson, D. L. (2019). NASA funds aviation research on a new fuel concept. *Physics.org*. Retrieved from https://phys.org/news/2019-05-nasa-funds-aviation-fuel-concept.html

Lee, B. L., & Worthington, A. C. (2010). *The Relative Efficiency of International, Domestic, and Budget Airlines: Nonparametric Evidence.* Research Gate.

Lee, C. H., Ling, H. Y., Yeow, J. A., & Hasan, M. A. (2012). Assessing Determinants of Green Purchase Intention. *International Proceedings of Economics Development and Research*, *44*, 25–29.

Lee, D. S. (2018). *The current state of scientific understanding of the non-CO2 effects of aviation on climate.* UK Department for Transport, Manchester Metropolitan University.

Lee, K., Carswell, J., & Allen, N. (2000). A meta-analytic review of occupational commitment: Relations with person and work-related variables. *The Journal of Applied Psychology*, *85*(5), 799–811. doi:10.1037/0021-9010.85.5.799 PMID:11055151

Leekha Chhabra, N., & Sharma, S. (2014). Employer branding: Strategy for improving employer attractiveness. *The International Journal of Organizational Analysis*, *22*(1), 48–60. doi:10.1108/IJOA-09-2011-0513

Lengnick-Hall, C., Beck, T., & Lengnick-Hall, M. (2011). Developing a capacity for organizational resilience through strategic human resource management. *Human Resource Management Review*, *21*(3), 243–255. doi:10.1016/j.hrmr.2010.07.001

Li, J., Wu, J., Wang, G., Wang, W., Wu, M., & Xu, J. (2020). Architecture and Key Technologies for Airline Operational Control System Based on Risk Preprocessing. *IEEE 4th Information Technology, Networking, Electronic and Automation Control Conference*. IEEE. 10.1109/ITNEC48623.2020.9085064

Linde, L., Sjödin, D., Parida, V., & Gebauer, H. (2020). Evaluation of digital business model opportunities: A framework for avoiding digitalization traps. *Research Technology Management*, *64*(1), 43–53. doi:10.1080/08956308.2021.1842664

Liu, H., Skibniewski, M. J., & Wang, M. (2016). Identification and hierarchical structure of critical success factors for innovation in construction projects: Chinese perspective. *Journal of Civil Engineering and Management*, *22*(3), 401–416. doi:10.3846/13923730.2014.975739

Li, Y., Liu, Z., Qin, K., Cui, J., Zeng, X., Ji, M., & Li, Y. (2021). Organizational trust and safety operation behavior in airline pilots: The mediating effects of organizational identification and organizational commitment. *Journal of Air Transport Management*, *92*, 102018. doi:10.1016/j.jairtraman.2021.102018

Ližbetinová, L., Lorincová, S., & Caha, Z. (2016). The Application of the Organizational Culture Assessment Instrument (OCAI) to Logistics Enterprises. *International Journal of Maritime Science & Technology*, *63*(3), 170–176.

Lloyd, S. (2002). Branding from the inside out. *Business Review Weekly - BRW, 24*(10), 64-66.

Locke, A. E., Feren, B. D., McCaleb, M. V., Shaw, N. K., & Denny, T. A. (1980). The relative effectiveness of four methods of motivating em-ployee performance. *Changes in working life*, 363-388.

Lordan, O. (2014). Study of the full-service and low-cost carriers network configuration. *Journal of Industrial Engineering and Management*, *7*(5), 1112–1123. doi:10.3926/jiem.1191

Lufthansa Group. (2022). *Annual Report 2021*. LG.

Lusa, A. S. Agência. (2016). Portugália passa a TAP Express e ganha novos aviões [Supporting and Explaining Miles and Snow's Strategy Typology with Denison's Suggested Adaptability Capabilities and Examining Their Effects on Business Performance Outputs]. *Observador*. https://observador.pt/2016/01/14/pinto-anuncia-fim-da-portugalia-lancamento-da-tap-express-novos-avioes/

MacCallum, R. C., Browne, M. W., & Sugawara, H. M. (1996). Power analysis and determination of sample size for covariance structure modeling. *Psychological Methods, 1*(2), 130–149. doi:10.1037/1082-989X.1.2.130

MacCallum, R. C., Widaman, K. F., Zhang, S., & Hong, S. (1999). Sample size in factor analysis. *Psychological Methods, 4*(1), 84–99. doi:10.1037/1082-989X.4.1.84

Ma, G., Jia, J., Ding, J., Shang, S., & Jiang, S. (2019). Interpretive structural model based factor analysis of BIM adoption in Chinese construction organizations. *Sustainability, 11*(7), 1–16. doi:10.3390u11071982

Magdalina, A., & Bouzaima, M. (2021). An empirical investigation of European airline business models: Classification and hybridization. *Journal of Air Transport Management, 93*, 102059. 10.1016/j.jairtraman.2021.102059

Mahfuza, J., Nawrasb, N., Jassimc, A. G., Hafazd, N. A., Hayatule, S., & Imanf, B. (2020). The effect of after COVID-19 human resource management approaches on organizational entrepreneurship. *Systematic Reviews in Pharmacy, 11*(10).

Mahmood, A.-L., Ammar, A.-B., & Roxani, A. (2016). Improving_Workforce_Planning_of_Aircraft. *Proceedings of the 3rd International Aviation Managemnt Conference, IAMC*, (pp. 34–43). EAU.

Ma, J., & Ma, C. (2011). Factors Analysis Based on the COSO Framework and the Government Audit Performance of Control Theory. *Procedia Engineering, 15*, 5584–5589. doi:10.1016/j.proeng.2011.08.1036

Maloni, M., Hiatt, M. S., & Campbell, S. (2019). Understanding the work values of Gen Z business students. *International Journal of Management Education, 17*(3), 100320. doi:10.1016/j.ijme.2019.100320

Mannheim, K. (1952). The Problem of Generations. In P. Kecskemeti (Ed.), *Essays on the Sociology of Knowledge* (pp. 276–320). Routledge and Kegan Paul.

Markatos, D. N., & Pantelakis, S. G. (2022). Assessment of the Impact of Material Selection on Aviation Sustainability, from a Circular Economy Perspective. *Aerospace (Basel, Switzerland), 9*(2), 2–14. doi:10.3390/aerospace9020052

Massey, J. E. (2005). Public relations in the airline industry: The crisis response to the september 11th attacks. *Journal of Hospitality & Leisure Marketing, 12*(1–2), 97–114. doi:10.1300/J150v12n01_07

Massukado, M. S. (2008). Análise comparativa de estratégias qualitativas de investigação: Possibilidades para a pesquisa em turismo. [Comparative analysis of qualitative research strategies: possibilities for tourism research]. *Turismo e Sociedade, 1*(1), 9–27.

Mathurin, P., Aliaj, O., & Fontanella-Khan, J. (2020). Pandemic triggers wave of billion-dollar US bankruptcies. *Financial Times*.

Matson, A. (1985). Service for the Disabled at United Kingdom Airports. *Nınnescah Internatıonal Servıce, 3*(4). *Upational Therapy in Health Care, 1*(4), 55–67.

Maxwell, R., & Knox, S. (2009). Motivating employees to "live the brand": A comparative case study of employer brand attractiveness within the firm. *Journal of Marketing Management, 25*(9–10), 893–907. doi:10.1362/026725709X479282

Mayer, R., Ryley, T., & Gillingwater, D. (2012). Passenger perceptions of the green image associated with airlines. *Journal of Transport Geography, 22*, 179–186. doi:10.1016/j.jtrangeo.2012.01.007

Mazareanu, E. (2021, October 26). *Low-cost carrier market share in Europe 2001-2020.* Statista. https://www.statista.com/statistics/1117218/low-cost-carrier-market-share-europe/

MD80.it. R. (2008). Aeroporto di torino: 2007 da record, oltre 3 milioni e mezzo i passeggeri [turin airport: 2007 a record, over 3 and a half million passengers]. *MD80.it.* https://www.md80.it/2008/01/15/aeroporto-di-torino-2007-da-record-oltre-3-milioni-e-mezzo-i-passeggeri/

Merkert, R. (2022). Quo vadis air transport management research? *Journal of Air Transport Management, 100*, 102205. doi:10.1016/j.jairtraman.2022.102205

Meyer, A. D. (1982). Adapting to environmental jolts. *Administrative Science Quarterly, 27*(4), 515–537. doi:10.2307/2392528 PMID:10257768

Meyer, J. P., & Allen, N. J. (1991). A three component conceptualization of organizational commitmen. *Human Resource Management Review, 1*(1), 61–89. doi:10.1016/1053-4822(91)90011-Z

Meyer, J. P., Allen, N. J., & Smith, C. A. (1993). Commitment to organizations and occupations: Extension and test of a three component conceptualization. *The Journal of Applied Psychology, 78*(4), 538–551. doi:10.1037/0021-9010.78.4.538

Meyer, J. P., & Espinoza, J. A. (2016). Occupational commitment. In J. P. Meyer (Ed.), *Handbook of employee commitment* (pp. 135–149). Edward Elgar Publishing. doi:10.4337/9781784711740.00021

Meyer, J. P., & Herscovitch, L. (2001). Commitment in the workplace toward a general model. *Human Resource Management Review, 11*(3), 299–326. doi:10.1016/S1053-4822(00)00053-X

Mhalla, M. (2020). The impact of novel coronavirus (COVID-19) on the global oil and aviation markets. *Journal of Asian Scientific Research, 10*(2), 96–104. doi:10.18488/journal.2.2020.102.96.104

Miller, K. (2020, December 8). Blog: Harward Business School. *Harward Business School.* https://online.hbs.edu/blog/post/what-is-the-triple-bottom-line

Ministry of Interior. (2021). *Coronavirus Disease 2019 Epidemic Administrative Center.* MOI. https://www.moicovid.com/

Ministry of Tourism and Sport. (2021). *Tourist Statistic 2020.* MOTS. https://www.mots.go.th/more_news_new.php?cid=592

Mishal, A., Dubey, R., Gupta, O. K., & Luo, Z. (2017). Dynamics of environmental consciousness and green purchase behaviour: An empirical study. *International Journal of Climate Change Strategies and Management, 9*(5), 682–706. doi:10.1108/IJCCSM-11-2016-0168

Mohd Suki, N. (2016). Green product purchase intention: Impact of green brands, attitude, and knowledge. *British Food Journal, 118*(12), 2893–2910. doi:10.1108/BFJ-06-2016-0295

Molchanova, K. M., Trushkina, N. V., & Katerna, O. K. (2020). Digital platforms and their application in the aviation industry. *Intellectualization of logistics and Supply Chain Management*, (3), 83-98.

Molenaar, D.-M., Bosch, F., Guggenheim, J., & Jhunjhunwala, P. Hean-Ho Loh, & Wade, B. (2021, March 29). *The Post-COVID-19 Flight Plan for Airlines.* Boston Consulting Group. https://www.bcg.com/publications/2020/post-covid-airline-industry-strategy

Mondal, J., & Chakrabarti, S. (2021). The abandonment behaviour of the branded app consumer: A study using interpretive structural modelling approach. *Journal of Retailing and Consumer Services, 63*, 102695. doi:10.1016/j.jretconser.2021.102695

Mowday, R. T. (1998). Reflections on the study and relevance of organizational commitment. *Human Resource Management Review, 8*(4), 387–401. doi:10.1016/S1053-4822(99)00006-6

Mrázová, M., & Kazda, A. (2021). The direction of airlines business models - up or down? (Post Covid-19 crisis - case study). *INCAS Bulletin, 13*(1), 263–274. doi:10.13111/2066-8201.2021.13.1.27

Muller, H., & Rolen, G. (1977). *Airlines And Disabled Travellers*. Department of Transportation.

Mundfrom, D. J., Shaw, D. G., & Ke, T. L. (2005). Minimum Sample Size Recommendations for Conducting Factor Analyses. *International Journal of Testing, 5*(2), 159–168. doi:10.120715327574ijt0502_4

Munisamy, S., Tahir Jan, M., Asraf Abdullah, M., & Mohd Satar, N. (2013). *A Critical Review of Literature on Performance of Airlines Antecedents of Loyalty in the Airline Industry of Malaysia:*Academia.

Nayeri, M. D., & Moradi, M. (2019). Green Purchase Determinants Based on Interpretive Structural Modeling: Case of Iran'S Green Marketing. *ASEAN Marketing Journal, 11*(1), 1–21. doi:10.21002/amj.v11i1.11479

Naz, F., Oláh, J., Vasile, D., & Magda, R. (2020). Green purchase behavior of university students in Hungary: An empirical study. *Sustainability, 12*(23), 1–21. doi:10.3390u122310077

Neufville, R., & Odoni, A. (2003). *Airport Systems Planning, Design and Management*. McGraw Hill.

Ng, E. S. W., Schweitzer, L., & Lyons, S. T. (2010). New Generation, Great Expectations: A Field Study of the Millennial Generation. *Journal of Business and Psychology, 25*(2), 281–292. doi:10.100710869-010-9159-4

Nguyen, T. H., & Vu, D. C. (2020). Food delivery service during social distancing: Proactively preventing or potentially spreading coronavirus disease–2019? *Disaster Medicine and Public Health Preparedness, 14*(3), e9–e10. doi:10.1017/dmp.2020.135 PMID:32367789

Nguyen, T. T. M., Phan, T. H., Nguyen, H. L., Dang, T. K. T., & Nguyen, N. D. (2019). Antecedents of purchase intention toward organic food in an Asian emerging market: A study of urban Vietnamese consumers. *Sustainability, 11*(17), 1–16. doi:10.3390u11174773

Nisada, W. (2011). *Strategic Human Resource Management*. Golden Time Printing.

Nižetić, S. (2020). Impact of coronavirus (COVID-19) pandemic on air transport mobility, energy, and environment: A case study. *International Journal of Energy Research, 44*(13), 10953–10961. doi:10.1002/er.5706

Nunnally, J. (1978). *Psychometric Methods* (2nd ed.). McGraw-Hill.

O'Callaghan, J. (2020). *Quiet and green: Why hydrogen planes could be the future of aviation*. Horizon, The EU Research and Innovation Magazine.

O'Connell, J. F., & Williams, G. (2005). Passengers' perceptions of low cost airlines and full service carriers: A case study involving Ryanair, Aer Lingus, Air Asia and Malaysia Airlines. *Journal of Air Transport Management, 11*(4), 259–272. doi:10.1016/j.jairtraman.2005.01.007

Öge, S. (2004). Elektronik İnsan Kaynakları Yönetimi (E-HRM)'nde İnsan Kaynakları Enformasyon Sistemi (HRIS)'nin Önemi ve Temel Kullanım Alanları. [The Importance of Human Resources Information System (HRIS) in Electronic Human Resources Management (E-HRM) and Main Usage Areas.] Ulusal Bilgi, Ekonomi ve Yönetim Kongresi, (s. 109-117). Eskişehir.

Ogochukwu, U., Chinonye, U., & Callistus, I. (2008). Benchmarking Human Capital Strategies of Domestic and Foreign Airlines: Evidence from Nigeria. *International Journal of Development and Management Review., 13*(1), 36–55.

Onianwa, I. O. (2022). COVID-19 Pandemic and Air Transport Business in Nigeria. In K. Kankaew (Ed.), *Global Air Transport Management and Reshaping Business Models for the New Era.* IGI Global Publishing. doi:10.4018/978-1-6684-4615-7.ch002

Oriade, A., Osinaike, A., Aduthene, K., & Wang, Y. (2021). Sustainability Awareness, Management Practices and Organizational Culture in Hotels: Evidence from Developing Countries. *International Journal of Hospitality Management, 92*, 102699. doi:10.1016/j.ijhm.2020.102699

Osterwalder, A., & Pigneur, Y. (2010). *Business Models Generation.* John Wiley and Sons.

Ou, J., & Wong, I. K. A. (2021). Strategic crisis response through changing message frames: A case of airline corporations. *Current Issues in Tourism, 24*(20), 2890–2904. doi:10.1080/13683500.2020.1849051

Özmen, Ö. T., Özer, P. S., & Saatçioğlu, Ö. Y. (2005). Akademisyenlerde örgütsel ve mesleki bağliliğin incelenmesine ilişkin bir örnek araştirma. [An exemplary study of organizational and professional commitment in academics.] *Dokuz Eylül Üniversitesi İşletme Fakültesi Dergisi, 6 (2)* 1-14. https://dergipark.org.tr/en/pub/ifede/issue/4603/62878

Öztürk, H., & Uslu, T. (2021). The effect of stress and burnout levels of employees on their occupational commitment during the covid-19 pandemic process: Aviation industry example: Covid-19 pandemisi sürecinde çalışanların stres ve tükenmişlik düzeylerinin mesleki bağlılıklarına etkisi: Havacılık sektörü örneği. *Journal of Human Sciences, 18*(4), 728–736. doi:10.14687/jhs.v18i4.6246

P., D. B., J., H., & M, D. S. (1975). Dynamique de la recherche en sciences sociales [Dynamics of social science research]. *Revue Française de Sociologie, 16*(3), 421–421. https://www.persee.fr/doc/rfsoc_0035-2969_1975_num_16_3_5811

Padhra, A. (2018). Emissions from auxiliary power units and ground power units during intraday aircraft turnarounds at European airports. *Transportation Research Part D, Transport and Environment, 63*, 433–444. doi:10.1016/j.trd.2018.06.015

Padrón, S., Guimarans, D., Ramos, J. J., & Fitouri-Trabelsi, S. (2015). *A bi-objective approach for scheduling ground-handling vehicles in airports.* Science Direct.

Padrón, S., & Guimarans, D. (2019). An Improved Method for Scheduling Aircraft Ground Handling Operations from a Global Perspective. *Asia-Pacific Journal of Operational Research, 36*(4), 1950020. doi:10.1142/S0217595919500209

Pangarkar, N. (2016). A framework for effective crisis response. *Journal of Organizational Change Management, 29*(4), 464–483. doi:10.1108/JOCM-04-2015-0063

Park, K., Qin, J., Seidel, T., & Zhou, J. (2021). Determinants and Consequences of noncompliance with the 2013 COSO Framework. *Journal of Accounting and Public Policy, 40*(6), 106899. doi:10.1016/j.jaccpubpol.2021.106899

Parviainen, P., Tihinen, M., Kääriäinen, J., & Teppola, S. (2017). Tackling the digitalization challenge: how to benefit from digitalization in practice. *International journal of information systems and project management, 5*(1), 63-77.

Patton, M. Q. (2002). *Qualitative research and evaluation methods* (3rd ed.). Sage Publications, Inc.

Payne, A., Frow, P., & Eggert, A. (2017). The customer value proposition: Evolution, development, and application in marketing. *Journal of the Academy of Marketing Science, 45*(4), 467–489. doi:10.100711747-017-0523-z

Persons, W. (1995). Crisis Management. *Career Development International*, 27.

Pestana, J. F. (2022, June 16th). *EasyJet ganha 18 novos slots em Lisboa e torna-se a companhia aérea n.º 2 no aeroporto de Lisboa.* [EasyJet wins 18 new slots in Lisbon and becomes the #2 airline at Lisbon airport]. Diário de Notícias. https://www.dnoticias.pt/2022/6/16/315790-easyjet-ganha-18-novos-slots-em-lisboa-e-torna-se-a-companhia-aerea-no-2-no-aeroporto-de-lisboa/

Petit, J. R., Jouzel, J., Raynaud, D., Barkov, N. I., Barnola, J. M., Basile, I., Bender, M., Chappellaz, J., Davis, M., Delaygue, G., Delmotte, M., Kotlyakov, V. M., Legrand, M., Lipenkov, V. Y., Lorius, C., PÉpin, L., Ritz, C., Saltzman, E., & Stievenard, M. (1999). Climate and atmospheric history of the past 420,000 years from the Vostok ice core, Antarctica. *Nature, 399*(6735), 429–436. doi:10.1038/20859

PGA Homepage. (n.d.). PGA. https://www.portugalia-airlines.pt/

Pickvance, C. G. (2001). Four varieties of comparative analysis. *Journal of Housing and the Built Environment, 16*(1), 7–28.

Polishchuk, O., Kulinich, T., Martynovych, N., & Popova, Y. (2022). Digitalization and Sustainable Development: The New COVID-19 Challenge Requires Non-Standard Solutions. *Problemy Ekorozwoju, 17*(2).

Popova, E. (2015). *Staff Scheduling in Ground Handling Company*. MathMods. https://www.mathmods.eu/resources/downloads/master-theses/finish/37-master-s-theses/251-elena-popova-thesis

Pordata (2022). *Tráfego de passageiros nos principais aeroportos: Lisboa, Porto e Faro*. [Passenger traffic at the main airports: Lisbon, Porto and Faro]. Pordata. https://www.pordata.pt/portugal/trafego+de+passageiros+nos+principais+aeroportos+lisboa++porto+e+faro-3248

Porter, M. E. (1980). *Competitive strategy: techniques for analyzing industries and competitors*. Harvard Business School.

Porto (2021). *TAP "não tem nenhuma utilidade para o Porto", critica Rui Moreira*. [TAP "has no use for Porto", criticizes Rui Moreira]. Porto.pt. https://www.porto.pt/pt/noticia/tap-nao-tem-nenhuma-utilidade-para-o-porto-critica-rui-moreira

Prager, S. (2011, July). Pioneering Passengers' Rights: Legislation and Jurisprudence from the Aviation Sector. In *Era Forum* (*Vol. 12*, No. 2, Pp. 301-320). Springer-Verlag.

PTI. (2022). COVID-19 impact: 10% jobs in aviation sector lost between Apr 2020 and Dec 2021, says govt. *Business-Today*. https://www.businesstoday.in/industry/aviation/story/covid-19-impact-10-jobs-in-aviation-sector-lost-between-apr-2020-and-dec-2021-says-govt-327667-2022-03-28

Publico. (2006). TAP paga 140 milhões ao Grupo Espírito Santo pela Portugália [TAP pays 140 million to Grupo Espírito Santo for Portugália]. *PÚBLICO*. https://www.publico.pt/2006/11/07/jornal/tap-paga-140-milhoes-ao-grupo-espirito-santo-pela-portugalia-106011

Publituris (2021, August 5th). *Ryanair abre 22 novas rotas e coloca mais três aviões em Lisboa no inverno*. [Ryanair opens 22 new routes and places three more planes in Lisbon in winter]. Publituris. https://www.publituris.pt/2021/08/05/ryanair-abre-22-novas-rotas-e-coloca-mais-tres-avioes-em-lisboa-no-inverno

Publituris (2022, March 8th). *Ryanair retira três aviões de Lisboa e cancela 19 rotas devido a falta de slots*. [Ryanair withdraws three planes from Lisbon and cancels 19 routes due to lack of slots]. Publituris. https://www.publituris.pt/2022/03/08/ryanair-retira-tres-avioes-de-lisboa-e-cancela-19-rotas-devido-a-falta-de-slots

Qarnain, S. S., Muthuvel, S., & Sankaranarayanan, B. (2021). Analysis of Energy Conservation Factors in Buildings Using Interpretive Structural Modeling Methodology: An Indian Perspective. *Journal of The Institution of Engineers (India): Series A, 102*(1), 43–61. doi:10.1007/s40030-020-00483-z

Qiu, S., Yao, D., Wang, Z., & Zhao, G. (2020). Research on Airline Operational Risk Management and Control Method. *IOP Conf. Series. Materials Science and Engineering*. IOP. 10.1088/1757-899X/715/1/012053

Quivy, R., & Van Campenhoudt, L. (1998). Manual de Investigação em Ciências Sociais (Gradiva (ed.); 2nd ed.).

Ragbir, N. K., Rice, S., Winter, S. R., & Choy, E. C. (2021). Emotions and caring mediate the relationship between knowledge of sustainability and willingness to pay for greener aviation. *Technology in Society*, *64*, 1–7. doi:10.1016/j.techsoc.2020.101491

Ramayah, T., Samat, N., & Lo, M. (2011). Market orientation, service quality and organizational performance in service organizations in Malaysia. *Asia-Pacific Journal of Business Administration*, *3*(1), 8–27. doi:10.1108/17574321111116379

Rappeport, A., & Chokshi, N. (2020). Crippled Airline Industry to Get $25 Billion Bailout, Part of It as Loans. *The New York Times*. https://www.nytimes.com/2020/04/14/business/coronavirusairlines-bailout-treasury-department

Rathnaweera, T. (2010). *Do hrm practices impact employee satisfaction, commitment or retention?* Research Gate. https://www.researchgate.net/publication/345764147

Raut, R., Priyadarshinee, P., Jha, M., Gardas, B. B., & Kamble, S. (2018). Modeling the implementation barriers of cloud computing adoption: An interpretive structural modeling. *Benchmarking*, *25*(8), 2760–2782. doi:10.1108/BIJ-12-2016-0189

Reida, E. (1984, October). Available Research on Disabled and Elderly Air Passengers. In *Thırd Internatıonal Conference on Mobılıty and Transport of Elderly and Handıcapped Personsflorida State University.* Tallahasseeloughborough University of Technology.

Reis, G. G., & Braga, B. M. (2016). Employer attractiveness from a generation perspective: Implications for employer branding. *Revista ADM*, *51*(1), 103–116. doi:10.5700/rausp1226

Report, H. A. (1937). U.S. Commerce Department. *Airship.net, Airship Blogs*. https://www.airships.net/hindenburg/disaster/commerce-department-report/

Report, H. A. (1937). U.S. Commerce Department. *Airship.net*, Airship Blogs. Retrieved from https://www.airships.net/hindenburg/disaster/commerce-department-report/

Revilla, M. A., Saris, W. E., & Krosnick, J. A. (2014). Choosing the Number of Categories in Agree–Disagree Scales. *Sociological Methods & Research*, *43*(1), 73–97. doi:10.1177/0049124113509605

Reynolds-Feighan, A. (2010). Characterisation of airline networks: A North American and European comparison. *Journal of Air Transport Management*, *16*(3), 109–120. doi:10.1016/j.jairtraman.2009.07.009

Reynolds-Feighan, A. (2018). US feeder airlines: Industry structure, networks and performance. *Transportation Research Part A, Policy and Practice*, *117*, 142–157. doi:10.1016/j.tra.2018.08.009

Richards, P. R. (1968). Modern Air Transport and Invalids. *The Journal of the Royal College of General Practitioners*, *15*(5), 336–343. PMID:5689750

Ritz, A., & Waldner, C. (2011). Competing for Future Leaders. *Review of Public Personnel Administration*, *31*(3), 291–316. doi:10.1177/0734371X11408703

Robbins, S. P., Judge, T. A., & Millett, B. (2015). *OB: the essentials*. Pearson Higher Education AU.

Robertson, B. (2022). *Which part of a flight uses the most fuel?* OAG. https://www.oag.com/blog/which-part-flight-uses-most-fuel

Robinson, J., III, & Kamgarpour, M. (2010). Benefits of continuous descent operations in high-density terminal airspace considering scheduling constraints. In *10th AIAA aviation technology, integration, and operations (ATIO) conference* (p. 9115). 10.2514/6.2010-9115

Robinson, T. (2020). *High Time for Hydrogen*. Royal Aeronautical Society. https://www.aerosociety.com/news/high-time-for-hydrogen/

Roegiers, X., & Ketele, J.-M. (1993). *Metodologia da Recolha de Dados*. IPIA.

Rondinelli, S., Gardi, A., Kapoor, R., & Sabatini, R. (2017). Benefits and challenges of liquid hydrogen fuels in commercial aviation. *International Journal of Sustainable Aviation*, *3*(3), 200–216. doi:10.1504/IJSA.2017.086845

Rose, N. L., Barnett, A., Borenstein, S., Dionne, G., Fisher, F., Hausman, J., Kahn, A., Oster, C., Papke, L., Pickrell, D., Poterba, J., Reiss, P., & Rosen, S. (1990). Profitability and Product Quality: Economic Determinants of Airline Safety Performance. *Journal of Political Economy*, *98*(5), 944–964. https://about.jstor.org/terms. doi:10.1086/261714

Royal Thai Government. (2021, October 2021). Prime Minister's Statement "Open the Country for Tourists Entering without Detention". Royal Thai Government. www.thaigov.go.th

Roy, S. (2013). Flying Low: HR Challenges in the Airline Industry. *IPBJ*, *5*(1), 49–75.

Roy, S. K. (2008). Identifying the dimensions of attractiveness of an employer brand in the Indian context. *South Asian Journal of Management*, *15*(4), 110–130.

Rozenberg, R., Szabo, S., & Šebeščáková, I. (2014). Comparison of FSC and LCC and their market share in aviation. *International Review of Aerospace Engineering*, *7*(5), 149–154. doi:10.15866/irease.v7i5.4439

Ruangkanjanases, A., You, J. J., Chien, S. W., Ma, Y., Chen, S. C., & Chao, L. C. (2020). Elucidating the Effect of Antecedents on Consumers' Green Purchase Intention: An Extension of the Theory of Planned Behavior. *Frontiers in Psychology*, *11*, 1–13. doi:10.3389/fpsyg.2020.01433 PMID:32793023

Ruiz-Moreno, F., Mas-Ruiz, F. J., & Sancho-Esper, F. M. (2021). Strategic groups and product differentiation: Evidence from the Spanish airline market deregulation. *Research in Transportation Economics*, *90*(C), 101030. https://ideas.repec.org/a/eee/retrec/v90y2021ics0739885921000020.html. doi:10.1016/j.retrec.2021.101030

Russell, E. (2022). *Whats behind the airlines operational mess this summer*. Skift.com. https://skift.com/2022/06/27/whats-behind-the-airlines-operational-mess-this-summer/

Russian Aviation. (2021). News. Retrieved from https://www.ruaviation.com/news/2021/4/15/16105/

Russo, R. F. S. M., & Camanho, R.de Russo, F.S.M. (2015). Criteria in AHP: A systematic review of literature. *Procedia Computer Science*, *55*, 1123–1132. doi:10.1016/j.procs.2015.07.081

Ryanair Group. (2019). *Annual Report 2019*. Ryanair. https://www.annualreports.com/HostedData/AnnualReportArchive/r/NASDAQ_RYAAY_2019.pdf

Ryanair Group. (2020). *Annual Report 2020*. Ryanair. https://www.annualreports.com/HostedData/AnnualReportArchive/r/NASDAQ_RYAAY_2020.pdf

Ryanair Group. (2021). *Annual Report 2021*. Ryanair. https://investor.ryanair.com/wp-content/uploads/2021/08/FINAL_Ryanair-Holdings-plc-Annual-Report-FY21.pdf

Ryanair Group. (2022a). *Annual Report 2022*. Ryanair. https://investor.ryanair.com/wp-content/uploads/2022/07/Ryanair-2022-Annual-Report.pdf

Ryanair Group. (2022b). Ryanair reports Half-Year profits of €1.37BN. Ryanair. https://investor.ryanair.com/wp-content/uploads/2022/11/H1-FY23-Results.pdf

Ryley, T., Baumeister, S., & Coulter, L. (2020). Climate change influences on aviation: A literature review. *Transport Policy*, *92*, 55–64. doi:10.1016/j.tranpol.2020.04.010

Saaty, T. L. (2008). Decision making with the analytic hierarchy process. *International Journal of Services Sciences, 1*(1), 83–98. doi:10.1504/IJSSCI.2008.017590

Sabharwal, R., & Miah, S. J. (2022). An intelligent literature review: Adopting inductive approach to define machine learning applications in the clinical domain. *Journal of Big Data, 9*(1), 1–18. doi:10.118640537-022-00605-3

Saggar, S., Tomasella, M., Cattaneo, G., & Matta, A. (2021). *Enhanced Operational Management Of Airport Ground Support Equipment For Better Aircraft Turnaround Performance.* IEEE.

Saini, G. K., Gopal, A., & Kumari, N. (2015). Employer Brand and Job Application Decisions: Insights from the Best Employers. *Management and Labour Studies, 40*(1–2), 34–51. doi:10.1177/0258042X15601532

Saini, G. K., Rai, P., & Chaudhary, M. K. (2014). What do best employer surveys reveal about employer branding and intention to apply? *Journal of Brand Management, 21*(2), 95–111. doi:10.1057/bm.2013.10

Sanjog, J., Karmakar, S., Patel, T., & Chowdhury, A. (2015). Towards virtual ergonomics: aviation and aerospace. *Aircraft Engineering and Aerospace Technology: An International Journal.*

Santos-Reyes, J., & Badillo-Portillo, A. (2022). Planning for a Trip: Selecting the Mode of Air Transport. In K. Kankaew (Ed.), *Global Air Transport Management and Reshaping Business Models for the New Era.* IGI Global Publishing. doi:10.4018/978-1-6684-4615-7.ch001

Sarkar, A., Qian, L., Peau, A. K., & Shahriar, S. (2021). Modeling drivers for successful adoption of green business: An interpretive structural modeling approach. *Environmental Science and Pollution Research International, 28*(1), 1077–1096. doi:10.100711356-020-10490-z PMID:32829426

Satardien, M., Jano, R., & Mahembe, B. (2019). The relationship between perceived organisational support, organisational commitment, and turnover intention among employees in a selected organisation in the aviation industry. *SA Journal of Human Resource Management/SA Tydskrif vir Menslikehulpbronbestuur, 17*(0), a1123. doi:10.4102/sajhrm.v17i0.1123

Savaneviciene, A., & Stankeviciute, Z. (2011). *Human Resource Management Practices Linkage With Organizational Commitment And Job Satisfaction.* Research Gate.

Savenkovs, K., & Sztore, M. (2021). *Features Of Development Of The Market Of Airlines -Lowcosters At The Present Stage Institūts.* RePec.

Saxton, J. (2006). Iran's Oil and Gas Wealth. *Joint Economic Committee Research Report.*

Scarlett, G. H. (2021). Tourism recovery and the economic impact: A panel assessment. *Research in Globalization., 3*, 100044. doi:10.1016/j.resglo.2021.100044

Schäfer, A. W., & Waitz, I. A. (2014). Air transportation and the environment. *Transport Policy, 34*, 1–4. doi:10.1016/j.tranpol.2014.02.012

Scheiwiller, S., & Zizka, L. (2021). Strategic responses by European airlines to the Covid-19 pandemic: A soft landing or a turbulent ride? *Journal of Air Transport Management, 95*, 102103. doi:10.1016/j.jairtraman.2021.102103 PMID:36568570

Scheraga, C. A. (2004). Operational efficiency versus financial mobility in the global airline industry: A data envelopment and Tobit analysis. *Transportation Research Part A, Policy and Practice, 38*(5), 383–404. doi:10.1016/j.tra.2003.12.003

Schlager, T., Bodderas, M., Maas, P., & Luc Cachelin, J. (2011). The influence of the employer brand on employee attitudes relevant for service branding: An empirical investigation. *Journal of Services Marketing, 25*(7), 497–508. doi:10.1108/08876041111173624

Schmidberger, S., Bals, L., Hartmann, E., & Jahns, C. (2009). Ground handling services at European hub airports: Development of a performance measurement system for benchmarking. *International Journal of Production Economics*, *117*(1), 104–116. doi:10.1016/j.ijpe.2008.10.006

Schmidt, E. (1941). Die entstehung von eisnebel aus den auspuffgasen von flugmotoren. [The formation of ice mist from the exhaust gases of aircraft engines.] Schriften der Deutschen Akademie der Luftfahrtforschung, Verlag R. Oldenbourg, München, Heft 44, 5(44), 1-15.

Schmidtchen, U., Behrend, E., Pohl, H. W., & Rostek, N. (1997). Hydrogen Aircraft and Airport Safety. *Renewable & Sustainable Energy Reviews*, *1*(4), 239–269. doi:10.1016/S1364-0321(97)00007-5

Schmitt, D., & Gollnick, V. (2016). The Air Transport System. In *Air Transport System* (pp. 1–17). Springer Vienna. doi:10.1007/978-3-7091-1880-1_1

Schulman, R. P. (2020). Organizational Structure and Safety Culture: Conceptual and Practical Challenges. *Safety Science*, *126*, 104669. doi:10.1016/j.ssci.2020.104669

Schultz, M., & Fricke, H. (2011). *Managing Passenger Handling at Airport Terminals Managing Passenger Handling at Airport Terminals Individual-based Approach for Modeling the Stochastic Passenger Behavior.* Research Gate. https://www.researchgate.net/publication/262567357

Schultz, M., Evler, J., Asadi, E., Preis, H., Fricke, H., & Wu, C. L. (2020). Future aircraft turnaround operations considering post-pandemic requirements. *Journal of Air Transport Management*, *89*, 101886. doi:10.1016/j.jairtraman.2020.101886 PMID:33013010

Seanyen, T., Wentbap, W., Chuayprasit, W., Srayudh, K., & Hasoh, A. (2020). Analysis of Dealing with COVID-19: The Impact on Aviation Industry. *Journal of Multidisciplinary in Humanities and Social Sciences*, *3*(2), 209–220.

Seçgin, S. (2016). *Pozitif örgütsel davranışın örgütsel bağlılık üzerindeki etkilerini belirlemeye yönelik Eskişehir havacılık sektöründe yapılan bir araştırma [A research conducted in Eskişehir aviation sector to determine the effects of positive organizational behavior on organizational commitment.].* [Unpublished master dissertation, Anadolu Üniversitesi, Sosyal Bilimler Enstitüsü, Eskişehir, Türkiye]. earsiv.anadolu.edu.tr

Seddon, P., & Lewis, G. (2003). Strategy and Business Models: What's the difference? *In proceedings of the 7th Pacific Asia Conference on Information System*, (pp. 236-248). https://core.ac.uk/download/pdf/301340932.pdf doi10.1.1.200.7881&rep=rep1&ty pe=pdf

Sefain, M. (2005). *Hydrogen aircraft concepts & ground support.* [Ph.D. thesis, Cranfield University, School of Engineering, Cranfield, England].

Semenick Alam, I. M., & Sickles, R. C. (1998). The Relationship between Stock Market Returns and Technical Efficiency Innovations: Evidence from the US Airline Industry. *Journal of Productivity Analysis*, *9*(1), 35–51. doi:10.1023/A:1018368313411

Serrano, F., & Kazda, A. (2020). The future of airports post COVID-19. *Journal of Air Transport Management*, *89*, 101900. doi:10.1016/j.jairtraman.2020.101900 PMID:32834696

Sezgen, E., Mason, K. J., & Mayer, R. (2019). Voice of airline passenger: A text mining approach to understand customer satisfaction. *Journal of Air Transport Management*, *77*, 65–74. doi:10.1016/j.jairtraman.2019.04.001

Shah, D., Rust, R. T., Parasuraman, A., Staelin, R., & Day, G. S. (2006). The Path to Customer Centricity. *Journal of Service Research*, *9*(2), 113–124. doi:10.1177/1094670506294666

Shah, F. T., Syed, Z., Imam, A., & Raza, A. (2020). The impact of airline service quality on passengers' behavioral intentions using passenger satisfaction as a mediator. *Journal of Air Transport Management*, *85*, 101815. doi:10.1016/j.jairtraman.2020.101815

Shanker, M. (2019). Recruitment process and its impact on retention of commercial pilots in Indian aviation industry. *Business Process Management Journal*, *26*(3), 736–751. doi:10.1108/BPMJ-12-2018-0376

Shank, J. K., & Govindarajan, V. (1993). *Strategic Cost Management: The New Tool for Competitive Advantage*. The Free Press.

Sharma, A., & Foropon, C. (2019). Green product attributes and green purchase behavior: A theory of planned behavior perspective with implications for circular economy. *Management Decision*, *57*(4), 1018–1042. doi:10.1108/MD-10-2018-1092

Sheaffer, Z., & Rita, M.-N. (2003). Executives' Orientation As Indicators of Crisis Management Policies and Practices. *Journal of Management Studies*, *40*(2), 577. doi:10.1111/1467-6486.00351

Sheng, J., Amankwah-Amoah, J., Khan, Z., & Wang, X. (2020). COVID-19 pandemic in the new era of big data analytics: Methodological innovations and future research directions. *British Journal of Management*, *32*(4), 1164–1183. doi:10.1111/1467-8551.12441

Shin, Y., Choi, J. O., & Hyun, S. S. (2022). The Effect of Psychological Anxiety Caused by COVID-19 on Job Self-Esteem and Job Satisfaction of Airline Flight Attendants. *International Journal of Environmental Research and Public Health*, *19*(7), 4043. doi:10.3390/ijerph19074043 PMID:35409725

Shrivastava, N. K., & Shukla, A. V. (2021). Measuring an employer brand: a study towards valid scale development (as a second-order factor of a structural model). *International Journal of Organizational Analysis*. doi:10.1108/IJOA-02-2021-2615

Silva, B. (2022, July 8th). A cidadania brasileira da TAP. [TAP's Brazilian citizenship]. Isto é Dinheiro. https://www.istoedinheiro.com.br/a-cidadania-brasileira-da-tap/

Şimşek, H. (2021). Havayolu Taşımacılığında Kriz Yönetimi. [Crisis Management in Air Transport.]. *Journal of Aviation Research*, *3*(1), 21–40. doi:10.51785/jar.777273

Singh, J., Rana, S., Hamid, A. B., & Gupta, P. (2022). Who should hold the baton of aviation sustainability? *Social Responsibility Journal*.

Singh, M. D., & Kant, R. (2008). Knowledge management barriers: An interpretive structural modeling approach. *International Journal of Management Science and Engineering Management*, *3*(2), 141–150. doi:10.1080/17509653.2008.10671042

Singh, S., & Gupta, A. (2021). An ISM modeling for factors affecting the purchase of green products. *Journal of Modelling in Management*, *16*(1), 103–123. doi:10.1108/JM2-05-2019-0099

Sivertzen, A.-M., Nilsen, E. R., & Olafsen, A. H. (2013). Employer branding: Employer attractiveness and the use of social media. *Journal of Product and Brand Management*, *22*(7), 473–483. doi:10.1108/JPBM-09-2013-0393

Slaper, T. F., & Hall, T. J. (2011). The Triple Bottom Line: What Is It and How Does It Work? *Indiana Business Review*, 1-8.

Slåtten, T., Lien, G., & Svenkerud, P. J. (2019). The role of organizational attractiveness in an internal market-oriented culture (IMOC): A study of hospital frontline employees. *BMC Health Services Research*, *19*(1), 307. doi:10.118612913-019-4144-8 PMID:31088463

Snape, E., & Redman, T. (2003). An evaluation of a three-component model of occupational commitment: Dimensionality and consequences among United Kingdom human resource management specialists. *The Journal of Applied Psychology, 88*(1), 152–159. doi:10.1037/0021-9010.88.1.152 PMID:12675402

Sobre Nós [About Us]. (n.d.). PGA. https://www.portugalia-airlines.pt/sobre-nos#historia

Soltani, S. H. K., Sham, M., Awang, M., & Yaman, R. (2012). Accessibility for disabled in public transportation terminal. *Procedia: Social and Behavioral Sciences, 35*, 89–96. doi:10.1016/j.sbspro.2012.02.066

Somerville, A., Marino, M., Baxter, G., & Graham, W. (2016). Understanding Box Wing Aircraft: Essential Technology To Improve Sustainability In The Aviation Industry. *Aviation*, 129–136.

Song, K. H., Choi, S., & Han, I. H. (2020). Competitiveness Evaluation Methodology for Aviation Industry Sustainability Using Network DEA. *Sustainability, 12*(24), 1–16. doi:10.3390u122410323

Speidel, L. K. (1984). Air Transporation and the Disabled: Current and Future Trends. *Occupational Therapy in Health Care, 1*(4), 55-67.

Spicer, A. (2020). Organizational Culture and COVID-19. *Journal of Management Studies, 57*(8), 1737–1740. doi:10.1111/joms.12625

Sreen, N., Purbey, S., & Sadarangani, P. (2018). Impact of culture, behavior and gender on green purchase intention. *Journal of Retailing and Consumer Services, 41*, 177–189. doi:10.1016/j.jretconser.2017.12.002

Starling, M. (2020). Future of aviation: Airbus to develop world's first zero-emission aircraft. *The Week.* https://www.theweek.co.uk/aviation/108150/future-of-aviation-airbus-zero-emission-aircraft

Statista. (2022). *Market share of low cost carriers (LCCs) in selected regions from 2008 to 2020.* Statista. https://www.statista.com/statistics/1097337/low-cost-carrier-market-regional-share/

Statista. (2022). Worldwide air freight traffic from 2004 to 2022. *Statista.com.* https://www.statista.com/statistics/564668/worldwide-air-cargo-traffic/

Steenkamp, J.-B. E. M., & Maydeu-Olivares, A. (2021). An updated paradigm for evaluating measurement invariance incorporating common method variance and its assessment. *Journal of the Academy of Marketing Science, 49*(1), 5–29. doi:10.100711747-020-00745-z

Števárová, L., & Badánik, B. (2018). Performance of Hub and Spoke Networks of Selected Airlines. *Transportation Research Procedia, 35*, 240–249. doi:10.1016/j.trpro.2018.12.024

Stoanescu, C., & Gheorghe, C. (2017). "Hybrid" airlines – Generating value between low-cost and traditional. *Proceedings of the International Conference on Business Excellence*, (pp. 577-587). doi:10.1515/picbe-2017-0062

Strategy, A. (2008). *The full-blown fuel crisis.* Aviation Economics.

Suidan, J., Cicetti, M., & Fleming, P. (n.d.). *IATA Ground Operations & Audits.* IATA. www.iata.org/ground-operations

Suk, M., & Kim, W. (2021). COVID-19 and the airline industry: Crisis management and resilience. *Tourism Review, 76*(4), 984–998. doi:10.1108/TR-07-2020-0348

Summerill, C., Pollard, J. T. S., & Smith, A. J. (2010). The Role of Organizational Culture and Leadership Implementation for Improved Risk Management. *The Science of the Total Environment, 408*(20), 4319–4327. doi:10.1016/j.scitotenv.2010.06.043 PMID:20624635

Sun, X., Wandelt, S., & Zhang, A. (2020). How did COVID-19 impact air transportation? A first peek through the lens of complex networks. *Journal of Air Transport Management*, *89*, 101928. doi:10.1016/j.jairtraman.2020.101928 PMID:32952320

Sun, X., Wandelt, S., & Zhang, A. (2021). Technological and educational challenges towards pandemic-resilient aviation. *Transport Policy*, *114*, 104–115. doi:10.1016/j.tranpol.2021.09.010 PMID:34566279

Sun, X., Wandelt, S., Zheng, C., & Zhang, A. (2021). COVID-19 pandemic and air transportation: Successfully navigating the paper hurricane. *Journal of Air Transport Management*, *94*, 102062. doi:10.1016/j.jairtraman.2021.102062 PMID:33875908

Supporting European Aviation (Eurokontrol) (2020). Dashboard, Daily traffic variation – states. Eurocontrol.int.

Suspiro, A. (2022, July 27th). *Para onde vai a Easyjet voar com os slots entregues pela TAP: 13 novas rotas na Europa e Marrocos. [Where is Easyjet going to fly with the slots delivered by TAP: 13 new routes in Europe and Morocco].* Observador. https://observador.pt/2022/07/27/para-onde-vai-a-easyjet-voar-com-os-slots-entregues-pela-tap/

Szondy, D. (2019). Nasa backs development of cryogenic hydrogen system to power all-electric aircraft. *New Atlas.* Image retrieved from https://newatlas.com/nasa-cheeta-funding-aircraft-fuel-cell/59725/#gallery:3

Tabachnick, B. G., & Fidell, L. S. (2007). *Using multivariate statistics* (5th ed.). Allyn & Bacon/Pearson Education.

Tabachnick, B. G., & Fidell, L. S. (2012). Using multivariate statistics. 6h edition, Person Education, Boston. Tak, B. & Çiftcioglu Aydemir B. (2009). Testing three dimentional occupational commitment scale on Turkish sample: An empirical investigation. *Journal of Business Faculty*, *10*(1), 35–54.

Tabares, D. A., & Mora-Camino, F. (2017). Aircraft ground handling: Analysis for automation. *17th AIAA Aviation Technology, Integration, and Operations Conference*, 1–16. 10.2514/6.2017-3425

Tagod, M., Adeleke, A. Q., & Moshood, D. T. (2021). Coercive Pressure as a Moderator of Organizational Structure and Risk Management: Empirical Evidence from Malaysian Construction Industry. *Journal of Safety Research*, *77*, 139–150. doi:10.1016/j.jsr.2021.02.011 PMID:34092304

Tan, B. C. (2011). The role of perceived consumer effectiveness on Value-Attitude-Behaviour model in green buying behaviour context. *Australian Journal of Basic and Applied Sciences*, *5*(12), 1766–1771.

Tanrıverdi, G., Bakır, M., & Merkert, R. (2020). What can we learn from the JATM literature for the future of aviation post Covid-19? - A bibliometric and visualization analysis. *Journal of Air Transport Management*, *89*, 101916. doi:10.1016/j.jairtraman.2020.101916 PMID:32901183

Tanwar, K., & Prasad, A. (2017). Employer brand scale development and validation: A second-order factor approach. *Personnel Review*, *46*(2), 389–409. doi:10.1108/PR-03-2015-0065

TAP. (2022). *Relatórios Anuais*. [Annual Reports] TAP. https://www.tapairportugal.com/pt/sobre-nos/relatorios-anuais

Teoh, R., Schumann, U., & Stettler, M. E. (2020). Beyond contrail avoidance: Efficacy of flight altitude changes to minimise contrail climate forcing. *Aerospace (Basel, Switzerland)*, *7*(9), 121. doi:10.3390/aerospace7090121

Ter Mors, A., Mao, X., Roos, N., Witteveen, C., & Salden Almende, A. B. (2007). *Multi-Agent Support System for Airport De-icing Scheduling.*

Thairath. (2021). *Ang sīang sa'ư̄n čhāk phanakngān sāi kān bin wọ̄n rat hai ngœ̄nkū Soft Loan tọ̄ lomhāičhai* [In Thai: Listen to the sobbing from the airline staff. Ask the state to give soft loans per breath]. Thairath. www.thairath.co.th

The Guardian. (2020, Ekim). Major US airlines to lay off thousands of workers as Covid-19 support expires. *The Guardian.* https://www.theguardian.com/: https://www.theguardian.com/business/2020/aug/14/us-airlines-layoffs-covid-19-support-delta-united-american

The Hindenburg Disaster. (1997). Airships.net, Airship Blogs. https:// www.airships.net/hindenburg/disaster/

The Hindenburg Disaster. (2020). History.com. https://www.history.com/this-day-in-history/the-hindenburg-disaster

The World Bank. (2021). *Air Transport: Annual Report 2021.* World Bank. https://documents.worldbank.org/en/publication/documents-reports/documentdetail/099346308042217146/idu0a0f612fd04dd504b880866e0b851ba17ff30

Thokala, P., Scanlan, J., & Chipperfield, A. (2010). Life cycle cost modelling as an aircraft design support tool. *Journal of Aerospace Engineering*, 477–488.

Timmis, A. J., Hodzic, A., Koh, L., Bonner, M., Soutis, C., Schäfer, A. W., & Dray, L. (2015). Environmental impact assessment of aviation emission reduction through the implementation of composite materials. *The International Journal of Life Cycle Assessment, 20*(2), 233–243. doi:10.100711367-014-0824-0

Tkalac Verčič, A., Galić, Z., & Žnidar, K. (2021). The Relationship of Internal Communication Satisfaction With Employee Engagement and Employer Attractiveness: Testing the Joint Mediating Effect of the Social Exchange Quality Indicators. *International Journal of Business Communication, 232948842110538.* doi:10.1177/23294884211053839

Tnews (2022, October 25th). Ryanair prevê aumentar tráfego de passageiros para Portugal em 36% até 2027. [Ryanair expects to increase passenger traffic to Portugal by 36% by 2027]. Tnews. https://tnews.pt/ryanair-preve-aumentar-trafego-de-passageiros-para-portugal-em-36-ate-2027/

Tomová, A., & Kirschnerová, I. (2015). Network strategies of European low cost and hybrid(ized) airlines: Is Ryanair unique? *Ekonomiczne problemy usług, 890*, 120-137.

Tomová, A., & Materna, M. (2018). Miscellaneous "modi vivendi" of regional and network airlines: The tracks for future research. *Transportation Research Procedia, 35*, 305–314. doi:10.1016/j.trpro.2018.12.007

Tomová, A., & Ramajová, L. (2014). Frequent Flyer Programs and Low-cost Airlines: Ongoing Hybridization? *Procedia: Social and Behavioral Sciences, 110*, 787–795. doi:10.1016/j.sbspro.2013.12.923

Topham, G. (2016, February). EasyJet plans to cut carbon emissions with hydrogen fuel-cell trial. *The Guardians.*

Topham, G. (April, 6th). EasyJet secures £600m coronavirus loan from UK Treasury and Bank. *The Guardian.* https://www.theguardian.com/business/2020/apr/06/easyjet-secures-600m-coronavirus-loan-from-uk-treasury-and-bank

Triple Bottom Line: Evidence from Aviation Sector. (2020). *International Journal of Business Ethics and Governance (IJBEG)*, 71-80.

Tuchen, S., Arora, M., & Blessing, L. (2020). Airport user experience unpacked: Conceptualizing its potential in the face of COVID-19. *Journal of Air Transport Management, 89*, 101919. doi:10.1016/j.jairtraman.2020.101919 PMID:32874022

Tucker, R. (2001). Strategy Innovation takes Imagination. *The Journal of Business Strategy, 22*(3), 23–27. https://dx.doi.org/10.1108/eb040168

Turban, D. B. (2001). Organizational Attractiveness as an Employer on College Campuses: An Examination of the Applicant Population. *Journal of Vocational Behavior, 58*(2), 293–312. doi:10.1006/jvbe.2000.1765

Türeli, N., Durmaz, V., Bahçecik, Y. S., & Akay, S. S. (2019). An Analysis of Importance of Innovatice Behaviors of Ground Handling Human Resources in Ensuring Customer Satisfaction. *Procedia Computer Science, 158*, 1077–1087. doi:10.1016/j.procs.2019.09.149

Turismo de Portugal. (2022). Turismo em números – Agosto 2022. [Tourism in numbers – August 2022]. Turismo de Portugal. https://travelbi.turismodeportugal.pt/turismo-em-portugal/turismo-em-numeros-agosto-2022/

Türk, A. (2017). 21. *Yüzyılda Terör Konsepti Havacılık Güvenliği. [Century Terror Concept Aviation Security.]* [Thesis of SHGM Havacılık Güvenliği ve Daire Başkanlığı Uzmanlık. Ankara].

Türk, A., Cevher, M. F., & Mizrak, K. C. (2021). The effect of informal relations and executive support on organizational commitment in the aviation cector. *International Journal of Innovative Science and Research Technology*, *6*(2), 243–253. http://bitly.ws/9nMw

Türker, G. Ö. (2020). Covid-19 Salgını Turizm Sektörünü Nasıl Etkiler? Turizm Akademisyenleri Perspektifinden Bir Değerlendirme. [How Does the Covid-19 Outbreak Affect the Tourism Sector? An Evaluation from the Perspective of Tourism Academics.] *International Journal of Social Sciences and Education Research*, 207-224.

Tutar, H. (2016). *Örgütsel davranış.1.baskı* [Organizational behavior.1st edition.]. Detay Yayıncılık.

UN (United Nations). (1992). United Nations Framework Convention on Climate Change. UN.

UNWTO. (2022). Word Tourism Barometer – May 2022: Vol. 20. *Issue 3*. World Tourism Organization.

Urbancová, H., & Hudáková, M. (2017). Benefits of Employer Brand and the Supporting Trends. *Economia e Sociologia*, *10*(4), 41–50. doi:10.14254/2071-789X.2017/10-4/4

Uslu, D. Yeter, & Kedikli, E. (2017). Sürdürülebilirlik Kapsamında Yeşil İnsan Kaynakları Yönetimine Genel Bir Bakış. [An Overview of Green Human Resources Management in the Context of Sustainability.] *Üçüncü Sektör Sosyal Ekonomi, 52*(3), 66-81.

Vaidya, O. S., & Kumar, S. (2006). Analytic hierarchy process: An overview of applications. *European Journal of Operational Research*, *169*(1), 1–29. doi:10.1016/j.ejor.2004.04.028

Vardarlıer, P. (2016). *Strategic approach to human resources management during crisis.* 12th International Strategic Management Conference, (pp. 464-472), Antalya, Turkey. 10.1016/j.sbspro.2016.11.057

Varshneya, G., Pandey, S. K., & Das, G. (2017). Impact of Social Influence and Green Consumption Values on Purchase Intention of Organic Clothing: A Study on Collectivist Developing Economy. *Global Business Review*, *18*(2), 478–492. doi:10.1177/0972150916668620

Veríssimo, A. (2022, October 11th). *Capital increase will remove Pedrosa and workers from TAP SGPS.* [Capital increase will remove Pedrosa and workers from TAP SGPS]. Eco. https://eco.sapo.pt/2022/10/11/aumento-de-capital-tira-humberto-pedrosa-da-tap-sgps/

Verstraete, D. (2009, April). *The Potential of Liquid Hydrogen for long range aircraft propulsion.* [Ph.D. thesis, Cranfield University, School of Engineering, Cranfield, England].

Vidovic, L. (2022). Industries Most and Least Impacted by COVID-19 from a Probability of Default Perspective - January 2022 Update. *S&P Global Market Intelligence*. https://www.spglobal.com/marketintelligence/en/news-insights/blog/industries-most-and-least-impacted-by-covid-19-from-a-probability-of-default-perspective-january-2022-update

Vidović, A., Štimac, I., & Vince, D. (2013). Development Of Business Models Of Low-Cost Airlines. *IJTTE. International Journal for Traffic and Transport Engineering*, *3*(1), 69–81. doi:10.7708/ijtte.2013.3(1).07

Vlaar, P., De Vries, P., & Willenborg, M. (2005). Why incumbents struggle to extract value from new strategic options: Case of the European Airline. *European Management Journal*, *23*(2), 154–169. doi:10.1016/j.emj.2005.02.009

Volker, C. E., Hoffmann, H., & Busch, T. (n.d.). *Airlines' Flexibility in Facing Regulatory Uncertainty: To Anticipate or Adapt?* WHO. https://www.who.int/emergencies/diseases/novel-coronavirus-2

Walker, C. (2018). Heathrow Airport 2017 Emission Inventory. *Ricardo Energy and Environment, Report number ED 11486* (1) 38. Cheshire, UK.

Wang, R., Shu-Li, Hsu, Lin, Y. H., & Tseng, M.-L. (2011). Evaluation of customer perceptions on airline service quality in uncertainty. *Procedia: Social and Behavioral Sciences*, *25*, 419–437. doi:10.1016/j.sbspro.2012.02.054

Warfield, J. N. (1974). Developing Subsystem Matrices in Structural Modeling. *IEEE Transactions on Systems, Man, and Cybernetics*, *4*(1), 74–80. doi:10.1109/TSMC.1974.5408523

Way, A. S., & Johnson, E. D. (2005). Theorizing about the impact of strategic human resource management. *Human Resource Management Review*, *15*(1), 1–19. doi:10.1016/j.hrmr.2005.01.004

Weber, J., & Alabi, L. O. (2022, August 9th). Ryanair warns days of cheap fares are coming to an end. *Financial Times*. https://www.ft.com/content/1da8eff7-3da8-45db-98de-866b1f68817c

Wege, L. G., Ngige, C. D., & Dieli, O. J. (2019). Electronic Human Resource Management (E-Hrm) and Organizational Performance In Nigeria. *International Journal of Management and Entrepreneurship*, *1*(1), 53–69.

Weng, Q., & McElroy, J. C. (2012). Organizational career growth, affective occupational commitment and turnover intentions. *Journal of Vocational Behavior*, *80*(2), 56–265. doi:10.1016/j.jvb.2012.01.014

Wenzel, M., Stanske, S., & Lieberman, M. B. (2020). Strategic responses to crises. *Strategic Management Journal*, *41*, V7–V18.

Werbel, J. D., & DeMarie, S. M. (2005). Aligning Strategic Human Resource Management and Person- Environment Fit. *Human Resource Management Review*, *15*(4), 247–262. doi:10.1016/j.hrmr.2005.10.001

Whelan, T., & Fink, C. (2016, October 21). Sustainable Business Practices. *Harvard Business Review*. https://hbr.org/2016/10/the-comprehensive-business-case-for-sustainability

Whyte, R., & Lohmann, G. (2016). Airline business models. In L. Budd & S. Ison (Eds.), *Air Transport Management: An international perspective* (pp. 107–121). Routledge.

Willard, B. (2012). The New Sustainability Advantage: Seven Business Case Benefits of a Triple Bottom Line. Gabriola Island: New Society Publishers.

Will, G. M., Al-Kfairy, M., & Mellor, B. R. (2019). How Organizational Structure Transforms Risky Innovations into Performance- A Computer Simulation. *Simulation Modelling Practice and Theory*, *94*, 264–285. doi:10.1016/j.simpat.2019.03.007

Williams, P. D. (2016). Transatlantic flight times and climate change. *Environmental Research Letters*, *11*(2), 024008. doi:10.1088/1748-9326/11/2/024008

Williams, P. D., & Joshi, M. M. (2013). Intensification of winter transatlantic aviation turbulence in response to climate change. *Nature Climate Change*, *3*(7), 644–648. doi:10.1038/nclimate1866

WizzAir. (n.d.). *About Wizz*. WizzAir. https://wizzair.com/en-gb/information-and-services/about-us/about-wizz

Wong, L. J., Sia, J. K. M., & Ling, T. H. Y. (2020). Airline passengers' perceived sacrifice and green practices adoption behaviours. *Asian Journal of Business Research*, *10*(3), 85–110. doi:10.14707/ajbr.200092

Woo, E., & Kim, Y. G. (2019). Consumer attitudes and buying behavior for green food products: From the aspect of green perceived value (GPV). *British Food Journal, 121*(2), 320–332. doi:10.1108/BFJ-01-2018-0027

Worasuwannarak, B., & Kankaew, K. (2022). The Image Value of Southeast Asia Airlines: A Study of Attribute that Led to Image Value of Choosing Southeast Asia Airlines by Mean-End Theory Approach. In K. Kankaew (Ed.), *Global Air Transport Management and Reshaping Business Models for the New Era.* IGI Global Publishing. doi:10.4018/978-1-6684-4615-7.ch009

World Bank. (2021). *Air transport, passengers carried.* World Bank. https://data.worldbank.org/indicator/IS.AIR.PSGR

World Health Organization (WHO). (2020). WHO. https://www.who.int/

WowEssays. (2020). Research Paper On HRM Challenges In Aviation. *WowEssays.* https://www.wowessays.com/free-samples/research-paper-on-hrm-challenges-in-aviation/

Wu, J., Zhang, P., Wang, Y., & Shi, J. (2022). Integrated aviation model and metaheuristic algorithm for hub-and-spoke network design and airline fleet planning. *Elsevier Enhanced Reader.* doi:10.1016/j.tre.2022.102755

Wu, H.-C., Cheng, C.-C., & Ai, C.-H. (2018). An empirical analysis of green switching intentions in the airline industry. *Journal of Environmental Planning and Management, 61*(8), 1438–1468. doi:10.1080/09640568.2017.1352495

Wu, S.-I., & Chen, Y.-J. (2014). The Impact of Green Marketing and Perceived Innovation on Purchase Intention for Green Products. *International Journal of Marketing Studies, 6*(5), 81–100. doi:10.5539/ijms.v6n5p81

Xiao, L. (2018). Analyzing consumer online group buying motivations: An interpretive structural modeling approach. *Telematics and Informatics, 35*(4), 629–642. doi:10.1016/j.tele.2018.01.010

Xiarewana, B., & Civelek, M. E. (2020). Effects of covid-19 on China and the world economy: Birth pains of the post-digital ecosystem. Journal of International Trade. *Logistics and Law, 6*(1), 147–157.

Xinhua.net. (2020). China to Support Air Transport Enterprises Amid Fight Against COVID19. *Xinhua.* http://www.xinhuanet.com/english/2020-05/27/c_139091729

Xiong, M., & Wang, H. (2022). Digital twin applications in aviation industry: A review. *International Journal of Advanced Manufacturing Technology, 121*(9-10), 1–16. doi:10.100700170-022-09717-9

Yager, T. J. (1983). *Factors Influencingaircraft Ground Handling Performance. Research Gate.*

Yaşlıoğlu, M. M. (2017). Sosyal bilimlerde faktör analizi ve geçerlilik : Keşfedici ve doğrulayıcı faktör analizlerinin kullanılması [Factor analysis and validity in social sciences: using exploratory and confirmatory factor analysis]. *Istanbul University Journal of the School of Business, 46* (Özel Sayı 2017), 74–85.

Yesudian, A. N., & Dawson, R. J. (2021). Global analysis of sea level rise risk to airports. *Climate Risk Management, 31*, 100266. doi:10.1016/j.crm.2020.100266

Yüksek, H., Çelik, M., & Keser, A. (2022). The mediator role of well-being in the effect of COVID-19 anxiety on occupational commitment: Research in the aviation sector. *International Journal of Occupational Safety and Ergonomics*, 1–17. doi:10.1080/10803548.2022.2057011 PMID:35322753

Zaremohzzabieh, Z., Ismail, N., Ahrari, S., & Abu Samah, A. (2021). The effects of consumer attitude on green purchase intention: A meta-analytic path analysis. *Journal of Business Research, 132*, 732–743. doi:10.1016/j.jbusres.2020.10.053

Zehir, C., Gurol, Y., Karaboga, T., & Kole, M. (2016). Human Resource Management and Firm Performance: The Mediating Role of Entrepreneurial Orientation. *Procedia: Social and Behavioral Sciences, 235*, 372–381. doi:10.1016/j.sbspro.2016.11.045

Zeithaml, V. A. (1988). Consumer Perceptions of Price, Quality, and Value: A Means-End Model and Synthesis of Evidence. *Journal of Marketing*, *52*(3), 2–22. doi:10.1177/002224298805200302

Zhao, Y., & Bacao, F. (2021). How does the pandemic facilitate mobile payment? An investigation on users' perspective under the COVID-19 pandemic. *International Journal of Environmental Research and Public Health*, *18*(3), 1016. doi:10.3390/ijerph18031016 PMID:33498863

Zhuang, W., Luo, X., & Riaz, M. U. (2021). On the Factors Influencing Green Purchase Intention: A Meta-Analysis Approach. *Frontiers in Psychology*, *12*, 1–15. doi:10.3389/fpsyg.2021.644020 PMID:33897545

About the Contributors

Salim Kurnaz is an Assistant Professor at Süleyman Demirel University, School of Civil Aviation with 22 years of experience in aviation. He worked as a maintenance technician, quality control technician and maintenance instructor in the Turkish Armed Forces between 1997-2020. He worked at Joint Force Command (JFCBS) Brunssum/ Netherlands between 2010-2013. He received bachelor's degree on Public Management in 2005 and on Aviation Management in 2022 from Anadolu University; master's degree in International Relations from Oklahoma University Oklahoma/USA in 2013 and in Business Management from Malatya Inonu University in 2018. He received his doctorate degree on Management Sciences in 2019 from Malatya Inonu University. He continues his studies in the fields of aviation management, contemporary management systems, strategic management, and behavioral sciences. He is currently working as Visiting Assistant Professor at Kazimiero Simonaviciaus University Vilnius, Lithuania for his Post-Doctorate Research on Crisis Management in Aviation Sector which is supported by The Scientific and Technological Research Council of Türkiye (TÜBİTAK).

António Rodrigues obtained a PhD in Management and Marketing at the University of Seville in Spain. He is currently a professor and researcher at the ISG-Business & Economics School (Lisbon, Portugal). He has teaching experience in undergraduate and master's degrees in Aeronautical Management. He is the author of several opinion articles and scientific articles. Professionally, he has experience in the financial, marketing and general management areas. He is certified in Sustainable Finance (ESG Advisor) by EFPA-The European Financial Planning Association.

Dorothea Bowyer is a lecturer at Western Sydney University (WSU) School of Business with a research focus on (a) the air transport industry and stakeholder accountability, (b) gender equity in higher education and (c) graduate employability. The scope of her industry related research focuses on the air transport industry and infrastructure asset management. In 2019 she was awarded the Western Sydney University (WSU) School of Business Priority Research Initiative grant to assist with infrastructure asset management and reporting in preparation for Western Sydney airport development. The dissertation topic of her PhD on airport stakeholders (completion 2009) and the development of a stakeholder performance assessment model (SAPAM) to assess the Sydney Airport Privatisation received recognition by the American Institute of Management Accountants (IMA) Foundation for Applied Research (FAR) and she received external recognition as Early Career Researcher (post doc). It ignited her passion for expanding her research to regional aviation and public sector infrastructure reporting. Dorothea has been involved in multiple Gender Equity Funded projects and extended her work to focus on investigating gender equity policies for student-parents in Australia and the availability and delivery of innovative, flexible childcare minding options in the higher education sector to reflect the needs of working parents in a post-Covid environment. She is also involved in work-integrated learning initiatives in her School and has implemented project-based and research-based learning and collaborative online international learning as part of innovative student-centric approaches to student engagement. Dorothea has held guest lectureships in Europe and is the European Academic Lead in her School. She has been the recipient of several grants, including the Endeavour Leadership Program and the New Colombo Plan.

***

Jorge Abrantes has a PhD in Tourism from the University of Lisbon (Institute of Geography and Spatial Planning - IGOT). More than 30 years of professional experience in tourism and aviation, mostly in business management and administration. He is an invited professor at the Estoril Higher School of Hotel and Tourism (ESHTE) and Universidade Aberta. Professor in tourism university degrees since 1992, his research areas are related to tourism and transport, innovation in tourism and new models of accommodation.

Savaş Selahattin Ateş graduated from Eskişehir Anadolu University School of Civil Aviation, Air Transport Management in 2004. He worked at Onur Airlines as dispatcher assistant for 1 year between 2004-2005. He had continued to work as vice manager of Eskişehir Hasan Polatkan Airport between 2005-2007. He started academic career as a research assistant at Anadolu University School of Civil Aviation end of the 2007. He graduated from Anadolu University Institute of Social Sciences Civil Aviation Management Master's Program with a thesis titled "Airport master planning approaches and research on different applications" in 2008. He graduated from the doctorate program of the same department in 2013 with his thesis titled "Decision support system model proposal for reducing delays in the operation process of aviation companies and its application at Atatürk Airport". He was appointed to Anadolu University Faculty of Aeronautics and Astronautics, Department of Aviation Management as Assistant Professor Dr. in 2014. He was the Erasmus Coordinator of the School of Civil Aviation Between 2014-2015. He worked as the Deputy Director of Aviation Museum at Anadolu University. He worked as the Vice Dean of the Faculty of Aviation and Aeronautics and SHY147 Basic Education Maintenance Organization 2016-2020. He has been working as Assistant of Academic Coordinator in Aviation Management Program at Open Education Faculty since 2015 and a Board Member of Eskişehir Technical University Open and Distance Education Application Center since 2020. Savaş S. ATEŞ continues his academic career as an Associate Professor in the Department of Aviation Management.

Mahmut Bakır graduated from the Department of Civil Air Transport Management, Mustafa Kemal University, Turkey. He obtained his master's and PhD degrees in Civil Aviation Management Programme from the Graduate School of Social Sciences, Anadolu University. He is currently working as a faculty member at the Department of Aviation Management, Samsun University. His primary research interests include airline business, service marketing, consumer behavior, multi-attribute decision analysis, and multivariate statistics. His previous research work has been published in several international journals and book projects.

Ana Barqueira, has a degree in Applied Mathematics and Computation (Probabilities and Statistics area) from Instituto Superior Técnico, a master's in applied mathematics for Economics and Management from Instituto Superior de Economia e Gestão and a PhD in Stochastic Processes and Statistics from Instituto Superior Técnico. She is an adjunct professor at Instituto Superior de Educação e Ciências since 2004, teaching in several degrees and masters. She is a researcher at CEMAT (Center for Computational and Stochastic Mathematics) at Instituto Superior Técnico (Lisbon Technical University) and Senior Researcher at the Centro de Estudos e Investigação Aplicada (CEIA) at ISEC Lisboa. Her research interests include Education, Tourism, Health, Genetics and some Engineering.

K. Gulnaz Bulbul was graduated from the Department of Industrial Engineering, Faculty of Engineering, Başkent University. She received her master's and PhD degrees in Civil Aviation Management at Anadolu University. She is currently a faculty member of Faculty of Aeronautics and Astronautics of Eskisehir Technical University. Her researches are mainly on Operations Research in Air Transportation.

Kanittha Charennit is a lecturer in airline business program at College of hospitality industry Management; Suan Sunandha Rajabhat University, Thailand.

Kübra Nur Cingöz was born in Kayseri. She graduated from Erzincan Binali Yıldırım University as honored student and worked as operation officer in Antalya Airport. She started her master in 2019 and completed in 2021. She is currently having her PhD in Kocaeli University, Aviation Studies and Management Department. She has been working for about 2,5 years as a Research Assistant in Gaziantep University. She has been attending national and international conferences and studying on the field of management and aviation related to the subject of aviaition pschology, climate change, human factor, sustainability.

Mehmet Durak works at Department of Aviation Management, Lüleburgaz Faculty of Aeronautics and Astronautics in Kırklareli University. He graduated from the department of civil air transport management at Kocaeli University in 2011. He earned his master's and Ph.D. degrees in the civil aviation management field at Anadolu University in 2016 and 2020, respectively. His studies are mainly on strategic management in the airline business. He was also a part of works on various topics in the field of aviation management. He uses multi-criteria decision-making methods and quantitative methods in his studies.

Vildan Durmaz was born in Eskişehir. She graduated from Anadolu University, Educational Faculty, English Teaching Department in 1988. She had studied as an instructor at Anadolu University, School of Civil Aviation. She had her master degree on Management and Organization and doctorate on Civil Aviation Management at the same university. She has been teaching undergraduate, graduate and PhD courses such as; Air Transportation Management, Airport Management, Airports and Environment, Aviation English at Eskişehir Technical University. She has been attending national and international conferences and studying on the field of management and aviation related to the subject of environmental management systems, airport sustainability, team management, current management applications. She wrote several course and scientific books on strategic management.

Ender Gerede earned an associate degree in avionics from Anadolu University's Civil Aviation Vocational School in 1988 and a DEST degree in "Aviation Maintenance Management" from Bordeaux Technical University (France) in 1993. In 1996, he graduated from the Faculty of Business Administration's Department of Business Administration with a bachelor's degree. Anadolu University's Institute of Social Sciences, Department of Civil Aviation Management awarded him a master's degree in 1998 and a doctorate in 2002. In 2013, Gerede was named associate professor of management and strategy, and in 2019, he was named professor of aviation safety and air transport management. He teaches safety management, airline management, human factors, Crew Resource Management, and aircraft maintenance management at the undergraduate and graduate levels. Gerede has done scientific research and written academic papers in the domains of aviation safety and airline management, as well as book and book chapters. He is Turkey's first instructor accredited by the Turkish Directorate General of Civil Aviation to teach Safety Management Systems and Airport Emergency Response Planning. He has obtained numerous sector-specific trainings in the area of aviation safety. He has conducted numerous training sessions for industry stakeholders on how to create and successfully implement a Safety Management System in Turkey.

Kannapat Kankaew is currently entitled as Assistant Professor and Deputy Dean in Research and Academic Services at College of Hospitality Industry Management; Suan Sunandha Rajabhat University. He received Doctor of Public Administration majoring in Human Resources Management. Prior his academic professional, he used to work in the aviation industry for more than 13 years. He has published and participated in International Conferences more than 20 papers in repute journals. And he dedicated as an editorial board and reviewer for various journals both domestic and international. He is one of the working committee for Thailand Professional Qualification Institutes (Public Organization).

Mustafa Kaya is a Research Assistant who graduated from Eskisehir Technical University, Department of Aviation Management in 2021. In 2021, he started his master's degree in Aviation Management at the same university. In this process, he began to be interested in human resources and unemployment in aviation. He continued his studies on the subject and made publications. He started to work as a research assistant at Istanbul Rumeli University in September 2022. He continues to work on Human Resources in Aviation.

Nadjim Mkedder holds his PhD. in Marketing from Anadolu University, Turkey. He received a Master of Finance degree in 2011 and a Bachelor's degree of Banking and Insurance in 2009 from the University of Tlemcen, Algeria. His research interests focus on international marketing, consumer behavior, and digital marketing.

Saurav Negi holds a Ph.D. degree in supply chain and is currently working as faculty at, MCBS, Muscat Oman. He has more than 15 years of experience. He has published more than 20 research papers and articles in international journals of repute.

Emircan Özdemir is an Assistant Professor in the Faculty of Aeronautics and Astronautics, Eskisehir Technical University in Turkey. After graduating with a double major in industrial engineering and aviation management, he received his MA degree in civil aviation management in 2016. In 2022, he received his PhD degree in civil aviation management. His research interests center around service marketing, marketing research, marketing communication, and data mining applications.

Anil Padhra is an Associate Professor in Sustainable Aviation at the University of West London, UK and a Senior Associate at the University of Cambridge, UK helping to lead the Aviation Impact Accelerator Project. He has a background in engineering and atmospheric physics and has published widely on topics including aircraft and airport emissions and the climate impacts of aviation. He has previously worked in air traffic management, airline flight operations and operational research. He is an expert member at the UN World Meteorological Organisation assessing the impacts of climate change and variability on aviation. Anil's current research interests are in airport infrastructure and operational impacts for future aircraft, the climate impact of aircraft contrails due to alternative fuels and aviation planning and adaptation due to a changing climate.

Rui Castro e Quadros, more than 25 years of professional experience, currently he is a BSc Aviation Management Program Director at ISEC Lisbon (and professor) and invited professor at Estoril Higher School of Hotel and Tourism (ESHTE). Research interests focus on air transport, airports, and tourism. He was Executive Director and Member of the Board of Directors of Grupo SATA, Managing Director for Italy Portugália Airlines, where he was responsible for the Italian market and Sales Executive at Iberia, Linhas Aéreas de España, in Lisbon. BSc/MSc in Communication and Post-Graduate in Marketing and Sales; Specialist Title in Strategic Management; and Doctoral Student in Tourism.

Gökhan Tanrıverdi works at Department of Aviation Management, Ali Cavit Çelebioğlu School of Civil Aviation in Erzincan Binali Yıldırım University. He graduated from the department of civil air transport management at Erciyes University in 2014. He earned his master's and Ph.D. degrees in the civil aviation management field at Anadolu University in 2016 and 2021, respectively. His studies are mainly on strategic management in the airline business. He was also a part of works on various topics in the field of aviation management. He uses qualitative methods, multi-criteria decision-making methods, social network analysis, and quantitative methods in his studies.

Saurabh Tiwari is currently working as an associate professor at, UPES, Dehradun, He has more than 15 years of experience. He has published many research papers and articles in international journals.

Shantanu Trivedi is working as a faculty at the Center for Continuing education, UPES, Dehradun. He has more than 10 Years of experience. He has published many research papers and articles in international journals of repute and won many research awards at international conferences.

Umran Under is a research assistant at Eskisehir Technical University and received her PhD on the subject of airline strategic pricing decisions and their effects on consumer behavior. Her areas of research include airline and airport marketing.

Index

www.ingramcontent.com/pod-product-compliance
Lightning Source LLC
LaVergne TN
LVHW062143300926
849161LV00010B/2609